Wearable Sensors for Supporting Diagnosis, Prognosis, and Monitoring of Neurodegenerative Diseases

Wearable Sensors for Supporting Diagnosis, Prognosis, and Monitoring of Neurodegenerative Diseases

Editors

Gabriella Olmo
Florenc Demrozi
Luigi Borzì

MDPI • Basel • Beijing • Wuhan • Barcelona • Belgrade • Manchester • Tokyo • Cluj • Tianjin

Editors

Gabriella Olmo
Deparment of Control and
Computer Engineering
Politecnico di Torino
Turin
Italy

Florenc Demrozi
Department of Electrical
Engineering and Computer
Science
University of Stavanger
Stavanger
Norway

Luigi Borzì
Department of Control and
Computer Engineering
Politecnico di Torino
Turin
Italy

Editorial Office
MDPI
St. Alban-Anlage 66
4052 Basel, Switzerland

This is a reprint of articles from the Special Issue published online in the open access journal *Electronics* (ISSN 2079-9292) (available at: www.mdpi.com/journal/electronics/special_issues/Wearable_Diseases).

For citation purposes, cite each article independently as indicated on the article page online and as indicated below:

LastName, A.A.; LastName, B.B.; LastName, C.C. Article Title. *Journal Name* **Year**, *Volume Number*, Page Range.

ISBN 978-3-0365-7225-3 (Hbk)
ISBN 978-3-0365-7224-6 (PDF)

Contents

About the Editors

Gabriella Olmo

Gabriella Olmo (IEEE Senior Member) received her M.E. and Ph.D. degrees in Electronic Engineering from Politecnico di Torino, Italy, in 1986 and 1992, respectively. In 2016, she received her Master's degree in Medicine and Surgery from Università di Torino, Italy. She is currently a full professor in the Department of Control and Computer Engineering, Politecnico di Torino, Italy. Her main research interests are in the fields of wearable sensors, signal processing, and machine learning techniques for medical applications. She is coauthor of more than 250 publications in international journals and proceedings in international conferences.

Florenc Demrozi

Florenc Demrozi (IEEE Member), Ph.D. in Computer Science, received his B.S. and M.E. degrees in Computer Science and Engineering from the University of Verona, Italy, in 2014 and 2016, respectively, and completed his Ph.D. degree in Computer Science from the University of Verona, Italy, in 2020. He is currently an associate professor in Biomedical Engineering in the Department of Electrical Engineering and Computer Science, University of Stavanger, Norway, where he is researching human activity recognition (HAR), ambient assisted living (AAL), the Internet of Medical Things (IoMT), and sensors and measurements.

Luigi Borzì

Luigi Borzì (IEEE Member) received his B.S. and M.E. in Biomedical Engineering from Politecnico di Torino, Italy, in 2015 and 2018, respectively. In 2020, he was a research assistant in the Department of Control and Computer Engineering, Politecnico di Torino. He received his Ph.D. in Computer Engineering in 2023. He is currently a postdoctoral researcher in the Department of Control and Computer Engineering, Politecnico di Torino. His research interests include digital health, wearable technology, biomedical data processing, human activity recognition, machine and deep learning, and movement analysis.

electronics

Editorial

Wearable Sensors for Supporting Diagnosis, Prognosis, and Monitoring of Neurodegenerative Diseases

Florenc Demrozi [1], Luigi Borzì [2,*] and Gabriella Olmo [2]

[1] Department of Electrical Engineering and Computer Science, University of Stavanger, 4021 Stavanger, Norway
[2] Department of Control and Computer Engineering, Politecnico di Torino, 10129 Turin, Italy
* Correspondence: luigi.borzi@polito.it

Citation: Demrozi, F.; Borzì, L.; Olmo, G. Wearable Sensors for Supporting Diagnosis, Prognosis, and Monitoring of Neurodegenerative Diseases. *Electronics* **2023**, *12*, 1269. https://doi.org/10.3390/ electronics12061269

Received: 3 March 2023
Accepted: 6 March 2023
Published: 7 March 2023

1. Introduction

The incidence of neurodegenerative disorders (NDs) is increasing in an aging population. NDs encompass a wide range of disorders characterized by the progressive deterioration of the central or peripheral nervous system, affecting millions of individuals worldwide. Despite the clinical significance of monitoring ND's symptoms, this can be challenging in current practice due to the difficulty of accurately remembering and describing symptoms and the infrequency of clinical appointments. Moreover, individuals with NDs may experience difficulties in objectively assessing their symptoms, and these may be perceived differently by their care partners. Thus, there is an unmet need for more objective and continuous monitoring of symptoms in NDs.

To address this challenge, new technological solutions are required for computerized diagnosis, evaluation of the effectiveness of therapy, and continuous monitoring of disease progression. In such a context, wearable technology has emerged as a revolutionary approach to healthcare, offering a more personalized approach to diagnosis and disease management. For example, in the field of neurological diseases, wearable technology has the potential to improve diagnosis, provide inexpensive and non-invasive assessment tools, monitor disease progression, and inform ongoing disease management. Recent advances in wearable and portable sensors, information, and communication technologies have enabled continuous monitoring of NDs. The use of wearable technology allows the collection of high-dimensional data from different domains during daily activities. In addition, signal processing and machine learning (ML) approaches have provided powerful methods for analyzing large amounts of multimodal data, facilitating the obtaining of detailed, objective, and accurate information on disease manifestations.

Wearable technology offers several advantages in monitoring NDs, such as continuous monitoring, objective measurements, and remote monitoring, which can lead to earlier diagnosis, more accurate treatment decisions, and improved outcomes. Wearable technology can also be used to measure various parameters, such as heart rate, blood pressure, movement, sleep patterns, and brain activity, providing insights into cognitive function and facilitating the diagnosis of NDs. In addition, the data collected from wearable technology can be analyzed using ML algorithms to identify patterns and develop predictive models, supporting clinicians in making informed decisions about treatment and care. In conclusion, wearable technology has excellent potential in NDs, providing continuous and objective monitoring and enabling ML analysis of high-dimensional data. As wearable technology continues to advance, it is likely to play an increasingly important role in diagnosing and managing NDs.

2. The Present Special Issue

The present Special Issue comprises eleven research and review articles that propose wearable solutions and explore signal processing, ML, and deep learning (DL) approaches for the computerized diagnosis and monitoring of NDs. The following is a brief summary of each of the articles.

Masi et al. [1] provided an overview of non-intrusive approaches to sleep monitoring for NDs. The authors reviewed twenty-six articles to gather information on the proposed solutions in terms of technologies, methods, and fields of application. The results showed that wearable sensors were mainly used for automatic sleep staging and movement analysis, while non-wearable solutions were used for home monitoring. In addition, inertial sensors were the most commonly used technology, followed by environmental cameras and bedside sensors. The authors concluded that, despite the wide variety of proposed solutions, these need further validation before being applied in clinical practice and in patients' daily lives.

Sigcha et al. [2] proposed a wearable system to estimate the severity of bradykinesia (i.e., slowness of movement) in Parkinson's disease (PD). Six subjects with PD and seven age-matched healthy controls (HCs) were equipped with a consumer smartwatch and asked to perform a series of motor exercises for 6 weeks. Inertial data were processed using different data representations, data augmentation techniques, feature sets, and ML models. The combination of convolutional neural network (CNN) and random forest (RF) classifier provided the best performance, with an accuracy of 0.86. Furthermore, a Pearson's correlation coefficient (r) of 0.94 and a mean square error of 0.46 were obtained between the system output and the clinical severity score.

Carvajal-Castaño et al. [3] collected inertial data from forty-five subjects with PD and eighty-nine HCs, including forty-four young and forty-five elderly people. Participants were asked to perform various gait tasks while wearing inertial measurement units attached to their shoes. Different data representations and DL models were used to process the data. The CNN fed with the short-time Fourier transform provided comparable results to the gated recurrent unit fed with raw data. The further combination of both models did not significantly improve performance. Finally, discrimination of persons with PD from elderly people proved more difficult (0.93 accuracy) than discrimination from younger persons (0.83 accuracy).

Pau et al. [4] employed a single inertial sensor on the lower back to analyze the subjects' gait. Specifically, 449 elderly HCs were recruited and divided into three groups according to age. Acceleration signals were recorded while participants walked in a straight line. Spatial and temporal gait parameters and harmonic ratio were calculated. Finally, statistical analysis (i.e., two-way multivariate analysis of variance) was used to assess significant differences. Older subjects showed a reduction in gait speed, stride length, and cadence ($p < 0.001$), compared to younger participants. Furthermore, the harmonic ratio analysis revealed a general trend of linear decrease with age.

Pietrosanti et al. [5] used wearable inertial sensors to analyze the swinging movement of the forearms during walking. Fifty-eight PD patients and thirty-one age-matched HCs were enrolled and asked to wear sensors on each arm and upper back while performing a timed up-and-go test. The fast Fourier transform of the inertial data was generated and used to extract a series of harmonic features. The two-sample t-test was used to assess the differences between PD and HC subjects. In addition, Spearman's test was used to calculate the correlation between features and clinical scores. The results showed significant differences in arm swing characteristics between subjects with PD and HCs. Furthermore, the harmonic amplitude features correlated significantly with the clinical gait ($r = -0.64$), body bradykinesia ($r = -0.67$), and overall score ($r = -0.57$).

Casadei et al. [6] developed a systolic blood pressure monitoring system based on a wearable device. First, a public data set comprising photoplethysmographic (PPG) recordings of forty-seven subjects was used to train a DL algorithm. Subsequently, data from eight subjects were recorded using both a small wearable PPG sensor and a sphygmomanometer, which was used as a reference. The results showed that the performance of the system was up to standard, with an average absolute error of 3.85 mmHg.

Cesari et al. [7] investigated how wearable devices can be assembled and used to provide feedback to human subjects to improve gait and posture. This can be applied to the rehabilitation of motor disabilities of patients suffering from NDs. Twelve subjects were asked to perform certain postural and motor tasks on a proprioceptive board while being monitored via electromyographic sensors, a force platform, motion capture cameras, and wearable inertial sensors. From the pre-processed multimodal data, several time- and frequency-domain features were extracted and input to different ML models. Preliminary analysis showed that using the inertial sensor system in addition to the other data sources significantly improved performance. Furthermore, using only wearable motion sensors and an RF classifier, an F-score of 0.90 was obtained in the detection of the different phases of motor tasks.

Rana et al. [8] proposed a processing pipeline based on voice analysis for the computerized diagnosis of PD. The data set consisted of voice features extracted from twenty-three PD patients and eight HCs. The authors used different feature selection strategies and different ML classifiers. The proposed DL algorithm provided the best results, with an accuracy of 0.87.

Calvo-Ariza et al. [9] analyzed facial expressions (happiness, surprise, and anger) to discriminate between thirty-one PD patients and twenty-three HCs. The face was extracted from each video frame using a multi-task CNN cascade. Subsequently, two different feature sets, namely local binary patterns and histograms of oriented gradients, were extracted and given as input to a support vector machine for binary classification. The first feature set provided the best performance, achieving an accuracy of 0.80 for the happiness expression.

Sethuraman et al. [10] proposed a system for aiding the diagnosis of Alzheimer's disease (AD) from resting-state functional magnetic imaging (rs-fMRI). The data set comprised 152 patients, in which subjects with AD, mild cognitive impairment, and HCs were equally represented. The images were digitally processed and various frequency levels of the rs-fMRI time series were extracted. Finally, data transformation was applied to convert the time series into images to be input into the DL model. Two CNNs (AlexNet and Inception V2) were used for classification, which were then fine-tuned and optimized. The results showed excellent discrimination ability, with an accuracy of 0.97 and 0.83 in differentiating subjects with AD from HCs and subjects with MCI, respectively.

Besides the mere utilization of wearables for monitoring purposes, the integration of healthcare with the Internet of Things (IoT) presents numerous opportunities for patient monitoring. Nevertheless, a major challenge in the era of Healthcare 4.0 is identifying compromised and malicious nodes, which can threaten network security and user privacy. On such aim, Awan et al. [11] proposed a trust management approach for edge nodes based on ML to identify nodes with malicious behavior. The trust calculation was based on characteristics such as friendliness, trustworthiness, and cooperation. Data were pre-processed using feature selection and scaling and input into a naive Bayes classifier. The experiments were performed in different scenarios and attacks, varying the number of nodes in the network. The results showed that the proposed EdgeTrust system is able to recognize possible IoT attacks to maintain a robust environment. Furthermore, the low power consumption makes the system suitable for real-world scenarios.

3. Future Directions

In recent decades, the advancement of technologies and methodologies has facilitated scientific research in wearable sensors and data processing techniques for health monitoring, leading to a proliferation of wearable solutions for objective assessment, computer-aided diagnosis, and continuous monitoring of chronic disorders. However, challenges in the clinical validation of these solutions and patient compliance for long-term passive monitoring in daily life still persist. To address these challenges, the development of tiny sensors that can be attached to the body or smart textiles with embedded sensors has emerged as a promising solution. Additionally, while research on widely prevalent neurodegenerative disorders such as Parkinson's disease is extensive, there has been limited exploration of rare disorders such as ataxia, Huntington's disease, and progressive supranuclear palsy.

To develop effective, scalable, and clinically validated wearable sensor systems for human health monitoring, further research is necessary.

Acknowledgments: We would like to thank all the authors who submitted their excellent work to this Special Issue. We are grateful to all reviewers for their helpful comments and feedback that helped improve the quality of this Special Issue.

Conflicts of Interest: The authors declare no conflict of interest.

References

1. Masi, G.; Amprimo, G.; Priano, L.; Ferraris, C. New Perspectives in Nonintrusive Sleep Monitoring for Neurodegenerative Diseases—A Narrative Review. *Electronics* **2023**, *12*, 1098. [CrossRef]
2. Sigcha, L.; Domínguez, B.; Borzì, L.; Costa, N.; Costa, S.; Arezes, P.; López, J.M.; De Arcas, G.; Pavón, I. Bradykinesia Detection in Parkinson's Disease Using Smartwatches' Inertial Sensors and Deep Learning Methods. *Electronics* **2022**, *11*, 3879. [CrossRef]
3. Carvajal-Castaño, H.A.; Pérez-Toro, P.A.; Orozco-Arroyave, J.R. Classification of Parkinson's Disease Patients—A Deep Learning Strategy. *Electronics* **2022**, *11*, 2684. [CrossRef]
4. Pau, M.; Bernardelli, G.; Leban, B.; Porta, M.; Putzu, V.; Viale, D.; Asoni, G.; Riccio, D.; Cerfoglio, S.; Galli, M.; et al. Age-Associated Changes on Gait Smoothness in the Third and the Fourth Age. *Electronics* **2023**, *12*, 637. [CrossRef]
5. Pietrosanti, L.; Calado, A.; Verrelli, C.M.; Pisani, A.; Suppa, A.; Fattapposta, F.; Zampogna, A.; Patera, M.; Rosati, V.; Giannini, F.; et al. Harmonic Distortion Aspects in Upper Limb Swings during Gait in Parkinson's Disease. *Electronics* **2023**, *12*, 625. [CrossRef]
6. Casadei, B.C.; Gumiero, A.; Tantillo, G.; Della Torre, L.; Olmo, G. Systolic Blood Pressure Estimation from PPG Signal Using ANN. *Electronics* **2022**, *11*, 2909. [CrossRef]
7. Cesari, P.; Cristani, M.; Demrozi, F.; Pascucci, F.; Picotti, P.M.; Pravadelli, G.; Tomazzoli, C.; Turetta, C.; Workneh, T.C.; Zenti, L. Towards Posture and Gait Evaluation through Wearable-Based Biofeedback Technologies. *Electronics* **2023**, *12*, 644. [CrossRef]
8. Rana, A.; Dumka, A.; Singh, R.; Rashid, M.; Ahmad, N.; Panda, M.K. An Efficient Machine Learning Approach for Diagnosing Parkinson's Disease by Utilizing Voice Features. *Electronics* **2022**, *11*, 3782. [CrossRef]
9. Calvo-Ariza, N.R.; Gómez-Gómez, L.F.; Orozco-Arroyave, J.R. Classical FE Analysis to Classify Parkinson's Disease Patients. *Electronics* **2022**, *11*, 3533. [CrossRef]
10. Sethuraman, S.K.; Malaiyappan, N.; Ramalingam, R.; Basheer, S.; Rashid, M.; Ahmad, N. Predicting Alzheimer's Disease Using Deep Neuro-Functional Networks with Resting-State fMRI. *Electronics* **2023**, *12*, 1031. [CrossRef]
11. Awan, K.A.; Ud Din, I.; Almogren, A.; Khattak, H.A.; Rodrigues, J.J. EdgeTrust: A Lightweight Data-Centric Trust Management Approach for IoT-Based Healthcare 4.0. *Electronics* **2023**, *12*, 140. [CrossRef]

Review

New Perspectives in Nonintrusive Sleep Monitoring for Neurodegenerative Diseases—A Narrative Review

Giulia Masi [1,*] , Gianluca Amprimo [2,3] , Lorenzo Priano [1,4,*] and Claudia Ferraris [2]

1 Department of Neurosciences, University of Turin, Via Cherasco 15, 10100 Torino, Italy
2 Institute of Electronics, Information Engineering and Telecommunication, National Research Council, Corso Duca degli Abruzzi 24, 10129 Torino, Italy
3 Department of Control and Computer Engineering, Politecnico di Torino, Corso Duca degli Abruzzi 24, 10129 Torino, Italy
4 Istituto Auxologico Italiano, IRCCS, Department of Neurology and Neurorehabilitation, S. Giuseppe Hospital, 28824 Piancavallo, Italy
* Correspondence: giulia.masi@unito.it (G.M.); lorenzo.priano@unito.it (L.P.)

Abstract: Good sleep quality is of primary importance in ensuring people's health and well-being. In fact, sleep disorders have well-known adverse effects on quality of life, as they influence attention, memory, mood, and various physiological regulatory body functions. Sleep alterations are often strictly related to age and comorbidities. For example, in neurodegenerative diseases, symptoms may be aggravated by alterations in sleep cycles or, vice versa, may be the cause of sleep disruption. Polysomnography is the primary instrumental method to investigate sleep diseases; however, its use is limited to clinical practice. This review aims to provide a comprehensive overview of the available innovative technologies and methodologies proposed for less invasive sleep-disorder analysis, with a focus on neurodegenerative disorders. The paper intends to summarize the main studies, selected between 2010 and 2022, from different perspectives covering three relevant contexts, the use of wearable and non-wearable technologies, and application to specific neurodegenerative diseases. In addition, the review provides a qualitative summary for each selected article concerning the objectives, instrumentation, metrics, and impact of the results obtained, in order to facilitate the comparison among methodological approaches and overall findings.

Keywords: neurodegenerative diseases; sleep monitoring; sleep disorders; Parkinson disease; dementia; Alzheimer Disease; wearable sensors; inertial sensors; video analysis; Internet of Things

Citation: Masi, G.; Amprimo, G.; Priano, L.; Ferraris, C. New Perspectives in Nonintrusive Sleep Monitoring for Neurodegenerative Diseases—A Narrative Review. *Electronics* **2023**, *12*, 1098. https://doi.org/10.3390/electronics12051098

Academic Editor: Hyun Jae Baek

Received: 3 January 2023
Revised: 6 February 2023
Accepted: 18 February 2023
Published: 22 February 2023

1. Introduction

Sleep plays a fundamental role in the lives of many animals, from some invertebrates to humans. It has both physiological and behavioral connotations and, although its functions and evolutionary significance are not yet fully known, its fundamental role in the maintenance of homeostasis and the adverse effects due to its sub-optimality are well-known in humans. Indeed, it influences attention, memory, mood, blood pressure, immune and inflammatory response, and stress response [1–3]. Under physiological conditions, a sleep phase and a wakefulness phase alternate in a regular manner, constituting the sleep–wake circadian rhythm. The sleep phase is a dynamic process aimed at obtaining the required neurophysiological states at certain times, according to circadian and homeostatic needs and despite external or internal interfering stimuli. Moreover, the so-called macrostructure of sleep, as recorded by electroencephalography (EEG) during polysomnography (PSG), is characterized by a chain of regular and predictable events (cyclic alternation of rapid eye movements (REM) and non-REM (NREM) sleep stages). The process shows an intrinsic variability and has to finely modulate itself in order to maintain the maximum adaptability while preserving sleep macrostructure. In this context, peculiar transient EEG patterns (sleep microstructure) are supposed to play the main role in the building up of

EEG synchronization and in the flexible adaptation against perturbations. Alterations in sleep macro- or microstructure provoke sleep disruption, sleep instability and loss of sleep quantity and quality [4,5]. Sleep and wakefulness influence each other; therefore, sleep quality degradation, when persisting over time, may translate into severe and irreversible symptoms, taking the form of a pathological framework. Therefore, it is very important to create the best possible sleeping conditions and to intervene promptly when sleep disturbances occur, both in their diagnosis and eventual treatments. Even though sleep time and quality lessen with age, sleep disorders are related to comorbidities rather than age [6]. In particular, sleep disorders have a high incidence in neurodegenerative diseases (ND) and are known to influence well-being and quality of life [7]. Indeed, the symptoms of the NDs may be worsened by the sleep disorders, but, at the same time, the latter may be caused or augmented by the neurodegenerative disease, creating a more complex clinical picture. Optimized, sometimes individualized, treatments are being developed in clinical practice [8]. The relationship between sleep abnormalities/disorders and NDs is so close that sleep disorders can be used as criteria for the diagnosis of specific NDs [9]. As an example, stridor co-occurs with multiple system atrophy (MSA), while a REM-sleep behavior disorder may discriminate between Alzheimer's disease (AD) and dementia with Lewy body (DLB). The most interesting discovery in the field is that, in some cases, especially in Parkinson's disease (PD), the onset of sleep disturbance could reflect early alterations in the neural pathways involved, thus constituting a prodromal symptom [10]. This allows earlier intervention in treatment and follow-up; moreover, it will be crucial when neuroprotective drugs become available [11]. The assessment of sleep macro- and microstructure, movements, respiratory pattern or other neurophysiological changes that occur during sleep is essential to verify the quality of sleep and detect sleep disorders. For clinical purposes, PSG is the gold standard for the assessment of sleep disorders, and guidelines are available for recommended uses. In PSG, selected electrophysiological signals are recorded along with other biological signals of interest, such as airflow, oxygen saturation, chest movements or snoring. The type and number of signals that are recorded depends on the reported symptoms and the aim of the PSG. EEG, plectrooculography (EOG), electrocardiography (ECG), and electromyography (EMG) are required for sleep staging, whereas in the detection of sleep apnea, for instance, the primary focus is on oxygen saturation, airflow, and thorax and abdominal movements [12]. Complete polysomnographic examinations are very complex and invasive; they need cumbersome instrumentation, a proper location, night-time assistance by experienced personnel, time, money and they bring discomfort for the patient as well. The medical inspection of the signals (many hours of recording) needs to be performed by qualified experts and it is, however, subjected to inter- operator variability [13,14]. For these reasons, PSG can only be performed in proper settings and usually for in-patients, mainly when precise diagnosis is essential for targeting therapy. Therefore, many alternatives have been proposed in the research to cope with this limitation, in particular for screening or monitoring purposes. They exploit, in general, new technologies and automatic algorithms to reduce the invasiveness of the instrumentation required and the intervention of specialized personnel. This would allow a much more frequent, if not continuous, assessment of the patients' condition with reduced cost and discomfort, providing the conditions for optimized diagnosis and treatments. Research in this area has several objectives:

- To update and simplify the work of medical staff by automating or semi-automating certain procedures—such as sleep staging or sleep disorders diagnosis—through new instrumentation.
- To verify medical treatment efficacy and, eventually, to optimize it, through sleep monitoring.
- To ensure frequent or continuous follow-up by providing instrumentation and protocols to be used in non-hospital settings.

This review wants to explore the available new technologies for minimally invasive sleep monitoring, specifically applied to the field of the NDs, focusing on wearable and non-wearable solutions. The paper is organized as follows. The next Sections 1.1 and 1.2 provides a general background on clinical aspects of sleep monitoring and an overview on the use of technological approaches in NDs. Section 2 provides the description of the methodology employed for the paper selection in this review, Section 3 illustrates results and, lastly, Sections 4 contains discussion and conclusions.

1.1. Background of Sleep Monitoring in Neurodegenerative Diseases

In NDs, the progressive loss of neurons in particular structures of the central nervous system (CNS) causes dysfunctions of neural pathways, leading to the symptoms typical of each disease. In some cases, treatments are available for symptoms relief, but the neurodegeneration process is unstoppable and irreversible. AD and PD are among the most common neurodegenerative disorders worldwide, with a high incidence in the elderly population [15]. In fact, aging is one of the main risk factors in developing NDs, even though their etiology can vary, and are not completely understood. Moreover, genes and environment are believed to be together responsible of these diseases' onset. Other less common NDs are Huntington disease, DLB, amyotrophic lateral sclerosis (ALS), Friedreich ataxia, and MSA. A brief description of the principal symptoms and characteristics is provided in Table 1, with a focus on the diseases' effects on sleep. In fact, these pathologies have a complex relationship with the sphere of sleep. Sleep disruption and disorders can be commonly found in patients with ND and may constitute an early biomarker. Iranzo in [11] highlights the frequent occurrence of the subsequent sleep disorders in ND:

- Insomnia.
- Excessive daytime sleepiness (EDS).
- Rapid eye movement (REM) sleep behavior disorder (RBD).
- Periodic leg movements in sleep (PLMS).
- Restless legs syndrome (RLS).
- Central or obstructive sleep apnea (CSA, OSA).
- Sleep disordered Breathing (SDS).
- Nocturnal stridor.
- Circadian rhythm disorders.

Further, sleep-quality impairment, sleep-time reduction, and presence of abnormal movements (both excessive and impaired) are other typical features. Sleep symptoms derive from multifactorial causes, including the deterioration of sleep–wake regulatory circuitries caused by the neurodegeneration itself and altered neural pathways, movement or respiratory symptoms specific to each pathology or several indirect mechanisms [16]. Sleep has, in turn, an influence on the neurodegeneration process, realizing a complex bi-directional relationship that could lead to new targeted interventions [17]. For instance, sub-optimal sleep—e.g., lack of sleep, disturbed sleep, sleep disorders—was found correlated to cognitive-impairment severity in AD patients and in the elderly, thus constituting a possible risk factor for the onset of cognitive impairment [18,19]. Lately, the discoveries regarding this relationship have been translated in the clinical practice, renovating disease diagnostic criteria and treatments [20]. However, sleep-related symptoms are still under-reported by patients and under-diagnosed by healthcare professionals. This is a flaw in optimized diagnosis and intervention, because of the reduced descriptive power of a complete clinical framework that considers these aspects. The result is a reduced quality of life for patients, sub-optimal treatments, and, sometimes, late diagnosis or misdiagnosis. In clinical practice, these sleep disruptions and disorders, including abnormal movements, are assessed through different tools, such as individual interviews (anamnesis), sleep diaries, sleep questionnaires, clinical scales, reduced or complete PSG, sleep diaries, and clinical scales; moreover, clinical protocols establish assessing procedures [21,22]. Typical sleep symptoms and main clinical assessing protocols are described in Table 2. PSG is the most complete clinical examination, able to evaluate every aspect of sleep and derive quantitative measures, constituting the gold standard in assessment and diagnosis of sleep-related problems. Sleep staging, REM sleep without atonia, apneas, oxygen saturation, sleep microstructure including the cyclic alternating pattern (CAP), and sleep parameters computation can be investigated by PSG. Some of the typical sleep parameters employed, besides sleep-stages descriptors, are total sleep time (TST), sleep latency, sleep efficiency, wake after sleep onset (WASO), and REM latency [23]. Standardized semiquantitative evaluation of symptom severity and quality-of-life reduction is provided by clinical rating scales, such as those shown in Table 2. The latter are employed for various sleep disturbances and disorders, including restless legs syndrome (RLS), insomnia, nocturia, breathing disorders, and daytime sleepiness [24]. It must be considered that each subject's clinical history deeply influences the sleep evaluation tools; in fact, perception of symptoms is subjective and can be influenced by the clinical framework. As an example, in dementia, cognitive impairment can make it difficult to obtain a subject's collaboration in clinical interviews and physical exams [25]. In synucleinopathies—such as PD, DBL, and MSA—RBD assessment is particularly relevant because its idiopathic occurrence is known to be a prodromal symptom that can anticipate any other symptom by decades [26]. In contrast, RBD developing after the onset of other symptoms may indicate a particular disease phenotype. For this reason, RBD screening and diagnosis have attracted much clinical attention in the last years.

Table 1. Neurodegenerative diseases (ND) and sleep-related symptoms, and sleep disorders incidence (sleep disorders incidence (SD)) [11].

ND	Symptoms	Sleep Symptoms	SD
Parkinson's disease [27,28]	Motor: tremors, postural issues, bradykinesia, ON/OFF states, dystonia, rigidity, dyskinesias. Non-motor: orthostatic hypotension, depression, gastrointestinal symptoms, speech and writing change	RBD (also prodromal), sleep-disordered breathing, EDS, Insomnia, RLS, PLMS	60–90%
Multiple system atrophy [29]	Parkinsonism, breathing problems	RBD, fragmented sleep, insomnia, stridor, EDS	80–100%
Dementia with Lewy body [30]	Dementia, parkinsonism, fluctuations, and visual hallucinations	Insomnia, circadian rhythm disorder, RBD [1] (also prodromal), confusional awakenings, EDS	80%
Alzheimer's Disease [31]	Cognitive impairment, dementia. Altered behavior, confusion, aggressiveness	Frequent daytime napping, difficulty in falling asleep and early wakeups, sleep fragmentation, reduced deep and REM sleep amounts, OSA, circadian rhythm alterations, slowdown of sleep EEG rhythms.	45%
Huntington Disease [32] (genetic)	Dementia, psychiatric disturbances	Sleep quality loss, insomnia, sleep fragmentation, EDS, circadian rhythm sleep disorders, reduced NREM and REM sleep.	87%
Amyotrophic lateral sclerosis [33]	Weakness, muscle atrophy, spasticity, respiratory dysfunction	Sleep-disordered breathing, nocturnal hypoventilation, nocturia, cramps, insomnia, EDS	17–76%
Friedreich ataxia [34] (genetic)	Impaired gait, balance, coordination, and speech	RBD, RLS, OSA	50%

[1] OSA: obstructive sleep apnea; EEG: electroencephalography; RBD: REM behavior disorder; EDS: excessive daytime sleepiness; RLS: restless leg syndrome. PLMS: periodic limb movements during sleep.

Table 2. Clinical assessing methods in sleep investigation.

Sleep Investigation	Clinical Assessing Methods
Sleep quality [35]	Anamnesis, diaries such as Consensus Sleep Diary (CSD), clinical scales such as Pittsburgh Sleep Quality Index (PSQI) for sleep disturbances, sleep duration, sleep latency, sleep efficiency, use of sleep medication, daytime dysfunction, and sleep-quality subjective evaluation in the past months.
SD: restless leg syndrome (RLS) [36]	Anamnesis, PSG for detecting associated PLMS, International Restless Legs Scale (IRLS).
SD: REM behavior disorder (RBD) [37]	Anamnesis; PSG [1] with sleep staging and REM sleep without atonia scorings; Video-PSG; screening questionnaires; rating scales: RBD Screening Questionnaire (RBDSQ), RBD Single-Question Screen (RBD1Q).
Sleep-related problems severity in PD	Rating scales: Parkinson's disease sleep scale (PDSS), ESS, SCOPA-SLEEP; PSG.
Nocturnal movements in PD [24]	Anamnesis; PSG; Video-PSG, Actigraphy; rating scales: • PDSS for leg or arm restlessness when resting, urgency to move when resting, getting out of bad for urination, nocturnal hypokinesia, painful posturing of the arms or legs, fidgeting. • MDS-UPDRS for turning in bad, getting out of bad. • NMSS for urgency to move when resting. • NMSQ for getting out of bed for urination, acting out dreams, urgency to move when resting. • PSQI, RLS and RBD scales.
Sleep disturbances in AD [25]	Anamnesis (manifestations of the sleep disorders can be atypical, cognitive impairment can make it difficult); RLS and breathing-disorders assessment; PSG; Actigraphy.
EDS [38]	Anamnesis, PSG, Multiple sleep latency test (MSLT), Maintenance of wakefulness test (MWT), Epworth sleepiness scale (ESS)

[1] PSG: polysomnography; SD: sleep disorder; SCOPA-Sleep: Scales for Outcomes in Parkinson's Disease-Sleep Disturbances; MDS-UPDRS: Movement Disorder Society-sponsored revision of the Unified Parkinson's Disease Rating Scale; NMSQ: Non-Motor Symptoms Questionnaire; NMSS: Non-Motor Symptoms Scale

1.2. Overview of Technologies for Neurodegenerative Diseases

Thanks to the progression of technology, many new-generation devices are available to the medical field. Reduction in costs and dimension for greater computational performances is the main followed trend in the hardware technology. This trend is influencing every aspect of medicine, from in-vitro studies to surgery, passing through virtual reality and robotics [39–42]. In particular, the development of good-quality low-cost sensors determined the development of new possible applications. In addition, the growing world population and the increase in life expectancy created new challenges that technologies, sensors, devices, and algorithms may help to resolve. Technologies in this field are being used to guarantee objectivity, continuity of care and massive screening for lower prices, employing wearable sensors, sensors networks, wireless communication, and automatic algorithms [43].

Companies are also riding the wave, in fact, many consumer products, including smartphones and smartwatches, integrate health monitoring tools and are available at affordable prices for the general population, providing new means for screening and the optimization of self-care. The information provided by this kind of technology does not usually have the aim of substituting standard clinical practice and is targeted to healthy population use; therefore, it is rare for these devices to comply with medical regulations. Nevertheless, some applications, such as heart-rate monitoring and movement analysis, have been proposed as medical tools and obtained American Food and Drug Administration (FDA) approval [44,45]. The gaming industry followed, as well, with the introduction of exergames for physical- and cognitive-health assistance and rehabilitation in neurodegenerative pathologies [46–48]. Sleep monitoring tools are also usually included in smartphones, smartwatches, and consoles, due to the well-known effects of sleep in cognitive and physical performances, as well as quality of life. However, the reliability of these devices in this field is not well-known yet [49,50]. Nevertheless, sleep monitoring is a wide field, where many aspects must be considered depending on the required observation (e.g., movements, sleep staging) and the final aim (e.g., diagnosis, screening) and it is influenced by many factors. Hence, it is very difficult to generalize results from general-purpose instrumentation, especially in the presence of diseases altering sleep characteristics. The latter is the case for NDs, for which sleep disturbances and disorders are important to consider, as presented in the previous section, but which manifest themselves through physical and cognitive symptoms which could influence monitoring tools in unknown and often unpredictable ways. From this perspective, a smaller portion of the research explores this declination, both for single-symptom assessment and generic care of the eldery or frail people. activity of daily living (ADL) recognition and assessment is one of the most interesting topics, because they allow continuous monitoring beyond clinic facilities and provide a multi-potential tool in the wide field of smart homes and assisted living. This is the main objective of Internet of Things (IoT) applications for the elderly. Indeed, due to the incidence of comorbidities in the elderly and their constantly growing number, management of their multiple needs will be possible only through new technologies. In the case of dementia and other NDs, this is one of the followed paths [51–56]. Besides ADLs monitoring, sleep patterns, disease diagnosis and progression assessment, vital signs, agitation, social interactions, compliance with medication intake, movement and fall detection/prevention are interests of these applications. Smart-home applications use a wide range of technological aids—such as radio frequency identification (RFID), wireless communication protocols, global positioning system (GPS), sensors, and cameras— frequently organized in a mixed architecture including wearable and non-wearable sensors. Studies on smart-home monitoring for NDs are reviewed in the dedicated results section (see Section 3.2).

Another trending topic in new technologies for sleep monitoring is the simplification of PSG. PSG is the gold-standard sleep-monitoring exam in clinical facilities. In its conventional set up, it involves multiple high-quality signals recordings. However, the instrumentation is cumbersome and uncomfortable for the subject to wear. In addition, the examination is long to carry out and to analyze, since clinicians have to deal with hours of recordings. This creates an imperative need for an intervention to simplify the whole procedure through new technologies. Moreover, the polysomnogram evaluation involves anomaly identification (REM sleep without atonia, arousals, apneas) and sleep staging which are subjected to intra- and inter-rater variability [13]. Simpler devices and methods are widely proposed in the literature: sleep staging through single-channel physiological recordings, actigraphy, respiratory dynamics and video were attempted [57–60]. Automatic sleep-staging solutions for NDs are reviewed in the dedicated results section (see Section 3.1).

Moreover, a wide range of unobtrusive sensors is employed in the literature for other aspects related to sleep monitoring, including wearables [61–65] or camera-based [66–69] systems. In PD and AD, sensors are widely used in symptoms management and assessment, also with a view to early diagnosis [70–80]. In these disorders, sleep is frequently investigated, especially in studies that focus on motor symptoms, such as the bradykinesia (BK) or dystonia in PD, which can lead to pain or create problems in changing positions or turning in bed. Actigraphy, which provides acceleration recordings from a wrist-worn unit, is already approved by the FDA in the medical field since it enables continuous monitoring (beyond single PSG evaluation). This approach is suitable for the evaluation of excessive daytime sleepiness (EDS), insomnia, and circadian-rhythm sleep disorders, where analysis of time spent in bed and asleep is more relevant. However, its boundary of use in sleep studies is still to be drawn and still a hot topic in the literature, such as in the assessment of NDs' sleep symptoms. In this framework, studies dedicated to NDs compliant with the inclusion criteria of this review are reported in the results section.

2. Materials and Methods

To provide a general overview of the main recent technological approaches used for the analysis of sleep disorders in NDs, an extensive search of the literature was performed through the online databases Web of Science and PubMed over the last 12 years. The search focused on published studies concerning the NDs listed in Table 1 and on the more exploited unobtrusive approaches for sleep monitoring. To this end, the following search criteria were set through:

- Customized queries using keywords and Boolean operators in the form "(Neurodegenerative Disorder OR Parkinson OR Alzheimer OR Huntington OR Lewy Body OR amyotrophic lateral sclerosis OR Ataxia OR Dementia OR Tremor) AND (sleep monitoring) AND (sensor OR IoT OR smart sensor OR environmental sensor OR inertial sensor OR wearable sensor OR optical sensor OR camera OR bed sensor)".
- Year range restriction to 2010–2022.
- Exclusion of pharmacology, veterinary and construction engineering categories.
- Writing language limitation to English.

No criterion was applied on the characteristics of studies participants, as long as the application proposed was explicitly aimed at use on ND-affected subjects.

3. Results

The total records found on Web of Science and PubMed were 142, of which 43 duplicates were excluded. Screening of the titles and abstracts reduced the records to 58. In the end, the full-text analysis of the remaining records led to a total of 26 articles. The selection procedure is shown in Figure 1.

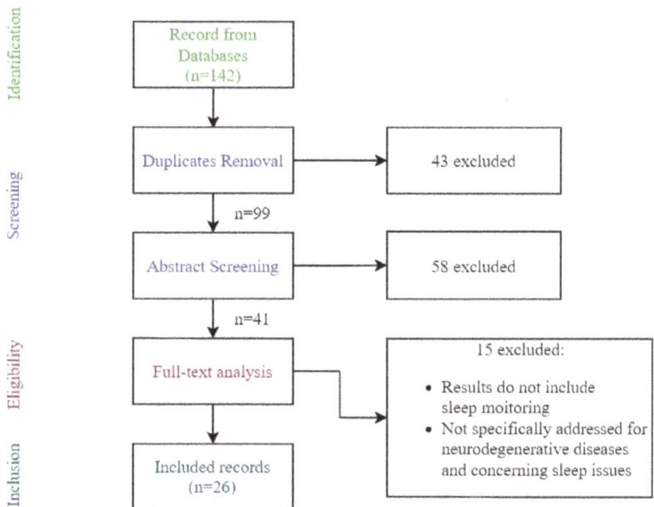

Figure 1. Article selection process.

The selected articles were then categorized considering the three main application domains: automatic sleep staging, at-home sleep monitoring and sleep-quality and movement analysis tools. The papers' distribution according to this categorization is shown in Figure 2a. Moreover, a qualitative synthesis is provided for each article, containing the main aim of the article, the instrumentation, the metrics and obtained results. The instrumentation employed in selected papers largely depended on the application and aims. Figure 2b shows the distribution of articles according to the use of wearable and non-wearable approaches, as well as the tested-sample-size type (e.g., PD-affected patients, healthy subjects). In addition, the collection of the sensors used in the reviewed paper was assessed; it includes: bed sensors, 3D cameras, infrared cameras, inertial sensors, smartwatches, headbands and novel tattooed electrodes. A pie chart summarizing sensors' employment is shown in Figure 3.

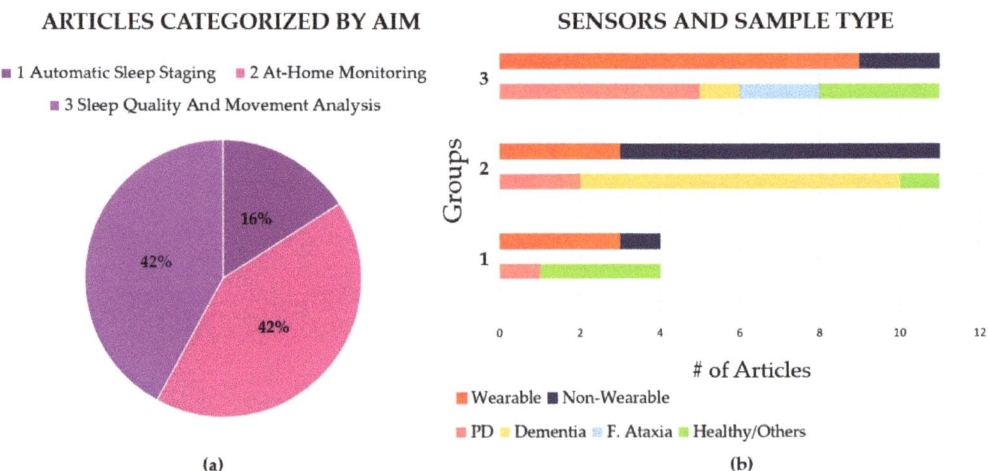

Figure 2. Distribution of selected papers according to chosen categorization. (**a**) Pie chart reporting the percentage of articles for the three mainly investigated categories in the literature; (**b**) bar plots of the distribution of the articles in the three categories of aim, considered sensor type (wearable or non-wearable) and type of targeted population.

Figure 3. Pie chart reporting the distribution of sensors employed in the reviewed articles.

3.1. Automatic Sleep-Staging Techniques

Various systems for simplifying the sleep-staging procedure are proposed in the literature, whether based on PSG or innovative instrumentation; however, few of these studies consider the peculiar condition of NDs, which, as already mentioned, can have a strong influence on the feasibility of the proposals and the generalizability of the results. Moreover, these diseases, together with their associated sleep disruptions, often require the observation of peculiar phenomena, to which the proposed new systems need to provide sensitivity. The gold-standard PSG or video-PSG procedure is the most descriptive and complete exam used in these cases. The research challenge is to reduce the cumbersome instrumentation needed, without losing the fundamental information for sleep-stage recognition and abnormality identification (e.g., k-complexes, sleep spindles, delta burst, apneas, muscle tone, eyes movements). To do so and understand the best configuration, automatic sleep-staging algorithms are also needed. From this perspective, the literature search provided four articles. Their qualitative analysis is displayed in Table 3.

Table 3. Qualitative summary of the selected articles proposing automatic sleep staging.

Article	Subjects	Instrumentation	Methods	Results
Casciola et al. [81]	12 healthy subjects (12 nights)	(W [1]) two-channel EEG headband (HB)	DL approach to overcome low-quality signals from EEG HB in sleep staging. Manual and automatic corrupted-epoch recognition and discard. Data augmentation. DL training in CNN plus LSTM configuration.	Accuracy: 74 ± 10 % with EEG HB signals, 77 ± 10 % with PSG signals.
Shustak et al. [82]	9 healthy subjects (5 nights)	(W) temporary tattooed dry electrode array: two submental EMG, two EOG and four forehead EEG electrodes. The signals were acquired through a customized wireless recording system and Bluetooth connection. See Figure 4a.	Assessment of sensing performance in three ways: by observing signal behavior in typical facial expression; in comparison with standard video-PSG, through qualitative and correlation measures; and in-home settings for feasibility and electrode-stability evaluation. In addition, the opinions of sleep technicians were collected.	Signals recorded with the temporary tattoo and the 10–20 system were visually similar (e.g., eye blinking, k-complexes, sleep spindles), making them easily interpretable for sleep technicians. Amplitude signal parameters and noise were evaluated in the presence of artifacts such rolling in bed or blinking.
Yi et al. [83]	5 healthy subjects (1 night)	(NW) hydraulic bed sensor.	74 features extraction from cardiac and respiratory signals. Classification into awake, REM, and non-REM stages by SVM and k-NN. Accuracy referred to manual PSG scoring.	Accuracy 85% with 0.74 kappa, in the detection of awake, REM, and non-REM stages.
Ko et al. [84]	30 healthy subjects, 27 PD patients divided into two subgroups: 15 PD patients taking clonezepam (PDcC), 12 PD patients without clonezepam (PDnC)	(W) Smartwatch (PPG). See Figure 4b.	Quantification analysis of light sleep, deep sleep, REM, and abnormal REM sleep. Classification into sleep/awake, light/deep sleep and REM sleep using Cole–Kripke algorithm and k-means clustering. Definition of abnormal REM epochs. Comparison between control group and PD group was conducted in the quantitative analysis of sleep stages.	Statistically significant differences between PD and controls were measured in the percentage of deep sleep and abnormal REM. Abnormal REM sleep was also able to distinguish between PDcC and PDnC.

[1] W: wearable; NW: non-wearable; ML: machine learning; EEG: electroencephalography; EOG: electrooculogram; CNN: convolutional neural network, LSTM: long short-term memory; SVM: support vector machine; PSG: polysomnography; PD: Parkinson diseases; REM: rapid eye movements; PSG: polysomnography; PPG: photoplethysmogram.

Some potential solutions were explored by Casciola et al. in [81], Shustak et al. in [82] and Yi et al. [83], on healthy subjects, whereas Ko et al., in [84], tested the capability of the proposed system for abnormal REM detection on PD patients. Casciola, in [81], considered the condition of dementia in AD, where cumbersome instrumentation is a critical issue, due to the typical patient behavior (fear, confusion, aggressive behavior [85,86]). From this perspective, portable EEG headbands (HB) could provide a solution. The authors wanted to overcome the typical reduced signal quality in HB through a deep learning (DL) approach. Their approach was tested on EEG HB and simultaneous PSG recordings. Accuracies of their automatic scoring algorithm were calculated according to manual scoring of PSG in the two cases (HB and PSG signals). The signal processing of HB included band pass filtering and corrupted-epoch manual identification and removal. This cleaning procedure was further deepened through an automatic identification of corrupted epochs using correlation metrics between channels and amplitude values. Data were augmented exploiting windows overlapping, and a DL model, based on convolutional neural network (CNN) and long short-term memory (LSTM), was developed and applied. Authors also implemented traditional sleep-staging techniques for performance comparison. In the end, the proposed DL sleep-staging model achieved 74% accuracy on low-quality HB EEG data and 77% with gold-standard PSG with respect to manual scores. Moreover, the balanced accuracy of the proposed DL method increased by almost 20% compared to any other machine-learning sleep-staging method attempted by them. To better understand the power of their method in the NDs' framework, their approach should be tested on a bigger and differentiated population, comprehending pathological subjects. Yi et al., in [83], proposed an automatic sleep-staging algorithm that exploits bed-sensor recordings consisting of four hydraulic bed transducers under the mattress. Their method aimed to classify sleep in awake, REM and NREM stages by computing 74 features and classifying them usingk-nearest neighbour (k-NN) and support vector machine (SVM) classifiers. Features related to temporal and frequency domains of heartbeat and respiration were considered (ballistocardiography signal analysis). The SVM classifier provided the best performances (accuracy 85.3%) and was also used in a hierarchical fashion (binary asleep–wake classification plus binary REM or NREM classification). In contrast, the other classifiers considered in this study showed inferior and similar performance when compared to the PSG manual score.

Regarding instrumentation developments, Shustak et al., in [82], proposed a wearable setup for sleep staging composed of temporary tattooed dry electrodes: two submental EMG, and two EOG and four forehead EEG electrodes. Data amplification and transfer to a laptop exploited a compact wireless recording system (a customized printed circuit board, a Bluetooth low-energy chip, and a battery). The electrode array employed is shown in Figure 4a. Signals were classically band-pass filtered, and a notch filter was also applied. The authors tested their system in three ways: firstly, they validated effectiveness of EOG, EMG, and EEG recordings using typical facial movements (e.g., smiling, blinking swallowing); secondly, they compared their EEG recordings to the gold-standard systems and, lastly, they assessed the feasibility in home environments. The tattooed electrodes provided signals visually similar to the ones from an EEG system with 10–20 international standard. It was possible to observe sleep spindles and k-complexes, and the recordings were easily interpretable for sleep technicians. Stable recordings were achieved both in a hospital environment and in home settings, where subjects reported good reviews and no impairments in sleep.

Lastly, Ko et al. in [84], provided a method for sleep staging and abnormal REM recognition using cardiac and acceleration signals provided by a smartwatch, see Figure 4c. The authors applied a hierarchical classification through machine-learning techniques, classifying firstly sleep/awake conditions with the Cole–Kripke algorithm, then deep and light sleep based on the G-value and, lastly, identifying REM through k-means clustering. They also defined identification criteria of abnormal REM stages, to be sensitive to REM parasomnias such as EDS typical in PD and MSA. They verified sleep-staging results in a clinical trial, comparing sleep stages and abnormal REM percentages in healthy-control versus PD patients treated with therapy for REM sleep behavior disorder (RBD) versus untreated PD patients. Although the classification accuracies were not very high, the results showed statistically significant differences between healthy-control and PD patients in the percentage of deep sleep. In addition, abnormal REM was found to be significantly different between PD patients with and without RBD therapy (in particular, using clonazepam).

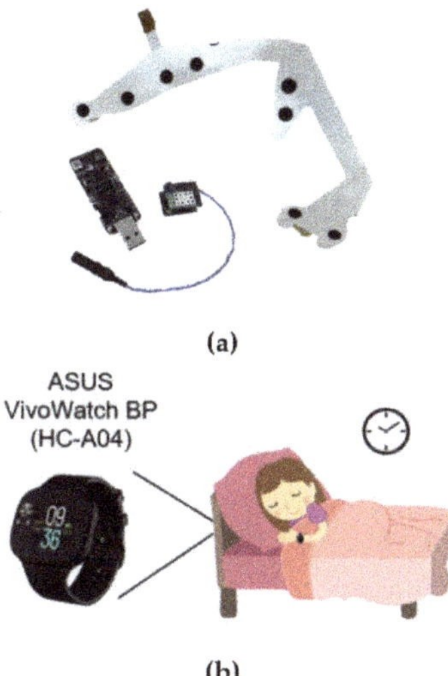

Figure 4. Examples of sensors employed for automatic sleep staging. (**a**) Electrodes array system, adapted from electromyography, electrooculography and electroencephalography, adapted from [82]. (**b**) Smartwatch for cardiac and inertial evaluation, adapted from [84].

3.2. At-Home Sleep Monitoring

The elderly population presents multiple needs simultaneously, since they are usually affected by several diseases with different symptoms. To cope with their conditions, more and more emphasis is being placed on wide-ranging monitoring over time within the home setting. In this way, various parameters can be monitored in a customized manner responding to multiple objectives: to verify health status; to assess the risks for the subject; to make preventive interventions; to diagnose diseases and observe their possible progression; ti check compliance with treatment and, finally, verify the effects of treatments. Such multi-approach monitoring is even more suitable in the presence of a diagnosed ND; in fact, significant efforts are focused on this line of research. Many of these studies include sleep monitoring in their set-up, given its importance for quality-of-life and symptom monitoring. Usually, these systems rely on a network of sensors, wearable and/or non-wearable, which transmit data to cloud services or platforms. In this way, subjects, caregiver, and clinicians can access the data and observe long-term results. Sometimes, these platforms provide custom-made analysis algorithms or they provide a summary of the outputs of the commercial/custom sensors employed. The literature search produced 11 articles in this framework; their summary description is presented in Table 4, where emphasis is placed on the advancement of the sleep-related study and instrumentation adopted.

Regarding cognitive impairment, a smart-home environment for continuous monitoring of elders with dementia is presented by Lazarou et al. in [87]. The article presents the architecture developed in the framework of the Dem@Care FP71 project [88,89]. In the Dem@Care project, the monitoring of sleep, physical activity and ADL were the main goals. In their setup, also used in [90], a commercial under-mattress sensor was employed that was able to determine sleep duration and stages. The proposed solution also involved the integration of the automatic evaluation of daily activity and anomalies by a wide range of sensors, with the assessment and final opinion of clinical experts to target the treatment. In Ref. [87], the authors wanted to verify that their system and the adapted clinical interventions could have positive effects on the physical and cognitive functions of participants. Results concerning sleep included reports of four use cases where, in general, a reduced number of sleep interruptions and increased deep sleep and REM phases were found. Detailed data of sleep patterns were presented for the four subjects (use cases). In Ref. [90], the long-terms effects of the use of the system were evaluated on a bigger subject sample (twelve mildly cognitive-impaired subjects and six subjects with AD): the results confirmed the previous observations: reporting better sleep quality. The effects of this system installation, along with a personalized non-pharmaceutical intervention suggested by the system, were compared with a control group that underwent traditional interventions and with a second control group that did not receive neither personalized nor traditional interventions. Thomas et al., in [91], proposed an at-home smart monitoring system able to assess treatment efficacy for AD. Part of the platform is shown in Figure 5. They considered sleep monitoring using a smartwatch, evaluating TST and compliance to wearing the watch. Specifically, they found that the watch was worn more during the day than at night (compliance 60%), and that subjects often forgot to put the watch back on their wrist when they put it away for some reason. This last result may suggest that wearable solutions, such wrist bands, may not be optimal for continuous sleep monitoring in elders, especially with any kind of memory impairment such as in mild cognitive impairment (MCI) or dementia. Kikhia et al., in [92], focused on nursing homes and proposed the DemaWare@NH monitoring framework system. The aim was to assess behavioral and psychological symptoms of dementia. Concerning sleep, they employed a smart clock connected with a smartphone able to detect respiration signals and movements. The system provided sleep staging in terms of awake, light-sleep and deep-sleep periods, and a 1–100 sleep score. The clinical staff accepted the system, but the smart-clock recordings were made difficult by patients who frequently interacted with the clock-phone system, moving it during the day or pulling cables. This forced the clinical stuff to set up the sensor only during the night. However, the clinical stuff considered the data provided informative on the status of the subjects. Rose et al., in [93], dealt with symptom assessment in AD. Specifically, they analyzed the correlation between nighttime agitation, sleep disturbances and urinary incontinence outside of the clinical setting. Even in this case, the authors designed a multiple-sensor network. To perform sleep monitoring, they used an under-mattress sensor, a microphone, and TEMPO nodes on wrists, i.e., a wireless inertial sensor net. They were able to detect the aforementioned symptoms and to find a correlation between them.

Table 4. Selected articles that present a system dedicated to neurodegenerative diseases in a smart-home monitoring framework, which includes sleep monitoring.

Article	Stage	Instrumentation	Subjects	Results
Dem@Care FP71 project [87,90]	Platform tested on patients	(NW [1]) Commercial under-mattress sensor providing sleep duration and stages	4 in [87]; 22 MCI + 4AD in [90];	Adaptation of treatment based on clinicians' observation of the platform output resulted in the improvement of the sleep quality, also comparing the results with subjects who received a standard intervention.
Thomas et al. [91]	System feasibility	(W) Smartwatch and automatic measures. See Figure 5.	30 AD + 30 spouses	Evaluation of feasibility, compliance in wearing watch, and total sleep-time extraction.
Kikhia et al. [92]	System feasibility and preliminary results	(NW) Smart clock with a smartphone (movement and respiration detection) able to provide sleep staging (awake, light sleep and deep sleep) and a sleep score.	4 subjects with Dementia	Good acceptability of the system by clinical staff, who were able to assess patients based on the output of the system.

Table 4. *Cont.*

Article	Stage	Instrumentation	Subjects	Results
Rose et al. in [93]	Platform tested on patients	(NW) Matress sensor, TEMPO nodes on wrists and a microphone, from which data are transmitted to an online platform where automatic event detection is performed and available for users' consultation.	12 AD subjects	Monitoring and correlation of symptoms, such as nighttime agitation and incontinence in AD, were performed. The correlation inference process showed a pattern for the time occurrence of symptoms.
Hayes et al. [94]	Platform tested on patients	(NW) Passive infrared sensors with custom automatic algorithm extracting sleep features (ORCATECH platform)	45 seniors, including 16 MCI (amnestic, aMCI, and non-amnestic MCI, naMCI) over 6 months	The comparisons of self-reported and platform measures in the three groups (healthy seniors, aMCI, naMCI) showed that movement in bed during the night, wake after sleep onset, and times up during the night were significantly different.
Au-Yeung et al. [95]	Case study with existing platforms	(NW) Aging & Technology (ORCATECH) platform + Emerald device	2AD, 1 frontotemporal dementia, and a major neurocognitive disorder affected subjects.	Sleep-score comparison in the presence/absence of drug administration. Night-time agitation and PLM assessment.
Rawtaer et al. [96]	Feasibility study	(NW) Bed-occupancy sensor based on fiberoptic technology, providing sleep duration and quality metrics (sleep duration, number of sleep interruptions)	28 MCI and 21 healthy controls (>65 years) subjects (HC)	Comparison of sleep duration and interruptions between MCI and HC subjects.
Abbate et al. [97]	Feasibility study	(W+NW) Bed sensor + EEG HB .	-	General discussion on the feasibility of sleep studies based on Enobio EEG HB and inference of risk of fall.
Branco et al. [98]	Feasibility study	(W) Inertial sensor included in the Datapark platform	22 PD subjects in rehabilitation center, for 2 months	Report of changes in sleep position and wakeups were provided to clinicians and patients along with other measures of general activity. Good acceptability of the system.
Silva de Lima [99]	Study presentation and beginning of recruiting	(W) Smartwatch + app	To be: 1000 PD subjects	The system aims to provide sleep-movement analyses.

[1] W: wearable; NW: non-wearable; MCI: mild cognitive impairment, AD: Alzheimer disease; EEG: Electroencephalography; HB: Headband; HC: healthy controls; PD: Parkinson disease; HB: headband.

Regarding continuous monitoring of AD, Oregon Center for Aging and Technology (ORCATECH) at the Oregon Health and Science University have been developing a home monitoring system since 2004. Their platform was meant to assess disease progression and intervention efficacy, relying on passive IR motion sensors and wireless magnetic-contact sensors. The project design and application are described in detail in [19]. Between the various activity recognition and evaluation, the findings regarding sleep by Heyes, in [94], are within the scope of this review. In this last study, the authors used a previously validated algorithm to automatically assess sleep, extracting sleep duration and permanence in bed features (e.g., WASO; TST; settling time: time from getting into bed until the start of the first 20 min period of no movement; times up at night: when the participant actually got out of bed; and total movement in bed at night). Authors also collected subjective sleep assessments and compared elderly volunteers with amnestic MCI and with non-amnestic MCI subjects. Passive sensing for dementia monitoring were also employed by

Au-Yeung in [95]. Their study evaluated only four subjects, two with the ORCATECH platform and two with the Emerald platform (Emerald Innovations Inc., Cambridge, MA, USA), which provides movement, location, and activity info from radio-wave sensors. They compared sleep scores, as provided by the two systems, in different pharmacological interventions. They were able to detect periodic leg movements, associated with drug side effects, providing a tool for modifying interventions and treatments.

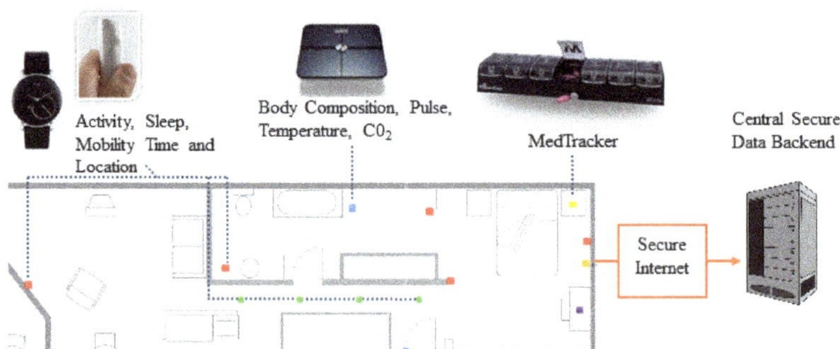

Figure 5. Example of an at-home smart platform for broad-spectrum assessment, including sleep. The Figure is adapted from [91], where a platform for the assessment of treatment efficacy in Alzheimer's disease is presented.

The works of Rawtaer et al. in Ref. [96] and of Abbate et al. in [97] were focused on the field of prevention and early detection. In Ref. [96], the authors evaluated the duration and quality of sleep with a sensor able to detect bed occupancy in terms of sleep duration and interruptions, both on healthy controls and subjects with MCI. The monitoring system reported a worse sleep quality in MCI subjects, in agreement with clinical questionnaires and almost all participants reported good acceptability (41 out of 49). In Ref. [97], the authors proposed a platform exploiting passive and physiological sensing. The study does not report any results on a specific group of subjects, but it claims the feasibility of sleep studies based on Enobio EEG HB (Starlab®, Neuroelectrics, Barcelona, Spain). From sleep data, they also intended to infer the risk of fall. Part of the presented platform architecture is shown in Figure 5a. Regarding Parkinson disease, Branco et al., in [98], presented a data platform (DataPark) able to collect continuous data from an accelerometer. The platform includes quantification algorithms of sleep and physical activity. They obtained preliminary results in a group of PD patients living in a rehabilitation clinic, observing sleep-position changes and wake-ups. In addition, authors reported that patient and personnel feedback were positive, especially regarding physical activity and sleep monitoring. Finally, Silva de Lima, in [99], presented their project and platform, feasibility study and recruiting procedure. Their system relied on a smartwatch connected to a smartphone to detect and analyze sleep movements.

3.3. Sleep Quality and Movement Analysis

In the literature, studies focused on sleep-quality evaluation and movements in sleep were found mainly addressed to PD, Friedreich ataxia and AD. The selected articles in this scope are shown in Table 5. Regarding PD, research focused on analyzing abnormal nocturnal movements during sleep. Those disturbances commonly affect PD patients because of disease-related symptoms or sleep disorders and are clinically assessed by PSG or video-PSG. Actigraphy is also commonly used for this purpose and is FDA-approved, while accelerometers and inertial sensors in various configurations have been gaining ground in this field in recent years [24,100].

Table 5. Selected articles that present a system dedicated to neurodegenerative diseases for sleep-quality and nocturnal-movements assessment.

Article	Subjects	Instrumentation	Methods	Results
Boroojerdi et al. in [101]	21 PD subjects	(W [1]) NIMBLE patch contains an accelerometer and an EMG	Tremor, postural instability, and sleep-quality-measures computation with different patch locations. Comparison with standard clinical scales. Feasibility evaluation.	No correlation between sleep measures and sleep diaries. General good usability and acceptability of the system.
Klingelhoefer in [102]	30 PD subjects with EDS and 33 PD subjects without EDS	(W) PKG (Parkinson's Kineti-Graph)	Bradykinesia and dyskinesia scores to determine disturbed nights. Comparison of the two groups by PKG and sleep-diary data (immobility, sleep duration, sleep interruptions).	In the PD-EDS group, correlation between subjective sleep reports and PKG parameters for quantity and quality of sleep. No correlation in the other group.
Xue in [103]	29 PD subjects, 17 with IBM	(W) multisite inertial sensors	Sleep-quality measure with traditional measures (total sleep time and sleep efficiency) and inertial sensors (acceleration, angular velocity, wakeups, turning in bad, limbs movements). Comparison between the two groups.	Negative correlation between turning-over events and disease duration. Positive correlation between TST and sleep-efficiency parameters and the number of turns in bed. Significant correlation between the number of turns and TST.
Bhidayasiri et al. in [104]	6 PD subjects and 6 spouses	(W) Inertial sensors	Night-time movement analysis, hypokinesia, rolling over description (degrees, duration, velocity, and acceleration) and wakeups	Impairment in turning in PD subjects (less frequent, slower, smaller).
Mirelman et al. in [105]	305 PD + 205 HC subjects	(W) Accelerometer	Nocturnal symptom assessment through lying, turning, and upright time.	Advanced PD subjects showed more upright periods, and a reduction in the number and velocity of their turns. Correlation between the reduction in nocturnal movements and increased PD motor severity, worse dysautonomia and cognition, and dopaminergic medication.
Gavriel et al. in [106,107]	9 F.Ataxia subjects	(W) 1 or 4 of wireless BSN nodes (inertial).	Extraction of biomarkers of Ataxia and Ataxia progression from segmentation of acceleration. They are based on movements and stillness intervals and were correlated to SARA (traditional Ataxia assessment method).	Correlation between the proposed biomarker and SARA assessment.
Wei et al. in [108]	10 healthy young subjects, 10 healthy elders, 8 subjects affected by Dementia	(W) Smartwatch (accelerometer) + actigraph and temperature sensors. See Figure 6.	Confront sleep diaries and accelerometer data. Sleep onset, sleep offset, and sleep duration and nighttime wakeups were calculated. Interday stability and intraday variability were calculated from temperature.	More movement during sleep, measured by actigraphy, in older adults than in the young, with an increasing trend in those with dementia. In addition, less temperature variation between night and day was measured in the elderly.

[1] W: wearable, NW: non-wearable; EMG: electromyogram; EDS: excessive daytime sleepiness; IBM: Impaired Bed Mobility; TST: Total Sleep Time; HC: Healthy Controls; BSN: Body Sensors Network; SARA: Scale for the Assessment and Rating of Ataxia; SAS: sleep apnea syndrome; iRBD: idiophatic behavior disorder.

Boroojerdi et al., in [101], and Klingelhoefer, in [102], focus on sleep-quality evaluation in PD, assessing movements during the night. In particular, Boroojerdi et al., in [101], studied PD motor symptoms with an EMG patch and an accelerometer, evaluating sleep quality in terms of time asleep and postural changes. The authors could not find a correlation between sleep-quality measures and the sleep-diary reports of the subjects. In contrast, Klingelhoefer et al., in [102], studied the effects of disturbed nights, such as daytime sleepiness, through scores for BK and dyskinesia (DK) during sleep computed from Parkinson's KinetiGraphTM (Global Kinectic Pty Ltd, Melbourne Victoria, Australia). The authors were able to correlate their algorithms for the definition of the quantity and quality of sleep, derived from immobility-period identification, to self-assessment reports, in the EDS affected group only. Nocturnal hypokinesia in PD was compared in [103] and [104]. Xue et al., in [103], compared standard clinical scores, such as Unified Parkinson's Disease Rating Scale (UPDRS), Hoehn and Yahr (HY), Pittsburg Sleep Quality Index (PSQI), Epworth

Sleepiness Scale (ESS), Parkinson's Disease Sleep Scale (PSS), with sleep-quality parameters extracted from inertial sensor analysis. They mainly considered TST, sleep efficiency and sleep turnings. In this way, they could find that sleep quality is influenced by turnings in bed and correlated to UPDRS or scores. Bhidayasiri et al., in [104], detected nocturnal movements with an inertial sensor as well. Specifically, the authors measured turning frequency and kinematic turnings parameters (e.g., degrees, velocities, accelerations). In addition, they compared turns in bed in PD patients and their spouses, finding significant impairment in PD subjects turnings (fewer, smaller, and slower turnings). The impact of PD on turning in bed was the main focus of [105] by Mirelman et al. as well. Specifically, the authors analyzed the influence of PD on sleep, obtaining information on sleep interruptions, turnings and laying from a single accelerometer, comparing data on 305 PD subjects and 205 healthy controls. In advanced PD, fewer turns, slower turns, and greater upright time were found, as expected. Moreover, newly diagnosed subjects were similar to controls in the number of turns, but differed in the speed and amplitude of turning, suggesting that this type of measurement can be used as a descriptive of disease progression and as a potential diagnostic tool.

Sleep quality and motion description were also considered relevant topics in Friedreich Ataxia by Gavriel et al. in ref. [106] and ref. [107], where a kinematic sensor network was used to assess disease progression and drug effect in an objective manner. Specific kinematic biomarkers were extracted from movement segmentation and compared with Scale for the Assessment and Rating of Ataxia (SARA) scores (standard assessing method). Finally, sleep quality was also explored in the field of dementia, where Wei et al., in [108], compared sleep-quality measures and outcomes in the presence of a dementia diagnosis and in subjects of different ages. They employed a commercial wristband together with a custom one equipped with actigraphy and temperature sensors, as shown in Figure 6a. The authors found significantly lower sleep and wake temperature difference in older adults with dementia. Furthermore, movements during sleep increased with age, and even more in the presence of dementia. Lastly, a group of innovative technologies related to RBD detection and evaluation were selected. In fact, RBD traditional assessment mainly relies on the identification of movements during the REM stage. Therefore, it requires the simultaneous identification of the REM stage and the analysis of EMG recordings, which constitutes one of the most complex procedures. Given its discovered importance in synucleinopathies, interest grew around prodromal RBD, also considering the difficulty in distinguishing it from mimics, i.e., other motor manifestations or parasomnias during sleep. An attempt at simplification was provided by Cesari and Waser in [109,110], respectively, which exploited 3D video analysis to evaluate limbs movements. They used custom algorithms to identify limb movements. The video analysis was based on the motion signal, corresponding to pixel-wise variation in the 3D video frames over time. Specifically, the authors grouped the automatically identified movements into three regions of interest (upper body, lower body, and full body) based on their duration, estimated movement features for each group and, finally, evaluated their accuracy. In addition, they correlated the estimated features, which could better discriminate isolated RBD- [111] from sleep-disordered breathing (SDB)-affected patients for each group regarding REM sleep without atonia episodes. Finally, Filardi et al., in [112], exploited the analysis of rest–wake-cycle analysis obtained from actigraphy to identify subjects with RBD and to compare their features with those of subjects presenting with symptoms that mimic RBD. A qualitative summary of these works is shown in Table 6.

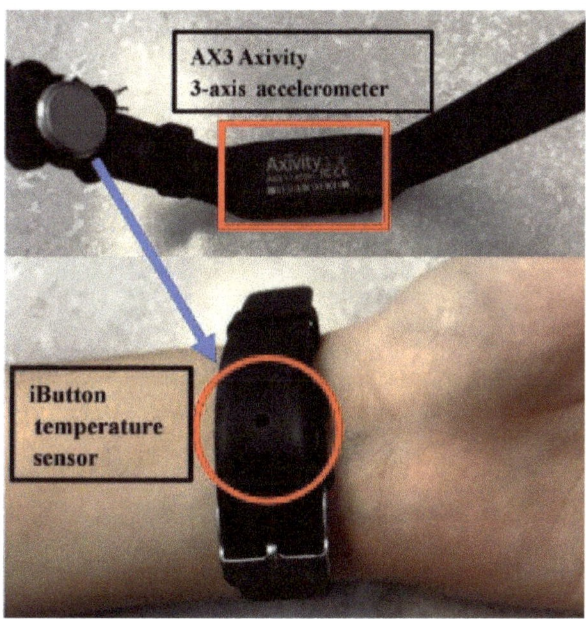

Figure 6. Custom-built wristband for actigraphy and temperature measures, employed for sleep quality assessment. The figure is adapted from [108].

Table 6. Selected articles that present a system dedicated to neurodegenerative diseases for nocturnal movements related to RBD.

Article	Subjects	Instrumentation	Methods	Results
Waser et al. in [110]	122 (40 iRBD, 18 prodromal RBD, 64 participants with mimic symptoms).	(NW[1]) 3D cameras	Custom algorithm for lower limb movement identification in REM. Feature extraction (movements rate, duration, extent, and intensity) and comparison with video-polysomnographic findings.	Significant increase in features analyzed among subjects with iRBD and prodromal RBD and mimic groups. In addition, leg movements with a duration <2 s discriminated iRBD with the highest accuracy (90.4%) from other motor activity during sleep.
Cesari et al. in [109]	20 RBD, 24 SDB subjects	(NW) 3D cameras	Custom algorithm for lower and upper limb movement identification in REM with a max. duration of 5s. Exclusion of breathing movements. Feature extraction (3D rate: the number of movements in REM sleep per hour of REM sleep, and 3D ratio: the total movement-duration time in seconds in REM sleep divided by the total REM-sleep time in seconds) and patient classification were performed (receiver operating characteristic curve to distinguish iRBD , positive class from SDB, negative class).	RBD vs. SDB classification provided an accuracy of 0.91 and F1-score of 0.90
Filardi et al. [112]	19 with iRBD, 19 RLS and 20 with untreated SAS and 16 healthy controls	(W) Micro Motionlogger® Actigraphy Watch (Ambulatory Monitoring, Inc.; NY) + light sensor.	Comparison of video-PSG and RBD-screening-questionnaires findings with the analysis of rest–activity cycles as derived from actigraphy. Features of rest–activity rhythm such as bedtime, wake-up time, midpoint of sleep, estimated wake after sleep onset (eWASO), estimated sleep efficiency (eSE) and activity bouts were extracted.	Lower sleep efficiency, augmented eWASO and increased frequency of prolonged activity bouts for subjects with iRBD compared with those with RLS and controls; no difference compared with SAS patients. In addition, features computed on 24h recording allowed to distinguish iRBD subjects better than screening questionnaires.

[1] NW: non-wearable; W: wearable; EMG: electromyogram; REM: Rapid Eye movement; SDB: sleep disordered breathing; iRBD: idiophatic behavior disorder.

4. Discussion and Conclusions

As discussed in the introduction section, sleep has an important role in guaranteeing a good quality of life, influencing cognitive and physical performances in healthy people and more extensively in the elderly, frail people or subjects with neurodegenerative disorders. Unobtrusive technologies for sleep monitoring are becoming the focus of many companies that develops health and well-being monitoring applications. The use of unobtrusive devices for sleep monitoring would also be of great value in the medical field, especially if applied to subjects affected by NDs, enabling more convenient and even continuous assessment of sleep-related disorders. However, analysis of the articles selected by this review showed that, in the latter area, the multiple proposed solutions still need further validation before application in clinical practice and in patients' daily lives. In fact, many different sensors were used in the reviewed works, showing the feasibility of different sleep monitoring tools, but, it was infrequently considered how these systems could fit into the complex consolidated clinical practice related to NDs.

First, the smart-home monitoring approach, even if interesting, requires the integration of sensors, data and interactions from many stakeholders: the house owner (who is also probably the end user), the company providing the system and the clinical facility that relies on the system and provides the medical service through it. At the moment, there are few healthcare facilities that actually provide these types of telemedicine services. Moreover, the literature search highlighted many smart-home monitoring solutions aimed at ND that included sleep monitoring, but most of them involved feasibility studies or only preliminary results about sleep. Ref. [90] and ref. [94] constitute exceptions, providing results on a moderate number of subjects with cognitive impairment and AD. However, the setup employed by these solutions, consisting of a network of several sensors, presents some drawbacks. For example, the large amount of data collected from all the sensors in continuous monitoring are very difficult (and expensive) to manage and analyze to obtain clinically meaningful results. In addition, custom algorithms should consider many use cases to be robust and subject-oriented, but structured guidelines for continuous home-monitoring applications are lacking in the literature. Moreover, the overall cost could be excessively expensive even in the validation phase of the solution, making these applications apparently suffer from the bottleneck effect typical of many telemedicine solutions [113,114].

Secondly, when there are multiple needs, as in multi-disease patients, it would not be feasible to employ a single device to assess each symptom. Therefore, patients and healthcare institutions need to rely on few trusted tools. From this perspective, actigraphy and inertial sensors are the main solutions for the movement analysis of daytime and nighttime symptoms, in addition or complementary to PSG. The wide applications of these types of sensors (e.g., gait analysis, limb movements, bradykinesia, tremor) make them suitable for integration in patients' daily life and hospitals. Indeed, they proved to be the most widespread and validated solutions. Actigraphy or "equivalent FDA approved devices that uses an accelerometer to measure limb activity associated with movement during sleep for physiologic applications" have already landed in the clinical sleep-monitoring field [115]. However, their use is always contingent on individual circumstances, such as the presence of ND. This is confirmed by the fact that the use of inertial sensors for sleep monitoring in ND is dominant between the reviewed articles, as shown in Figure 2, especially for sleep-quality assessment and movement analysis in a wearable configuration. The inertial sensors are mainly used to determine the permanence in bed, the number of sleep interruptions and the kinematic properties of the movements, such as the turning speed. This makes them good substitutes for sleep diaries, due to their ability to collect quantitative and objective information about sleep. In [102–105], inertial wearable sensors showed the ability of characterizing PD patients with respect to healthy subjects and disease progression; while in refs. [87,90], they were successfully used for AD treatment optimization and in refs. [106,107] for Ataxia characterization through the extraction of biomarkers correlated to standard scores. The feasibility and the importance of sleep evaluations in patients with

ND is, therefore, undoubtable, but a structured protocol of assessment that exploits these sensors has still to be established. For instance, hte optimized number and positioning of inertial sensors in the different disciplines is still to be defined. Fewer sensors would provide a cheaper and more convenient solution, but may not provide sufficient sensitivity to events of interest (e.g., the accelerometer on the arm may ignore foot/limb movements), not to mention that the events of interest depend on the analysis to be performed, which is not always completely defined a priori. Bed sensors are known to be able to provide information on bed occupancy and nighttime movements [116], but no articles presenting their use in ND other than AD were found in the literature search.

A separate discussion should be conducted on movement detection during REM phase to assess REM sleep without atonia for the diagnosis of RBD. In this literature search, two main approaches were found in this direction: 3D-video analysis [109,110] and actigraphy [112]. Both of them showed good performance and are cost-effective solutions. However, they need prior sleep-stage scoring (such as REM-stage recognition for 3D video analysis) or manual event tagging (such as day–night stage recognition for actigraphy). The potential of this type of screening is huge due to the possibility of observing other types of movements of clinical interest, such as thorax/abdomen movements during breathing, or turnings in bed. Therefore, these technologies are a promising line of research that should be further explored, while also considering mixed approaches. Lastly, the selected articles about automatic sleep staging showed interesting results using several types of sensors. However, the samples tested are not sufficient to evaluate a trend in this category. For example, in refs. [81–83], only healthy subjects were enrolled, with sample sizes ranging from 5 to 12 subjects. In contrast, ref. [84] included PD subjects but did not provide an accuracy comparison with PSG results.

To conclude, the literature research conducted in this review seems to demonstrate the feasibility of many different types of unobtrusive methods and technologies for sleep monitoring in ND, but further exploration needs to be performed to better establish the possibilities and limitations of these solutions in this specific scenario. Furthermore, a structured revision of the possible intersection with the actual clinical practice should be considered in order to select and adapt the possible solutions capable to cover, for each neurodegenerative disorder, the widest possible number of their clinical needs.

Author Contributions: Conceptualization, G.M., C.F. and G.A.; methodology, G.M. and C.F.; formal analysis, G.M. and C.F.; investigation, G.M.; resources, G.M.; writing—original draft preparation, G.M. and G.A.; writing—review and editing, G.M., C.F., G.A. and L.P.; visualization, G.M. and G.A.; supervision, C.F. and L.P.; project administration, L.P.; funding acquisition, L.P. All authors have read and agreed to the published version of the manuscript.

Funding: This work was supported by the Department of Excellence Grant of the Italian Ministry of Education, University and Research to the 'Rita Levi Montalcini' Department of Neuroscience, University of Torino, Italy.

Data Availability Statement: Not applicable.

Acknowledgments: G.M. would like to to thank Rechichi I. for her mentoring and inspiration in this field.

Conflicts of Interest: The authors declare no conflict of interest.

Abbreviations

The following abbreviations are used in this manuscript:

AD	Alzheimer's disease
ADL	activity of daily living
ALS	amyotrophic lateral sclerosis
BK	bradykinesia
CAP	cyclic alternating pattern
DK	dyskinesia
CNN	convolutional neural network
CNS	central nervous system
DL	deep learning
DLB	dementia with Lewy body
ECG	electrocardiography
EDS	excessive daytime sleepiness
EEG	electroencephalography
EMG	electromyography
EOG	electrooculography
ESS	Epworth Sleepiness Scale
FDA	American Food and Drug Administration
GPS	global positioning system
HY	Hoehn and Yahr
IoT	Internet of Things
k-NN	k-nearest neighbour
LSTM	long short-term memory
MCI	mild cognitive impairment
MSA	multiple system atrophy
ND	neurodegenerative diseases
NREM	non-REM
ORCATECH	Oregon Center for Aging and Technology
OSA	obstructive sleep apnea
PD	Parkinson's disease
PLMS	periodic leg movements in sleep
PSG	polysomnography
PSQI	Pittsburg Sleep Quality Index
PSS	Parkinson's Disease Sleep Scale
RBD	REM sleep behavior disorder
REM	rapid eye movements
RFID	radio frequency identification
RLS	restless legs syndrome
SARA	Scale for the Assessment and Rating of Ataxia
SD	sleep disorders incidence
SDB	sleep-disordered breathing
SVM	support vector machine
TST	total sleep time
UPDRS	Unified Parkinson's Disease Rating Scale
WASO	wake after sleep onset

References

1. Killgore, W.D.S. Effects of sleep deprivation on cognition. *Prog. Brain Res.* **2010**, *185*, 105–129.
2. Spiegel, K. Effect of sleep deprivation on response to immunizaton. *JAMA* **2002**, *288*, 1471–1472. [CrossRef]
3. Knutson, K.L.; Spiegel, K.; Penev, P.; Van Cauter, E. The metabolic consequences of sleep deprivation. *Sleep Med. Rev.* **2007**, *11*, 163–178. [CrossRef] [PubMed]
4. Halász, P. Hierarchy of micro-arousals and the microstructure of sleep. *Neurophysiol. Clin. Neurophysiol.* **1998**, *28*, 461–475. [CrossRef]
5. Parrino, L.; Ferrillo, F.; Smerieri, A.; Spaggiari, M.C.; Palomba, V.; Rossi, M.; Terzano, M.G. Is insomnia a neurophysiological disorder? The role of sleep EEG microstructure. *Brain Res. Bull.* **2004**, *63*, 377–383. [CrossRef] [PubMed]

6. Foley, D.; Ancoli-Israel, S.; Britz, P.; Walsh, J. Sleep disturbances and chronic disease in older adults: Results of the 2003 National Sleep Foundation Sleep in America Survey. *J. Psychosom. Res.* **2004**, *56*, 497–502. [CrossRef] [PubMed]
7. Iranzo, A.; Santamaria, J. Sleep in Neurodegenerative Diseases. In *Sleep Medicine*; Springer: New York, NY, USA, 2015; pp. 271–283.
8. Iranzo, A. Sleep in dementia and other neurodegenerative diseases. In *Sleep Disorders in Neurology*; John Wiley & Sons, Ltd: Chichester, UK, 2018; pp. 229–240.
9. Abbott, S.M.; Videnovic, A. Chronic sleep disturbance and neural injury: Links to neurodegenerative disease. *Nat. Sci. Sleep* **2016**, *8*, 55–61.
10. Priano, L.; Bigoni, M.; Albani, G.; Sellitti, L.; Giacomotti, E.; Picconi, R.; Cremascoli, R.; Zibetti, M.; Lopiano, L.; Mauro, A. Sleep microstructure in Parkinson's disease: Cycling alternating pattern (CAP) as a sensitive marker of early NREM sleep instability. *Sleep Med.* **2019**, *61*, 57–62. [CrossRef]
11. Iranzo, A. Sleep in neurodegenerative diseases. *Sleep Med. Clin.* **2016**, *11*, 1–18. [CrossRef]
12. Mendonca, F.; Mostafa, S.S.; Ravelo-Garcia, A.G.; Morgado-Dias, F.; Penzel, T. A review of obstructive sleep apnea detection approaches. *IEEE J. Biomed. Health Inform.* **2019**, *23*, 825–837. [CrossRef]
13. Younes, M.; Raneri, J.; Hanly, P. Staging sleep in polysomnograms: Analysis of inter-scorer variability. *J. Clin. Sleep Med.* **2016**, *12*, 885–894. [CrossRef] [PubMed]
14. Miladinović, Đ.; Muheim, C.; Bauer, S.; Spinnler, A.; Noain, D.; Bandarabadi, M.; Gallusser, B.; Krummenacher, G.; Baumann, C.; Adamantidis, A.; et al. SPINDLE: End-to-end learning from EEG/EMG to extrapolate animal sleep scoring across experimental settings, labs and species. *PLoS Comput. Biol.* **2019**, *15*, e1006968. [CrossRef] [PubMed]
15. Dorsey, E.R.; Sherer, T.; Okun, M.S.; Bloem, B.R. The emerging evidence of the Parkinson pandemic. *J. Parkinsons. Dis.* **2018**, *8*, S3–S8. [CrossRef] [PubMed]
16. Raggi, A.; Ferri, R. Sleep disorders in neurodegenerative diseases: Sleep in neurodegenerative diseases. *Eur. J. Neurol.* **2010**, *17*, 1326–1338. [CrossRef]
17. Fifel, K.; Videnovic, A. Circadian and sleep dysfunctions in neurodegenerative disorders-an update. *Front. Neurosci.* **2020**, *14*, 627330. [CrossRef]
18. Zhang, Y.; Ren, R.; Yang, L.; Zhang, H.; Shi, Y.; Okhravi, H.R.; Vitiello, M.V.; Sanford, L.D.; Tang, X. Sleep in Alzheimer's disease: A systematic review and meta-analysis of polysomnographic findings. *Transl. Psychiatry* **2022**, *12*, 136. [CrossRef] [PubMed]
19. Lyons, B.E.; Austin, D.; Seelye, A.; Petersen, J.; Yeargers, J.; Riley, T.; Sharma, N.; Mattek, N.; Dodge, H.; Wild, K.; et al. Corrigendum: Pervasive computing technologies to continuously assess Alzheimer's disease progression and intervention efficacy. *Front. Aging Neurosci.* **2015**, *7*, 232. [CrossRef] [PubMed]
20. Voysey, Z.J.; Barker, R.A.; Lazar, A.S. The treatment of sleep dysfunction in neurodegenerative disorders. *Neurotherapeutics* **2021**, *18*, 202–216. [CrossRef]
21. American Academy of Sleep Medicine. *International Classification of Sleep Disorders—Third Edition (ICSD-3)*; American Academy of Sleep Medicine (AASM): Darien, IL, USA, 2014.
22. Wood, E.A.; McCall, W.V. Assessment methodologies in sleep medicine clinical trials. *Clin. Investig.* **2013**, *3*, 791–800. [CrossRef]
23. Shrivastava, D.; Jung, S.; Saadat, M.; Sirohi, R.; Crewson, K. How to interpret the results of a sleep study. *J. Community Hosp. Intern. Med. Perspect.* **2014**, *4*, 24983. [CrossRef]
24. Zampogna, A.; Manoni, A.; Asci, F.; Liguori, C.; Irrera, F.; Suppa, A. Shedding light on nocturnal movements in Parkinson's disease: Evidence from wearable technologies. *Sensors* **2020**, *20*, 5171. [CrossRef]
25. Urrestarazu, E.; Iriarte, J. Clinical management of sleep disturbances in Alzheimer's disease: Current and emerging strategies. *Nat. Sci. Sleep* **2016**, *8*, 21–33. [CrossRef] [PubMed]
26. Claassen, D.; Josephs, K.; Ahlskog, J.; Silber, M.; Tippmann-Peikert, M.; Boeve, B. Rem sleep behavior disorder preceding other aspects of synucleinopathies by up to half a century. *Neurology* **2011**, *77*, 1155–1155. [CrossRef]
27. Magrinelli, F.; Picelli, A.; Tocco, P.; Federico, A.; Roncari, L.; Smania, N.; Zanette, G.; Tamburin, S. Pathophysiology of motor dysfunction in Parkinson's disease as the rationale for drug treatment and rehabilitation. *Parkinsons Dis.* **2016**, *2016*, 9832839. [CrossRef] [PubMed]
28. Lajoie, A.C.; Lafontaine, A.L.; Kaminska, M. The spectrum of sleep disorders in Parkinson disease: A review. *Chest* **2021**, *159*, 818–827. [CrossRef]
29. Lin, J.Y.; Zhang, L.Y.; Cao, B.; Wei, Q.Q.; Ou, R.W.; Hou, Y.B.; Liu, K.C.; Xu, X.R.; Jiang, Z.; Gu, X.J.; et al. Sleep-related symptoms in multiple system atrophy: Determinants and impact on disease severity. *Chin. Med. J.* **2020**, *134*, 690–698. [CrossRef]
30. Ferman, T.J.; Smith, G.E.; Dickson, D.W.; Graff-Radford, N.R.; Lin, S.C.; Wszolek, Z.; Van Gerpen, J.A.; Uitti, R.; Knopman, D.S.; Petersen, R.C.; et al. Abnormal daytime sleepiness in dementia with Lewy bodies compared to Alzheimer's disease using the Multiple Sleep Latency Test. *Alzheimer Res. Ther.* **2014**, *6*, 76. [CrossRef]
31. Duncan, M.J.; Veasey, S.C.; Zee, P. Editorial: Roles of sleep disruption and circadian rhythm alterations on neurodegeneration and Alzheimer's disease. *Front. Neurosci.* **2021**, *15*, 737895. [CrossRef]
32. Herzog-Krzywoszanska, R.; Krzywoszanski, L. Sleep disorders in Huntington's disease. *Front. Psychiatry* **2019**, *10*, 221. [CrossRef]
33. Boentert, M. Sleep disturbances in patients with amyotrophic lateral sclerosis: Current perspectives. *Nat. Sci. Sleep* **2019**, *11*, 97–111. [CrossRef]

34. Corben, L.A.; Ho, M.; Copland, J.; Tai, G.; Delatycki, M.B. Increased prevalence of sleep-disordered breathing in Friedreich ataxia. *Neurology* **2013**, *81*, 46–51. [CrossRef]

35. Mollayeva, T.; Thurairajah, P.; Burton, K.; Mollayeva, S.; Shapiro, C.M.; Colantonio, A. The Pittsburgh sleep quality index as a screening tool for sleep dysfunction in clinical and non-clinical samples: A systematic review and meta-analysis. *Sleep Med. Rev.* **2016**, *25*, 52–73. [CrossRef]

36. Kohnen, R.; Allen, R.P.; Benes, H.; Garcia-Borreguero, D.; Hening, W.A.; Stiasny-Kolster, K.; Zucconi, M. Assessment of restless legs syndrome—Methodological approaches for use in practice and clinical trials. *Mov. Disord.* **2007**, *22*, S485–S494. [CrossRef] [PubMed]

37. Skorvanek, M.; Feketeova, E.; Kurtis, M.M.; Rusz, J.; Sonka, K. Accuracy of rating scales and clinical measures for screening of rapid eye movement sleep behavior disorder and for predicting conversion to Parkinson's disease and other synucleinopathies. *Front. Neurol.* **2018**, *9*, 376. [CrossRef] [PubMed]

38. McWhirter, D.; Bae, C.; Budur, K. The assessment, diagnosis, and treatment of excessive sleepiness: Practical considerations for the psychiatrist. *Psychiatry* **2007**, *4*, 26–35.

39. Riek, L.D. Healthcare robotics. *Commun. ACM* **2017**, *60*, 68–78. [CrossRef]

40. Sivaparthipan, C.B.; Muthu, B.A.; Manogaran, G.; Maram, B.; Sundarasekar, R.; Krishnamoorthy, S.; Hsu, C.H.; Chandran, K. Innovative and efficient method of robotics for helping the Parkinson's disease patient using IoT in big data analytics. *Trans. Emerg. Telecommun. Technol.* **2020**, *31*, e3838. [CrossRef]

41. Moro, C.; Štromberga, Z.; Raikos, A.; Stirling, A. The effectiveness of virtual and augmented reality in health sciences and medical anatomy: VR and AR in Health Sciences and Medical Anatomy. *Anat. Sci. Educ.* **2017**, *10*, 549–559. [CrossRef]

42. Walper, S.A.; Lasarte Aragonés, G.; Sapsford, K.E.; Brown, C.W., 3rd; Rowland, C.E.; Breger, J.C.; Medintz, I.L. Detecting biothreat agents: From current diagnostics to developing sensor technologies. *ACS Sens.* **2018**, *3*, 1894–2024. [CrossRef] [PubMed]

43. Wang, Y.; Lai, F.; Vespa, P. Enabling technologies facilitate new healthcare delivery models for acute stroke. *Stroke* **2010**, *41*, 1076–1078. [CrossRef] [PubMed]

44. Guillodo, E.; Lemey, C.; Simonnet, M.; Walter, M.; Baca-García, E.; Masetti, V.; Moga, S.; Larsen, M.; HUGOPSY Network; Ropars, J.; et al. Clinical applications of mobile health wearable-based sleep monitoring: Systematic review. *JMIR mHealth uHealth* **2020**, *8*, e10733. [CrossRef] [PubMed]

45. Rune Labs Secures FDA Clearance for Parkinson's Disease Monitoring through StrivePD Ecosystem on Apple Watch. 2022. Available online: https://www.prnewswire.com/news-releases/rune-labs-secures-fda-clearance-for-parkinsons-disease-monitoring-through-strivepd-ecosystem-on-apple-watch-301566472.html (accessed on 16 December 2022).

46. Ben-Sadoun, G.; Manera, V.; Alvarez, J.; Sacco, G.; Robert, P. Recommendations for the design of serious games in neurodegenerative diseases. *Front. Aging Neurosci.* **2018**, *10*, 13. [CrossRef] [PubMed]

47. Alves, M.L.M.; Mesquita, B.S.; Morais, W.S.; Leal, J.C.; Satler, C.E.; Dos Santos Mendes, F.A. Nintendo Wii™ versus Xbox Kinect™ for assisting people with Parkinson's Disease. *Percept. Mot. Skills* **2018**, *125*, 31512518769204. [CrossRef]

48. Amprimo, G.; Masi, G.; Priano, L.; Azzaro, C.; Galli, F.; Pettiti, G.; Mauro, A.; Ferraris, C. Assessment tasks and virtual exergames for remote monitoring of Parkinson's disease: An integrated approach based on Azure Kinect. *Sensors* **2022**, *22*, 8173. [CrossRef]

49. Fino, E.; Mazzetti, M. Monitoring healthy and disturbed sleep through smartphone applications: A review of experimental evidence. *Sleep Breath.* **2019**, *23*, 13–24. [CrossRef] [PubMed]

50. Sun, X.; Qiu, L.; Wu, Y.; Tang, Y.; Cao, G. SleepMonitor: Monitoring respiratory rate and body position during sleep using smartwatch. In Proceedings of the ACM on Interactive, Mobile, Wearable and Ubiquitous Technologies, Association for Computing Machinery (ACM), New York, NY, USA, 2017; Volume 1, pp. 1–22.

51. Giannakopoulou, K.M.; Roussaki, I.; Demestichas, K. Internet of Things technologies and machine learning methods for Parkinson's disease diagnosis, monitoring and management: A systematic review. *Sensors* **2022**, *22*, 1799. [CrossRef]

52. Sheikhtaheri, A.; Sabermahani, F. Applications and outcomes of Internet of Things for patients with Alzheimer's disease/dementia: A scoping review. *Biomed. Res. Int.* **2022**, *2022*, 6274185. [CrossRef] [PubMed]

53. Simonet, C.; Noyce, A.J. Domotics, smart homes, and Parkinson's disease. *J. Parkinsons. Dis.* **2021**, *11*, S55–S63. [CrossRef] [PubMed]

54. Alberdi, A.; Weakley, A.; Schmitter-Edgecombe, M.; Cook, D.J.; Aztiria, A.; Basarab, A.; Barrenechea, M. Smart home-based prediction of multidomain symptoms related to Alzheimer's Disease. *IEEE J. Biomed. Health Inform.* **2018**, *22*, 1720–1731. [CrossRef] [PubMed]

55. ROSETTA. 2012. Available online: http://www.aal-europe.eu/projects/rosetta/ (accessed on 16 December 2022).

56. Home. 2014. Available online: https://www.neurodegenerationresearch.eu/ (accessed on 16 December 2022).

57. Cooray, N.; Andreotti, F.; Lo, C.; Symmonds, M.; Hu, M.T.M.; De Vos, M. Proof of concept: Screening for REM sleep behaviour disorder with a minimal set of sensors. *Clin. Neurophysiol.* **2021**, *132*, 904–913. [CrossRef]

58. Rechichi, I.; Zibetti, M.; Borzì, L.; Olmo, G.; Lopiano, L. Single-channel EEG classification of sleep stages based on REM microstructure. *Healthc. Technol. Lett.* **2021**, *8*, 58–65. [CrossRef] [PubMed]

59. Park, K.S.; Choi, S.H. Smart technologies toward sleep monitoring at home. *Biomed. Eng. Lett.* **2019**, *9*, 73–85. [CrossRef] [PubMed]

60. Nochino, T.; Ohno, Y.; Kato, T.; Taniike, M.; Okada, S. Sleep stage estimation method using a camera for home use. *Biomed. Eng. Lett.* **2019**, *9*, 257–265. [CrossRef] [PubMed]

61. Chen, Z.; Lin, M.; Chen, F.; Lane, N.; Cardone, G.; Wang, R.; Li, T.; Chen, Y.; Choudhury, T.; Cambell, A. Unobtrusive Sleep Monitoring using Smartphones. In Proceedings of the Proceedings of the ICTs for improving Patients Rehabilitation Research Techniques, Venice, Italy, 5–8 May 2013.
62. Ya-Ti Peng.; Ching-Yung Lin.; Ming-Ting Sun.; Landis, C.A. Multimodality sensor system for long-term sleep quality monitoring. *IEEE Trans. Biomed. Circuits Syst.* **2007**, *1*, 217–227. [CrossRef]
63. Kwon, S.; Kim, H.; Yeo, W.H. Recent advances in wearable sensors and portable electronics for sleep monitoring. *iScience* **2021**, *24*, 102461. [CrossRef]
64. Tran, V.P.; Al-Jumaily, A.A.; Islam, S.M.S. Doppler radar-based non-contact health monitoring for obstructive sleep apnea diagnosis: A comprehensive review. *Big Data Cogn. Comput.* **2019**, *3*, 3. [CrossRef]
65. Ma, L.; Liu, S.Y.; Cen, S.S.; Li, Y.; Zhang, H.; Han, C.; Gu, Z.Q.; Mao, W.; Ma, J.H.; Zhou, Y.T.; et al. Detection of motor dysfunction with wearable sensors in patients with idiopathic rapid eye movement disorder. *Front. Bioeng. Biotechnol.* **2021**, *9*, 627481. [CrossRef]
66. Febriana, N.; Rizal, A.; Susanto, E. Sleep monitoring system based on body posture movement using Microsoft Kinect sensor. In *AIP Conference Proceedings*; AIP Publishing LLC: New York, NY, USA, 2019.
67. Lee, J.; Hong, M.; Ryu, S. Sleep monitoring system using Kinect sensor. *Int. J. Distrib. Sens. Netw.* **2015**, *2015*, 1–9. [CrossRef]
68. Jakkaew, P.; Onoye, T. Non-contact respiration monitoring and body movements detection for sleep using thermal imaging. *Sensors* **2020**, *20*, 6307. [CrossRef]
69. Veauthier, C.; Ryczewski, J.; Mansow-Model, S.; Otte, K.; Kayser, B.; Glos, M.; Schöbel, C.; Paul, F.; Brandt, A.U.; Penzel, T. Contactless recording of sleep apnea and periodic leg movements by nocturnal 3-D-video and subsequent visual perceptive computing. *Sci. Rep.* **2019**, *9*, 16812. [CrossRef]
70. Ancona, S.; Faraci, F.D.; Khatab, E.; Fiorillo, L.; Gnarra, O.; Nef, T.; Bassetti, C.L.A.; Bargiotas, P. Wearables in the home-based assessment of abnormal movements in Parkinson's disease: A systematic review of the literature. *J. Neurol.* **2022**, *269*, 100–110. [CrossRef] [PubMed]
71. Breasail, M.Ó.; Biswas, B.; Smith, M.D.; Mazhar, M.K.A.; Tenison, E.; Cullen, A.; Lithander, F.E.; Roudaut, A.; Henderson, E.J. Wearable GPS and accelerometer technologies for monitoring mobility and physical activity in neurodegenerative disorders: A systematic review. *Sensors* **2021**, *21*, 8261. [CrossRef] [PubMed]
72. Channa, A.; Popescu, N.; Ciobanu, V. Wearable solutions for patients with Parkinson's disease and neurocognitive disorder: A systematic review. *Sensors* **2020**, *20*, 2713. [CrossRef]
73. Mughal, H.; Javed, A.R.; Rizwan, M.; Almadhor, A.S.; Kryvinska, N. Parkinson's disease management via wearable sensors: A systematic review. *IEEE Access* **2022**, *10*, 35219–35237. [CrossRef]
74. Woodberry, E.; Browne, G.; Hodges, S.; Watson, P.; Kapur, N.; Woodberry, K. The use of a wearable camera improves autobiographical memory in patients with Alzheimer's disease. *Memory* **2015**, *23*, 340–349. [CrossRef]
75. Lussier, M.; Lavoie, M.; Giroux, S.; Consel, C.; Guay, M.; Macoir, J.; Hudon, C.; Lorrain, D.; Talbot, L.; Langlois, F.; et al. Early detection of mild cognitive impairment with in-home monitoring sensor technologies using functional measures: A systematic review. *IEEE J. Biomed. Health Inform.* **2019**, *23*, 838–847. [CrossRef]
76. Saner, H.; Schütz, N.; Botros, A.; Urwyler, P.; Buluschek, P.; du Pasquier, G.; Nef, T. Potential of ambient sensor systems for early detection of health problems in older adults. *Front. Cardiovasc. Med.* **2020**, *7*, 110. [CrossRef]
77. Varatharajan, R.; Manogaran, G.; Priyan, M.K.; Sundarasekar, R. Wearable sensor devices for early detection of Alzheimer disease using dynamic time warping algorithm. *Cluster Comput.* **2018**, *21*, 681–690. [CrossRef]
78. Mc Ardle, R.; Del Din, S.; Galna, B.; Thomas, A.; Rochester, L. Differentiating dementia disease subtypes with gait analysis: Feasibility of wearable sensors? *Gait Posture* **2020**, *76*, 372–376. [CrossRef]
79. Kourtis, L.C.; Regele, O.B.; Wright, J.M.; Jones, G.B. Digital biomarkers for Alzheimer's disease: The mobile/wearable devices opportunity. *NPJ Digit. Med.* **2019**, *2*, 9. [CrossRef]
80. Sigcha, L.; Domínguez, B.; Borzì, L.; Costa, N.; Costa, S.; Arezes, P.; López, J.M.; De Arcas, G.; Pavón, I. Bradykinesia detection in Parkinson's disease using smartwatches' inertial sensors and deep learning methods. *Electronics* **2022**, *11*, 3879. [CrossRef]
81. Casciola, A.A.; Carlucci, S.K.; Kent, B.A.; Punch, A.M.; Muszynski, M.A.; Zhou, D.; Kazemi, A.; Mirian, M.S.; Valerio, J.; McKeown, M.J.; et al. A deep learning strategy for automatic sleep staging based on two-channel EEG headband data. *Sensors* **2021**, *21*, 3316. [CrossRef] [PubMed]
82. Shustak, S.; Inzelberg, L.; Steinberg, S.; Rand, D.; David Pur, M.; Hillel, I.; Katzav, S.; Fahoum, F.; De Vos, M.; Mirelman, A.; et al. Home monitoring of sleep with a temporary-tattoo EEG, EOG and EMG electrode array: A feasibility study. *J. Neural Eng.* **2019**, *16*, 026024. [CrossRef] [PubMed]
83. Yi, R.; Enayati, M.; Keller, J.M.; Popescu, M.; Skubic, M. Non-Invasive In-Home Sleep Stage Classification Using a Ballistocardiography Bed Sensor. In Proceedings of the IEEE EMBS International Conference on Biomedical and Health Informatics (BHI), Chicago, IL, USA, 19–22 May 2019.
84. Ko, Y.F.; Kuo, P.H.; Wang, C.F.; Chen, Y.J.; Chuang, P.C.; Li, S.Z.; Chen, B.W.; Yang, F.C.; Lo, Y.C.; Yang, Y.; et al. Quantification analysis of sleep based on smartwatch sensors for Parkinson's disease. *Biosensors* **2022**, *12*, 74. [CrossRef]
85. Mahoney, E.L.; Mahoney, D.F. Acceptance of wearable technology by people with Alzheimer's disease: Issues and accommodations. *Am. J. Alzheimer Dis. Other Demen.* **2010**, *25*, 527–531. [CrossRef]

86. Bate, G.; Richardson, S.; Taylor, J.P.; Burn, D.; Allan, L.; Yarnall, A.; Guan, Y.; Del-Din, S.; Lawson, R. Feasibility of using wearable sensors to monitor activity and sleep patterns in inpatients with delirium and Parkinson's disease. *Mov. Disord.* **2022**, *37*, S365–S366.

87. Lazarou, I.; Karakostas, A.; Stavropoulos, T.G.; Tsompanidis, T.; Meditskos, G.; Kompatsiaris, I.; Tsolaki, M. A novel and intelligent home monitoring system for care support of elders with cognitive impairment. *J. Alzheimer Dis.* **2016**, *54*, 1561–1591. [CrossRef]

88. Stavropoulos, T.G.; Meditskos, G.; Tsompanidis, T.; Andreadis, S.; Kompatsiaris, I. Dem@Home: Ambient Monitoring and Clinical Support for People Living with Dementia. In Proceedings of the 13th European Semantic Web Conference (ESWC) Crete, Greece, 29 May–2 June 2016; Volume 9989, pp. 26–29.

89. Andreadis, S.; Stavropoulos, T.G.; Meditskos, G.; Kompatsiaris, I. Dem@Home: Ambient Intelligence for Clinical Support of People Living with Dementia. In Proceedings of the 13th European Semantic Web Conference (ESWC), Crete, Greece, 29 May–2 June 2016.

90. Lazarou, I.; Stavropoulos, T.G.; Meditskos, G.; Andreadis, S.; Kompatsiaris, I.Y.; Tsolaki, M. Long-term impact of intelligent monitoring technology on people with cognitive impairment: An observational study. *J. Alzheimer Dis.* **2019**, *70*, 757–792. [CrossRef]

91. Thomas, N.W.D.; Beattie, Z.; Marcoe, J.; Wright, K.; Sharma, N.; Mattek, N.; Dodge, H.; Wild, K.; Kaye, J. An Ecologically Valid, longitudinal, and Unbiased Assessment of Treatment Efficacy in Alzheimer disease (the EVALUATE-AD trial): Proof-of-concept study. *JMIR Res. Protoc.* **2020**, *9*, e17603. [CrossRef]

92. Kikhia, B.; Stavropoulos, T.G.; Meditskos, G.; Kompatsiaris, I.; Hallberg, J.; Sävenstedt, S.; Melander, C. Utilizing ambient and wearable sensors to monitor sleep and stress for people with BPSD in nursing homes. *J. Ambient Intell. Humaniz. Comput.* **2018**, *9*, 261–273. [CrossRef]

93. Rose, K.M.; Lach, J.; Perkhounkova, Y.; Gong, J.; Dandu, S.R.; Dickerson, R.; Emi, I.A.; Fan, D.; Specht, J.; Stankovic, J. Use of body sensors to examine nocturnal agitation, sleep, and urinary incontinence in individuals with Alzheimer's disease. *J. Gerontol. Nurs.* **2018**, *44*, 19–26. [CrossRef] [PubMed]

94. Hayes, T.L.; Riley, T.; Mattek, N.; Pavel, M.; Kaye, J.A. Sleep habits in mild cognitive impairment. *Alzheimer Dis. Assoc. Disord.* **2014**, *28*, 145–150. [CrossRef] [PubMed]

95. Au-Yeung, W.T.M.; Miller, L.; Beattie, Z.; May, R.; Cray, H.V.; Kabelac, Z.; Katabi, D.; Kaye, J.; Vahia, I.V. Monitoring behaviors of patients with late-stage dementia using passive environmental sensing approaches: A case series. *Am. J. Geriatr. Psychiatry* **2022**, *30*, 1–11. [CrossRef] [PubMed]

96. Rawtaer, I.; Mahendran, R.; Kua, E.H.; Tan, H.P.; Tan, H.X.; Lee, T.S.; Ng, T.P. Early detection of mild cognitive impairment with in-home sensors to monitor behavior patterns in community-dwelling senior citizens in Singapore: Cross-sectional feasibility study. *J. Med. Internet Res.* **2020**, *22*, e16854. [CrossRef]

97. Abbate, S.; Avvenuti, M.; Light, J. MIMS: A minimally invasive monitoring sensor platform. *IEEE Sens. J.* **2012**, *12*, 677–684. [CrossRef]

98. Branco, D.; Bouça, R.; Ferreira, J.; Guerreiro, T. Designing free-living reports for Parkinson's disease. In Proceedings of the Extended Abstracts of the 2019 CHI Conference on Human Factors in Computing Systems—CHI EA '19 Glasgow, Scotland, UK, 4–9 May 2019; ACM Press: New York, NY, USA, 2019.

99. Silva de Lima, A.L.; Hahn, T.; de Vries, N.M.; Cohen, E.; Bataille, L.; Little, M.A.; Baldus, H.; Bloem, B.R.; Faber, M.J. Large-scale wearable sensor deployment in Parkinson's patients: The Parkinson@home study protocol. *JMIR Res. Protoc.* **2016**, *5*, e172. [CrossRef] [PubMed]

100. van Wamelen, D.J.; Sringean, J.; Trivedi, D.; Carroll, C.B.; Schrag, A.E.; Odin, P.; Antonini, A.; Bloem, B.R.; Bhidayasiri, R.; Chaudhuri, K.R.; et al. Digital health technology for non-motor symptoms in people with Parkinson's disease: Futile or future? *Parkinsonism Relat. Disord.* **2021**, *89*, 186–194. [CrossRef]

101. Boroojerdi, B.; Ghaffari, R.; Mahadevan, N.; Markowitz, M.; Melton, K.; Morey, B.; Otoul, C.; Patel, S.; Phillips, J.; Sen-Gupta, E.; et al. Clinical feasibility of a wearable, conformable sensor patch to monitor motor symptoms in Parkinson's disease. *Parkinsonism Relat. Disord.* **2019**, *61*, 70–76. [CrossRef]

102. Klingelhoefer, L.; Rizos, A.; Sauerbier, A.; McGregor, S.; Martinez-Martin, P.; Reichmann, H.; Horne, M.; Chaudhuri, K.R. Night-time sleep in Parkinson's disease - the potential use of Parkinson's KinetiGraph: A prospective comparative study. *Eur. J. Neurol.* **2016**, *23*, 1275–1288. [CrossRef]

103. Xue, F.; Wang, F.Y.; Mao, C.J.; Guo, S.P.; Chen, J.; Li, J.; Wang, Q.J.; Bei, H.Z.; Yu, Q.; Liu, C.F. Analysis of nocturnal hypokinesia and sleep quality in Parkinson's disease. *J. Clin. Neurosci.* **2018**, *54*, 96–101. [CrossRef]

104. Bhidayasiri, R.; Sringean, J.; Taechalertpaisarn, P.; Thanawattano, C. Capturing nighttime symptoms in Parkinson disease: Technical development and experimental verification of inertial sensors for nocturnal hypokinesia. *J. Rehabil. Res. Dev.* **2016**, *53*, 487–498. [CrossRef] [PubMed]

105. Mirelman, A.; Hillel, I.; Rochester, L.; Del Din, S.; Bloem, B.R.; Avanzino, L.; Nieuwboer, A.; Maidan, I.; Herman, T.; Thaler, A.; et al. Tossing and turning in bed: Nocturnal movements in Parkinson's disease: Nocturnal movement in pd. *Mov. Disord.* **2020**, *35*, 959–968. [CrossRef] [PubMed]

106. Gavriel, C.; Thomik, A.A.C.; Lourencco, P.R.; Nageshwaran, S.; Athanasopoulos, S.; Sylaidi, A.; Festenstein, R.; Faisal, A.A. Kinematic body sensor networks and behaviourmetrics for objective efficacy measurements in neurodegenerative disease drug

trials. In Proceedings of the 2015 IEEE 12th International Conference on Wearable and Implantable Body Sensor Networks (BSN), Cambridge, MA, USA, 9–12 June 2015.

107. Gavriel, C.; Thomik, A.A.C.; Lourenco, P.R.; Nageshwaran, S.; Athanasopoulos, S.; Sylaidi, A.; Festenstein, R.; Faisal, A.A. Towards neurobehavioral biomarkers for longitudinal monitoring of neurodegeneration with wearable body sensor networks. In Proceedings of the 2015 7th International IEEE/EMBS Conference on Neural Engineering (NER), Montpellier, France, 22–24 April 2015.

108. Wei, J.; Boger, J. Sleep detection for younger adults, healthy older adults, and older adults living with dementia using wrist temperature and actigraphy: Prototype testing and case study analysis. *JMIR mHealth uHealth* **2021**, *9*, e26462. [CrossRef] [PubMed]

109. Cesari, M.; Kohn, B.; Holzknecht, E.; Ibrahim, A.; Heidbreder, A.; Bergmann, M.; Brandauer, E.; Hogl, B.; Garn, H.; Stefani, A. Automatic 3D video analysis of upper and lower body movements to identify isolated REM sleep behavior disorder: A pilot study. *Annu. Int. Conf. IEEE Eng. Med. Biol. Soc.* **2021**, *2021*, 7050–7053. [PubMed]

110. Waser, M.; Stefani, A.; Holzknecht, E.; Kohn, B.; Hackner, H.; Brandauer, E.; Bergmann, M.; Taupe, P.; Gall, M.; Garn, H.; et al. Automated 3D video analysis of lower limb movements during REM sleep: A new diagnostic tool for isolated REM sleep behavior disorder. *Sleep* **2020**, *43*, zsaa100. [CrossRef]

111. Högl, B.; Stefani, A.; Videnovic, A. Idiopathic REM sleep behaviour disorder and neurodegeneration—An update. *Nat. Rev. Neurol.* **2018**, *14*, 40–55. [CrossRef]

112. Filardi, M.; Stefani, A.; Holzknecht, E.; Pizza, F.; Plazzi, G.; Högl, B. Objective rest-activity cycle analysis by actigraphy identifies isolated rapid eye movement sleep behavior disorder. *Eur. J. Neurol.* **2020**, *27*, 1848–1855. [CrossRef]

113. Hjelm, N.M. Benefits and drawbacks of telemedicine. *J. Telemed. Telecare* **2005**, *11*, 60–70. [CrossRef]

114. Dhanvijay, M.M.; Patil, S.C. Internet of Things: A survey of enabling technologies in healthcare and its applications. *Comput. Netw.* **2019**, *153*, 113–131. [CrossRef]

115. Smith, M.T.; McCrae, C.S.; Cheung, J.; Martin, J.L.; Harrod, C.G.; Heald, J.L.; Carden, K.A. Use of Actigraphy for the Evaluation of Sleep Disorders and Circadian Rhythm Sleep-Wake Disorders: An American Academy of Sleep Medicine Clinical Practice Guideline. *J. Clin. Sleep Med.*. **2018**, *14*, 1231–1237. [CrossRef]

116. Conti, M.; Orcioni, S.; Madrid, N.M.; Gaiduk, M.; Seepold, R. A review of health monitoring systems using sensors on bed or cushion. In *Bioinformatics and Biomedical Engineering*; Springer International Publishing: Cham, Switzerland, 2018; pp. 347–358.

Article

Bradykinesia Detection in Parkinson's Disease Using Smartwatches' Inertial Sensors and Deep Learning Methods

Luis Sigcha [1,2], Beatriz Domínguez [1], Luigi Borzì [3], Nélson Costa [2], Susana Costa [2], Pedro Arezes [2], Juan Manuel López [1], Guillermo De Arcas [1] and Ignacio Pavón [1,*]

[1] Instrumentation and Applied Acoustics Research Group (I2A2), ETSI Industriales, Universidad Politécnica de Madrid, Campus Sur UPM, Ctra. Valencia, Km 7, 28031 Madrid, Spain
[2] ALGORITMI Research Center, School of Engineering, University of Minho, 4800-058 Guimarães, Portugal
[3] Department of Control and Computer Engineering, Politecnico di Torino, 10129 Turin, Italy
* Correspondence: ignacio.pavon@upm.es; Tel.: +34-91-067-7222

Abstract: Bradykinesia is the defining motor symptom of Parkinson's disease (PD) and is reflected as a progressive reduction in speed and range of motion. The evaluation of bradykinesia severity is important for assessing disease progression, daily motor fluctuations, and therapy response. However, the clinical evaluation of PD motor signs is affected by subjectivity, leading to intra- and inter-rater variability. Moreover, the clinical assessment is performed a few times a year during pre-scheduled follow-up visits. To overcome these limitations, objective and unobtrusive methods based on wearable motion sensors and machine learning (ML) have been proposed, providing promising results. In this study, the combination of inertial sensors embedded in consumer smartwatches and different ML models is exploited to detect bradykinesia in the upper extremities and evaluate its severity. Six PD subjects and seven age-matched healthy controls were equipped with a consumer smartwatch and asked to perform a set of motor exercises for at least 6 weeks. Different feature sets, data representations, data augmentation methods, and ML models were implemented and combined. Data recorded from smartwatches' motion sensors, properly augmented and fed to a combination of Convolutional Neural Network and Random Forest model, provided the best results, with an accuracy of 0.86 and an area under the curve (AUC) of 0.94. Results suggest that the combination of consumer smartwatches and ML classification methods represents an unobtrusive solution for the detection of bradykinesia and the evaluation of its severity.

Keywords: Parkinson's disease; bradykinesia; wearables; inertial sensors; artificial intelligence; deep learning

Citation: Sigcha, L.; Domínguez, B.; Borzì, L.; Costa, N.; Costa, S.; Arezes, P.; López, J.M.; De Arcas, G.; Pavón, I. Bradykinesia Detection in Parkinson's Disease Using Smartwatches' Inertial Sensors and Deep Learning Methods. *Electronics* **2022**, *11*, 3879. https://doi.org/10.3390/electronics11233879

Academic Editor: Nicola Francesco Lopomo

Received: 31 October 2022
Accepted: 21 November 2022
Published: 24 November 2022

Publisher's Note: MDPI stays neutral with regard to jurisdictional claims in published maps and institutional affiliations.

1. Introduction

Parkinson's disease (PD) is one of the most common neurodegenerative diseases worldwide [1], affecting millions of people and impacting their quality of life (QoL) [2]. PD is a progressive disease with a slow and variable evolution. In the early stages, the symptoms are weak, and they increase in intensity as the disease progresses [3]. PD involves both motor and non-motor symptoms, with some of the latter (i.e., speech impairment and sleep disorders) manifesting up to 20 years before the clinical diagnosis [4]. Being primarily a movement disorder, several motor signs are associated with PD, including bradykinesia, tremor, and rigidity. As the disease progresses, postural instability and freezing of gait (FOG) manifest, increasing the risk of falls [5] and contributing to decreased mobility [6]. As the main biochemical abnormality in PD is dopamine deficiency [7], current treatments are mainly based on dopamine replacement, with Levodopa representing the most effective drug treatment for PD [8,9]. However, current treatments do not prevent disease progression, their effectiveness decreases with disease progression [10], and long-term therapy frequently leads to severe side effects [11]. Moreover, as the disease progresses and drug therapy is administrated, patients may experience fluctuations in the state of their motor system, between the so-called ON state, where symptoms are under control and the patient can move fluidly, and an OFF state, in which a lack of dopamine predominates and symptoms reappear when the effect of the medication vanishes.

Bradykinesia represents one of the earliest motor signs of PD, and it is one of the main aspects that specialists try to quantify to diagnose PD and optimize therapy. It is defined by the slowness and decrease in the amplitude or speed of movement in a body part [12]. Akinesia and hypokinesia refer respectively to poor spontaneous movements (i.e., in facial expression) or associated movement (i.e., arm swing during walking) and the low amplitude of movement [2,13]. Bradykinesia can vary throughout the day and its severity also vary depending on the timing and amount of the last medication. In addition, the symptom's severity also depends on the patient's emotional state and environment [2]. Bradykinesia is one of the key signs in the evaluation of PD, it is directly related to dopamine deficiency [14], and it shows an exceptional response to treatment [15]. Thus, objectively quantifying this symptom would provide relevant information for treatment adjustments and early diagnosis. Following the movement disorder society revised version of the unified Parkinson's disease rating scale (MDS-UPDRS), neurologists assess bradykinesia severity through the execution of rapid, repetitive, alternating hand and heel movements, and they observe the amplitude and slowness of the movement [16]. However, the assessment is performed sporadically, during brief follow-up visits, often without considering the effect of medication. Moreover, intra-rater and inter-rater variability affect the evaluation of the patient's motor performance [17,18].

Subjectivity and late diagnosis highlight the need for new, more objective methodologies allowing the early diagnosis of the disease, the continuous monitoring of its evolution, and the evaluation of the response to therapy [19]. In this context, digital technologies have demonstrated their potential to change the disease paradigm, providing unobtrusive yet efficient solutions for the diagnosis, assessment, monitoring, and treatment planning of PD patients [20,21]. Indeed, an objective measure of PD symptoms can help improve disease management and accelerate the development of new therapies [15]. Wearable sensors benefit from the current technological advances to provide lightweight, portable, easy-to-use, inexpensive devices which can provide accurate measurements of physical variables [22]. Wearable motion sensors and ML methods have been widely used for objectively and rigorously assessing motor symptoms, motor fluctuations, and other complications that are relevant to adjust treatment and remote assistance [15,23–26].

In this context, this paper evaluates the potential of consumer smartwatches for estimating bradykinesia severity in PD. To this end, upper limb motion data were recorded from a triaxial accelerometer and triaxial gyroscope placed on the patient's wrist. Then, signal processing, data augmentation, data transformation, and different ML and DL classification models were implemented to predict the bradykinesia severity following the standards of the MDS-UPDRS scale. The main contributions of this work are summarized as follows:

- This study evaluates the potential of accelerometer and gyroscope sensors embedded in commodity smartwatches to detect bradykinesia severity using a set of standardized exercises. This approach can present an unobtrusive solution for bradykinesia monitoring in ambulatory and non-supervised environments using low-cost devices instead of using proprietary monitoring devices or sensors.
- Different feature extraction methodologies proposed in the related literature are reproduced and evaluated with the data collected using commodity smartwatches. This task is performed to compare the predictive power of approaches based on ML and DL. In addition, the potential of different data representations and data augmentation techniques is evaluated with the aim of improving the performance of the systems for automatic bradykinesia severity scoring.
- This work also introduces the use of convolutional neural networks (CNN) with patch input (implemented with 1D-convolutional layers) for automatic temporal

window contextualization. The patch input strategy is proposed as a mechanism to automatically split and project the data of a single (multi-channel) sliding window into another dimension that can be exploited by classification algorithms. Additionally, the proposed approach is evaluated using an end-to-end neural network, and in combination with a Random Forest (RF) classifier located at the top of the neural network.

- Finally, a methodology for the aggregation of a set of predictions (severity ratings) obtained from the classifiers during a single clinical visit is proposed and evaluated. This methodology is carried out with the aim of improving the outcomes of the bradykinesia assessment by providing a single severity indicator of the motor function of the upper limbs.

The rest of this paper is organized as follows: An overview of the research studies focusing on bradykinesia detection using wearable sensors is provided in Section 2. Section 3 describes the data set used in this study, the implemented signal processing and ML methods, and the performance evaluation procedure. Results are reported in Section 4 and discussed in Section 5, together with conclusions.

2. Related Work

The quantification of bradykinesia using wearable technologies has been widely explored in the last several decades. Besides commercial solutions, such as Kinesia® (Great Lakes NeuroTechnologies Inc., Cleveland, OH, USA) [27] and PKG® (Global Kinetics Pty Ltd., Melbourne, Australia) [28], several research studies focused on the detection of bradykinesia by characterizing the movement of patients. In [29], 50 PD patients were monitored to quantify bradykinesia and hypokinesia. Two accelerometers on the wrist were used for data collection, obtaining sensitivities of 60–71% and specificities of 66–76%. In [30], seven gyroscopes and two accelerometers were placed on the forearms, shins, and trunk for diagnosing the presence or absence of bradykinesia, tremor, body posture, and gait parameters, obtaining a Pearson correlation coefficient r of 0.71 with the UPDRS scale. In [31], a combination of a flexible sensor placed on the hand (triaxial accelerometer and gyroscope) and a consumer smartwatch (triaxial accelerometer) was employed to monitor 13 PD subjects. By using an RF algorithm, the authors achieved an AUC of 0.65 in a multiclass classification (MDS-UPDRS). In [32], an inertial measurement unit (IMU) wristband with an accelerometer was used to monitor 31 PD patients and 50 healthy controls. The authors proposed a methodology to extract bradykinesia digital biomarkers, providing a strong correlation (Pearson $r = 0.67$) between hand motion measurements and the MDS-UPDRS scoring.

The leg agility task (MDS-UPDRS item 3.8) was addressed in different studies for the quantification of bradykinesia. In [33,34], 34 and 24 subjects were enrolled, respectively. Three IMUs were mounted on the patient's chest and each thigh. Time- and frequency-domain features were extracted and selected to feed classification algorithms, i.e., Support Vector Machine (SVM) and k-Nearest Neighbors (kNN). Bradykinesia severity (UPDRS score) was estimated with an accuracy of 43% in both studies. In [35], 19 subjects with PD were monitored with ankle-mounted IMUs for leg agility evaluation and treatment response. Time- and frequency-domain features were computed to feed different classifiers, i.e., SVM, Decision Tree, and Logistic Regression. Pearson correlation with the UPDRS bradykinesia score was found to be $r = 0.83$. Finally, in [17], smartphones' sensors and ML were used to detect bradykinesia using leg agility exercises, achieving an accuracy of 77.7% in a multi-class classification using the UPDRS scale.

In recent years, the research community has explored the use of DL techniques for the automatic analysis of motor symptoms. DL methods make it possible to process the recorded inertial signals without the need for additional processing techniques, reducing the effort in the design and selection of discriminative feature sets [36]. However, despite the advantages of technology in PD, the application of these techniques requires a high amount of quality data and high computational processing power [37].

Relevant works using DL methods and wearable technology to assess bradykinesia have been proposed in [38,39], where CNN and sensors placed on the upper limbs have been employed. The results of these works indicate that they can outperform (shallow) ML approaches achieving an accuracy of 90.9% [38], and an AUC of 0.926 [39]. In [40], 30 PD patients were monitored during different activities using a single accelerometer on the wrist. CNN was used to process raw data and predict bradykinesia severity, achieving an accuracy of 0.67, sensitivity of 0.65, and specificity of 0.89. In [41], six flexible wearable sensors were used for recording data from 20 individuals with PD throughout multiple clinical assessments. Raw inertial data were input to a CNN algorithm, providing an AUC of 0.77.

3. Materials and Methods

In this section, the methodology developed to obtain different bradykinesia detection methods is described. Section 3.1 describes the data used in this study, including information regarding subjects' characteristics, experimental procedures, and clinical assessment of bradykinesia. The preprocessing procedures, including filtering, feature extraction, data transformation, and data augmentation, are reported in Section 3.2. Section 3.3 describes the ML and DL classification algorithms employed in the present work, together with their implementation details. Finally, details regarding the performance evaluation methods are provided in Section 3.5.

3.1. Bradykinesia Dataset

The dataset employed in this study was collected using the Monipar application [42]. Monipar proposes a system based on wearable technology and artificial intelligence (AI) for monitoring motor activity in PD. The system consists of a mobile app that guides the user in performing 8 exercises of the MDS-UPDRS scale and a wearable module that records the subject's movement using the triaxial accelerometer and gyroscope embedded in a consumer smartwatch. Specifically, tasks consisted of a series of 8 exercises belonging to the MDS-UPDRS scale part III, concerning the examination of the motor aspects [16]. The selected exercises include rest tremor amplitude, postural tremor of the hands, movement of the hands to the chest, finger tapping, hand movements, pronation–supination movements of the hands, arising from a chair, and gait. The duration of the entire procedure is approximately 7 min.

3.1.1. Data Acquisition

Data were recorded from 6 subjects (3 females and 3 males, 64.2 ± 8.2 years) diagnosed with PD in the early stages of the disease, according to the Hoehn and Yarn scale [43] (H&Y = 1 in all subjects) and from 7 healthy control subjects (4 females and 3 males, 64.0 ± 5.4 years). The data collection process was carried out for 8 and 9 weeks, respectively, using the Monipar application. Each week, subjects performed the pre-defined motor tasks in a controlled environment. A total of 105 weekly sessions were collected during the experimentation (46 sessions for PD; 59 sessions for healthy controls). These data correspond to more than 13 h of movement data collected with a triaxial accelerometer and a triaxial gyroscope. However, only relevant data related to the movement of the upper limbs, i.e., that recorded during finger tapping, hand movement, and pronation–supination movement of the hands, were analyzed in this study. The data from the three hand exercises correspond to 80 min (10% of the entire data set) of movement data collected by each of the inertial sensors.

The smartwatch employed for data collection was available on the market in 2019. This device employs Android Wear operating system and an internal memory of 4 GB (2 GB of free space). The device has a calibrated triaxial accelerometer with a maximum amplitude set to ±2 g, and triaxial gyroscope with a measurement range set to ±2000 dps.

The smartwatch was placed on the wrist of the most affected side, according to the clinical indication of the physician attending to the patient and the dominant hand of healthy controls. Data were recorded using the accelerometer and gyroscope embedded in the smartwatch, with a sampling frequency of 50 Hz. Such a frequency is appropriate for human motion analysis, as the frequency content generated by common human movements lies in the 0–20 Hz band [44]. Figure 1 summarizes the data collection process carried out using the Monipar app.

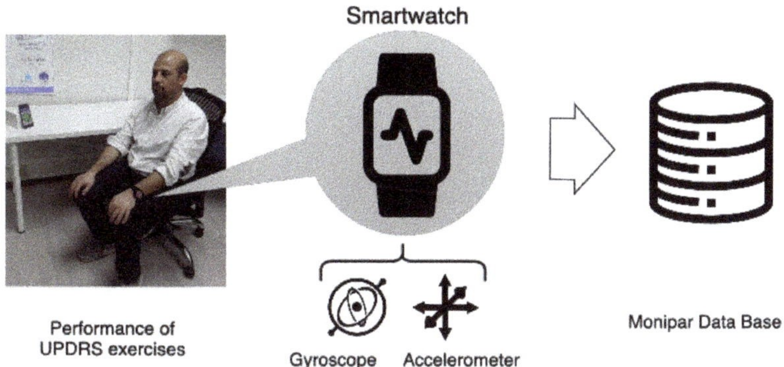

Figure 1. Data collection methodology to detect bradykinesia using smartwatches and MDS-UPDRS exercises.

3.1.2. Data Labeling

Training supervised ML and DL methods to automatically detect motor symptoms requires data to be labeled by expert clinicians, who recognize symptoms and evaluate their severity. The labeling of the Monipar data was performed by a trained expert neurologist, who reviewed the videos of the weekly trials performed by the subjects. For each motor task, the clinician identified the presence of bradykinesia and evaluated its severity. According to the MDS-UPDRS guidelines, a score between 0 (no bradykinesia) and 4 (severe bradykinesia) was assigned to each task. To assign a single severity value to the data of each weekly assessment, the sub-scores of the three upper limbs exercises were averaged and rounded, and finally used as the reference metric. The distribution of the severity of bradykinesia in the group of PD patients and control subjects is reported in Figure 2. It can be observed that the recorded bradykinesia severity corresponds to four UPDRS ratings, including normal (0), slight (1), mild (2), and moderate (3). As evident from Figure 2, the data distribution is unbalanced, with more than 57% of the data corresponding to the UPDRS 0 severity (no bradykinesia). Moreover, movements belonging to the class UPDRS 4 (severe bradykinesia) are not represented. This is likely due to the intrinsic composition of the PD sample, which encompasses patients in the early stages of the disease.

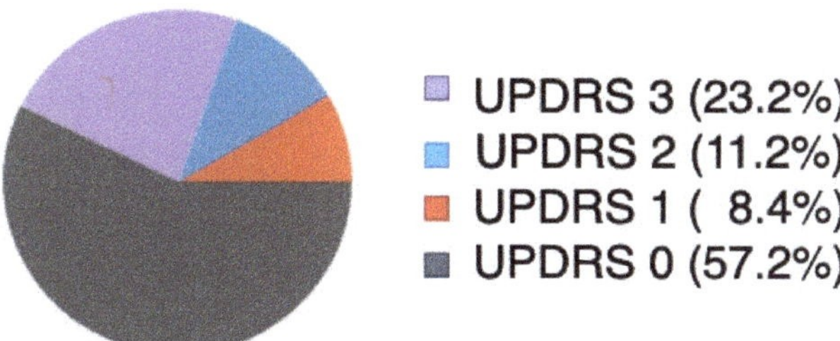

Figure 2. Distribution of the severity of bradykinesia in the dataset.

3.2. Signal Preprocessing

In order to prepare data for the subsequent classification step, some preprocessing procedures were performed. First, data were filtered and segmented (Section 3.2.1); then, different data transformation methods were applied (Section 3.2.2) to provide the input for ML and DL algorithms; finally, data augmentation was exploited to increase the data set size and provide a more balanced distribution of data (Section 3.2.3).

3.2.1. Filtering and Segmentation

The low-frequency components of the sensor readings are related to postural changes (gross movements), while the high-frequency components reflect the actual accelerations of the body segments, associated with rapid movements [45]. To remove the gravity effect and the noise produced by trembling or shaking, inertial data were filtered using a fourth-order zero-lag Butterworth band-pass infinite impulse response (IIR) digital filter, with cut-off frequencies of 0.25 Hz and 3.5 Hz. The advantage of the Butterworth-type filter is that it allows a nearly constant gain in the passband. Then, inertial signals were segmented using non-overlapping sliding windows of 5.12 s (i.e., 256 samples). Figure 3 shows a segment of the raw gyroscope signal (Figure 3a) and the corresponding filtered signal (Figure 3b).

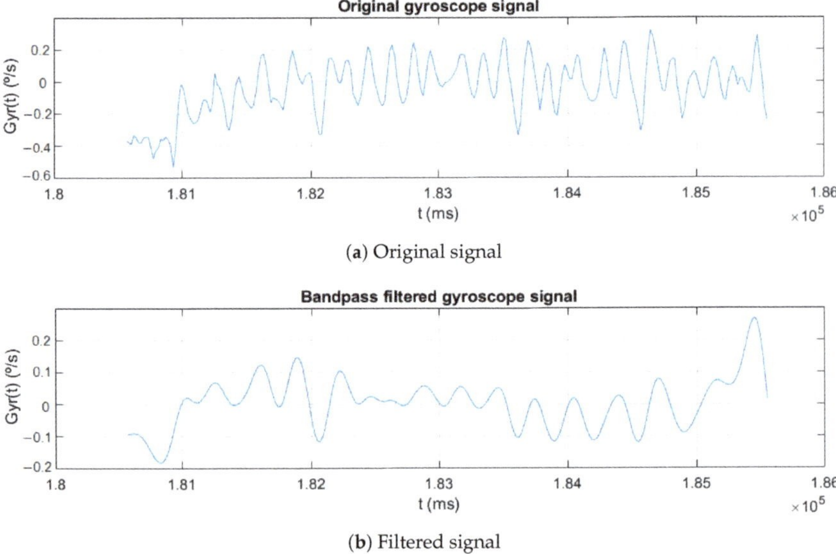

(a) Original signal

(b) Filtered signal

Figure 3. Filtering applied to the gyroscope signals. (a) sample of the original signal corresponding to exercise 4 (finger tapping); (b) gyroscope signal after applying a 0.25–3.5 Hz fourth-order Butterworth band pass filter.

3.2.2. Feature Extraction

Classic ML models such as RF require features to be extracted from recorded signals. Two feature sets proposed in the reference literature were reproduced in this study, belonging to both the time and frequency domains. This was carried out to establish a reference model for comparison with the proposed methods. The two sets of features [31,46] include a total number of 74 and 43 features, respectively.

As far as the input data for DL models are concerned, two different data representations were employed. The first consists of using the inertial readings, normalized in the range from -1 to 1. The second was created as follows. Every single window obtained from the segmentation process was divided into two consecutive windows of 2.56 s (i.e., 128 samples). Then, the signals' fast Fourier transform (FFT) was computed for both windows and used as an input feature set. This feature extraction method is based on contextual windows and will be referred to in the rest of the paper as Contextual FFT. The contextualization of adjacent FFT windows is based on methods proposed in the reference literature to improve the performance in FOG detection using accelerometers [47–49].

A summary of the feature set employed in this study is shown in Table 1.

Table 1. Summary of the data representations. FFT: fast Fourier transform.

Feature Set	Number of Features	Description of the Features
Shawen et al. [31]	74	- Time domain features (24) - Frequency domain features (24) - Features extracted from the derivatives of the signals (16) - Entropy (4) - Peak correlation between signals (3) - Cross-correlation delay (3)
Channa et al. [46]	43	- Time domain features (28) - Frequency domain features (12) - Peak correlation between signals (3)
Filtered signal (256-sample window)	768 (256 × 3)	Filtered signal obtained from the triaxial sensors (accelerometer or gyroscope).
Contextual FFT	384 (128 × 3)	Concatenated single-side FFT of two consecutive windows. A single window (256 samples) was divided into 2 windows of 128 samples before FFT computation.

3.2.3. Data Augmentation

The synthetic minority over-sampling technique (SMOTE) [50] was used to balance the data input to classic ML classifiers. Specifically, the classes with a minority number of sliding windows (i.e., UPDRS 1 = 768; UPDRS 2 = 1027; UPDRS 3 = 2132) were resampled to provide the same number of sliding windows as the majority class (UPDRS 0 = 5253). The number of nearest neighbors used to construct the synthetic samples was set to 5. This procedure produced an increase in the dataset size of 53%.

As far as the raw signals input to convolutional models are concerned, the application of signal permutation and magnitude warping [51] were employed to quadruple the amount of data. In the former case, the input data were sliced into four equal-length segments, and these segments were randomly permuted to create a new sliding window. As for the latter method, convolution between the input data and a smooth (randomly generated) curve was performed to change the magnitude of the samples of the sliding window.

All the described data augmentation techniques were applied only to the training subsets, while the testing subset remained unchanged. Figure 4 shows examples of the original data and the data augmentation techniques applied to the gyroscope signals. As shown in Figure 4b,e, portions of the signal were randomly permuted from the original signals (see Figure 4a,d), while, in Figure 4c,f, the amplitude of the original signals was modified by a randomly generated (smooth) curve.

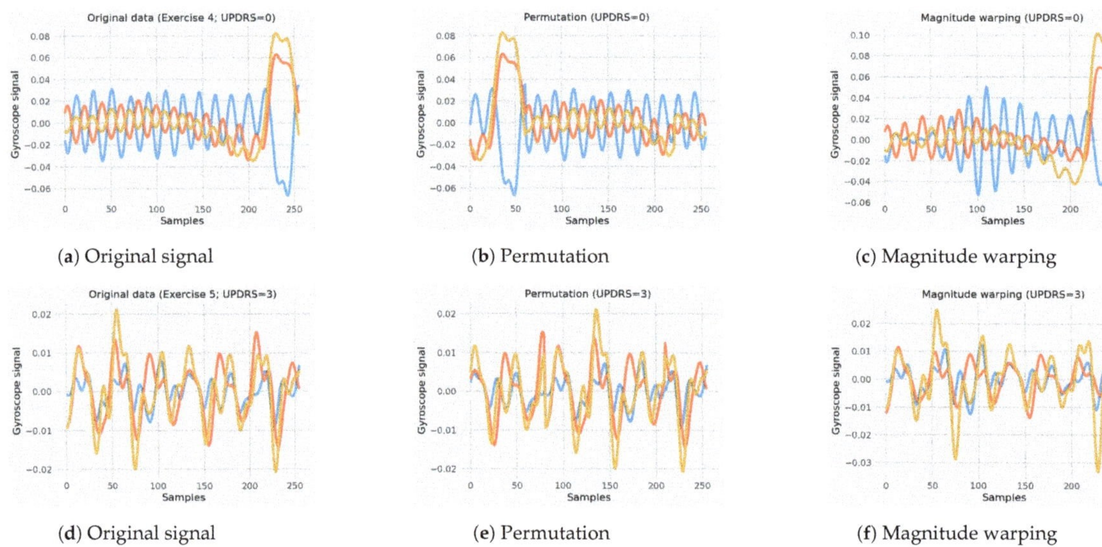

(**a**) Original signal   (**b**) Permutation   (**c**) Magnitude warping

(**d**) Original signal   (**e**) Permutation   (**f**) Magnitude warping

Figure 4. Data augmentation techniques applied in the gyroscope signals. (**a**) sample of the original signal corresponding to exercise 4 (finger tapping); (**b**) permutation of a sample signal of the exercise 4; (**c**) magnitude warping of a sample signal of exercise 4; (**d**) sample of the original signal corresponding to exercise 5 (hand movements); (**e**) permutation of a sample signal of the exercise 5; (**f**) magnitude warping of a sample signal of exercise 5.

3.3. Classification Algorithms

Different algorithms were implemented to predict the bradykinesia severity in PD patients and control subjects, resulting in a multi-class classification task. The output of the implemented models is a value between 0 and 3, according to the clinical bradykinesia score provided by the MDS-UPDRS scale.

For comparison proposes, two detection methods have been reproduced and evaluated to generate baseline metrics. The reproduced methods were the feature sets proposed in Shawen et al. [31] and Channa et al. [46]; these feature sets fed an RF classification model with 100 estimators, as proposed in [31]. Additional parameters of the RF classification algorithm were a minimum sample split equal to 2, a minimum sample leaf equal to 1, and the split criterion was Gini impurity [52].

The following DL algorithms were trained either using raw inertial signals or using the Contextual FFT data representation, as previously described in Section 3.2.2.

CNN. It consists of an input layer (256 features and 3 channels), connected to three one-dimensional convolutional layers of (1D-CNN), all three with 64 filters of size equal to 8 and rectified linear unit (ReLU) activation functions. Then, a global average pooling (GAP) layer was connected. For classification tasks, a multi-layer-perceptron (MLP) block composed of a densely connected layer with 260 units and ReLU activation was densely connected to a softmax layer with 4 units, corresponding to the number of output classes (i.e., bradykinesia severity score from 0 to 3).

Contextual CNN. The features extracted by the contextual windows method were evaluated. In this case, the architecture of the CNN used is composed of an input layer (256 features and 3 channels), connected to three 1D-CNN, the first one with 64 filters of size 8 and the next two with 20 filters of size 8, all of them with ReLU activation function. A GAP layer was then connected. For the classification tasks, an MLP block composed of a densely connected layer with 180 units and a ReLU activation function was connected to the classification layer, made of 4 units with a softmax activation function.

CNN (PI). As a novel approach, a CNN with patch input (PI) was proposed and evaluated. The patch extraction was implemented using a 1D-CNN. For this task, the kernel and stride parameters were set with the same value. In this way, the convolutional layers can act as an automatic patch extractor and bring equivalent results to patching extraction strategies such as those employed in Transformer-based models and isotropic computer-vision models [53,54], in which images are divided into non-overlapping square patches in raster-scan order. The proposed patching input strategy adapted to process multi-channel signals is shown in Figure 5.

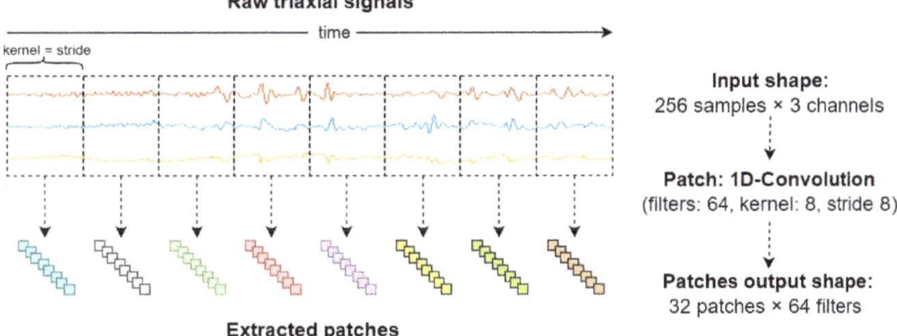

Figure 5. Patch input strategy with 1D-Convolution.

The training and evaluation of this latter model were performed using the filtered inertial signals. The architecture consists of an input layer implementing the PI strategy using 64 filters with kernel size and stride equal to 8 (in both cases). The input layer was connected to one 1D convolutional layer with 64 filters, a kernel size of 3, and a ReLU activation function. Then, a max pooling layer with a pool size of 2 and a subsequent GAP layer were connected. For the classification tasks, the MLP block included two densely connected layers, with 100 and 50 units, respectively, both with ReLU activation functions. Finally, these layers were connected to the final classification with 4 units and a softmax activation function. The architecture for the DNN with convolutional layers and PI is shown in Figure 6.

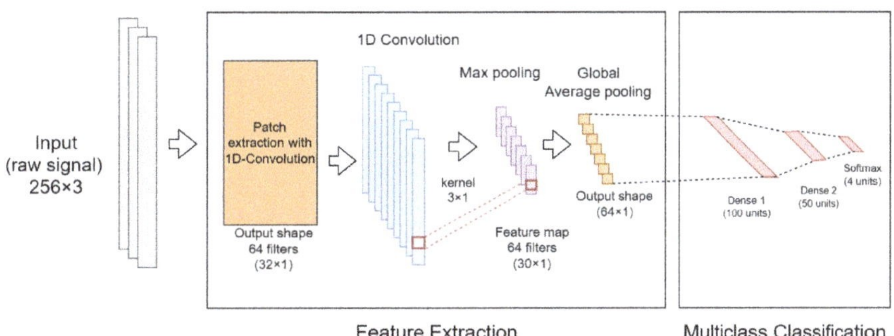

Figure 6. Proposed architecture for a CNN with patch input and MLP.

CNN (PI) + RF. The combination of CNN with PI and RF classification algorithm was evaluated. In this approach, the convolutional (with path input) block acts as a feature extractor, while the RF model (with 100 estimators) performs the classification tasks. While the CNN block allows the automatic extraction of features from the raw signal, the RF classifier takes advantage of a large number of individual decision trees that operate as an ensemble, providing good performance and high generalization capabilities.

In this classification algorithm, the parameters of the CNN block are similar to those used in the CNN architecture with path input (see Figure 6). The classification algorithm that combines convolutional layers with PI and RF is shown in Figure 7.

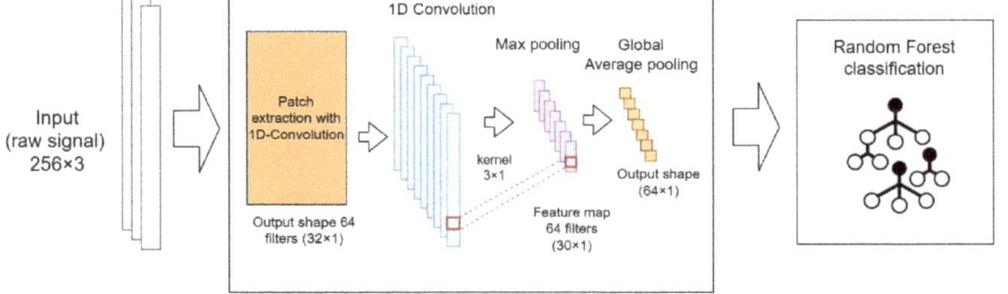

Figure 7. Proposed architecture for a CNN with patch input and Random Forest classification.

For the training of the DL algorithms, it was necessary to perform an initial hyperparameter tuning process. This adjustment was performed with the hyperband method [55], during which the learning rate, the number of filters, and the number of densely connected neurons were adjusted. A batch size of 64, a maximum number of epochs equal to 200, and a cross-entropy loss function were used in all cases to solve the multi-class classification problems. Moreover, an early stopping strategy was included, consisting of stopping the training process when the performance stops improving on the validation data set. The ADAM [56] backpropagation method was employed for optimizing the models' parameters. The learning rate was set to 2.3×10^{-3} for all DL architectures except for the Contextual CNN, where 5.9×10^{-3} was used.

3.4. Session-Based Analysis

For further evaluation, a session-based analysis was performed using the best approach for bradykinesia severity rating. Since the output of the classification algorithms is a specific class (UPDRS 0 to 3) for each single sliding window (hereinafter referred to as window-level detection), the aggregation of the predicted windows from a single (weekly) session was performed using statistical methods.

The aggregation of data from the three exercises was performed by calculating the 95th percentile value of the corresponding window-level predictions. This was accomplished to mimic the clinical assessment, which is based on the worst severity rating (i.e., maximum MDS-UPDRS rating) observed by the examiner during the assessment period. In addition, the 95th percentile was selected in agreement with the methodology proposed in [32] to derive bradykinesia digital biomarkers from hand movements using wrist-worn sensors. Finally, the predicted outcomes of a single session were compared with the reference metric (average of the MDS-UPDRS sub-scores) described in Section 3.1.2.

3.5. Evaluation Methodology

Stratified k-fold cross-validation (CV) with a k value equal to 5 (5-fold CV) was used to evaluate the performance of the algorithmic approaches. First, all the observations (sliding windows) of the data set were randomly shuffled; then, data were divided into 5 equal parts (folds) while preserving the percentage of samples for each class, as shown in Figure 8. At each interaction, ML and DL models were trained using four folds and tested on the final fold. The procedure was repeated 5 times, corresponding to the number of folds. Sliding windows of 256 samples with no overlap were used to avoid training and evaluation subsets sharing signal segments when using the 5-fold CV methodology. This validation approach was used to overcome the limited amount of bradykinesia data for each patient. Additionally, the performance metrics used to evaluate the bradykinesia detection models included accuracy, precision, recall, F1-score, area under the curve (AUC), Pearson r and root-mean-square error (RMSE) [57].

	Data set			
	Fold 1	Fold 2	Fold 3	Fold 4	Fold 5
Iteration 1	**Fold 1**	Fold 2	Fold 3	Fold 4	Fold 5
Iteration 2	Fold 1	**Fold 2**	Fold 3	Fold 4	Fold 5
Iteration 3	Fold 1	Fold 2	**Fold 3**	Fold 4	Fold 5
Iteration 4	Fold 1	Fold 2	Fold 3	**Fold 4**	Fold 5
Iteration 5	Fold 1	Fold 2	Fold 3	Fold 4	**Fold 5**

Figure 8. K-fold validation (k = 5) employed to evaluate the performance of the classification algorithms.

4. Experiments and Results

This section reports the results obtained in the present study. Several experiments were performed to evaluate the proposed approaches, to identify the combination of sensors (or combination of sensors), signal processing, input type, and DL algorithms that provide the best performance.

Section 4.1 reports the results of the RF classification model, fed with either accelerometer or gyroscope recordings, and using their combination. The experiments were performed using the two different feature sets proposed in the literature [31,46]. The results of the different DL approaches are reported in Section 4.2, evaluating the effect of different input types (raw data or FFT), data augmentation [50,51], and DL architectures. Finally, the results of the session-based analysis are reported in Section 4.3.

4.1. Baseline

Table 2 reports the performance of the RF classification model. The effect of different feature sets, sensors, and sensor combinations are evaluated. First, the feature set proposed by Channa et al. [46] provided better results than that used by [31], despite the smaller number of features extracted. This is reflected in all performance metrics for all types of sensor data. Specifically, the best performance in bradykinesia detection was obtained using features [46] extracted from the gyroscope recordings, achieving an AUC value up to 0.909 and a corresponding accuracy of 0.783. Moreover, from Table 2, it can be observed that the combination of accelerometer and gyroscope does not provide better performance than that obtained by using only the gyroscope data. This suggests that is possible to implement robust bradykinesia detection systems using a single inertial sensor.

In this study, the reproduction of Shawen et al. features [31] in conjunction with an RF algorithm achieved better performance than that reported by the authors (0.65 AUC) by using the gyroscope data. This behavior is expected because, in the work of Shawen et al. [31], the data employed corresponds to a set of activities of daily living (ADLs) in addition to the clinical assessment tasks (i.e., finger-to-nose).

Moreover, according to Table 2, the reproduction of both approaches [31,46] using the gyroscope data presents competitive results. These results are in line with the ones reported in similar studies, where an AUC of 0.926 [39] and accuracy up to 0.909 [38] were achieved.

Based on these results, the subsequent experiments were performed using only the gyroscope data. In addition, the feature set proposed by Channa et al. [46] in conjunction with an RF classifier with 100 estimators was selected as a baseline.

Table 2. Baseline methods. Accel: Accelerometer; Gyro: Gyroscope.

Feature Set	Sensor Data	Accuracy	Precision	Recall	F1-Score	AUC
Shawen et al. [31]	Accel	0.731	0.721	0.731	0.692	0.889
	Gyro	0.762	0.753	0.762	0.727	0.907
	Accel + Gyro	0.720	0.705	0.720	0.673	0.886
Channa et al. [46]	Accel	0.749	0.745	0.749	0.716	0.905
	Gyro	0.783	0.773	0.783	0.755	0.909
	Accel + Gyro	0.749	0.746	0.749	0.718	0.893

4.2. Classification Methods

In this section, the performance of six algorithmic approaches is reported and compared—specifically, the baseline model identified in the first experiment; the baseline model fed with data augmented using the SMOTE algorithm; the CNN model trained with the filtered inertial signals; the CNN model trained with the contextual FFT windows; the CNN model with the proposed PI and data augmentation; and, finally, the convolutional model (with PI) combined with the RF classification model.

In more detail, for the first approach, the performance metrics of the baseline model (hand-crafted features with an RF classification algorithm) were reported without additional processing. Second, to improve the predictive power of the first method, the SMOTE technique was employed to increment the data used to train the algorithm. Third, for comparison purposes, the performance of a (standard) CNN trained with a raw signal was evaluated. This approach was selected to take advantage of the capability of the CNNs to handle raw signals. Fourth, in an attempt to improve the CNN's results, the performance of a similar (three-layer) CNN was evaluated in conjunction with features extracted by the contextual windows method. Fifth, the CNN with PI and MLP (see Figure 6) was evaluated in conjunction with the data augmentation techniques (permutation and magnitude warping). This method was proposed as an end-to-end solution for the detection of bradykinesia which does not require feature extraction. In addition, sixth, to improve the predictive power of the CNN with PI, the classification block at the top of the network was changed to an RF classification algorithm (see Figure 7). RF classifiers can provide good performance and high generalization capabilities even when handling unbalanced data [58], in this case, by using the features extracted by the CNN block. This latter approach was trained similarly to the fifth approach to facilitate the comparison of results.

The performance of these models is reported in Table 3. The best results for each metric are bold in Table 3.

Table 3. Results of different bradykinesia detection methods using data collected for a single triaxial gyroscope. SMOTE: synthetic minority over-sampling technique; RF: random forest; CNN: convolutional neural network; FFT: fast Fourier transform; MLP: multi-layer perceptron; PI: patch input; DA: data augmentation.

Method	Data Representation	Classifier	Accuracy	Precision	Recall	F1-Score	AUC
Baseline	Channa et al. [46]	RF (100)	0.783	0.773	0.783	0.755	0.909
Baseline + SMOTE	Channa et al. [46]	RF (100)	0.792	0.702	0.667	0.680	0.912
CNN	Raw signal	CNN + MLP	0.675	0.661	0.675	0.618	0.687
Contextual CNN [1]	FFT contextual	CNN + MLP	0.629	0.561	0.629	0.563	0.706
CNN(PI) + DA [2]	Raw signal	CNN (PI) + MLP	0.826	0.733	**0.751**	0.738	**0.943**
CNN(PI) + RF + DA [2]	Raw signal	CNN (PI) + RF (100)	**0.835**	**0.750**	0.748	**0.746**	0.939

[1] Proposed method which employs contextualization of adjacent windows (Contextual FFT) in the input data.
[2] Proposed method which implements the patch input strategy in the CNN feature extraction block. The best results for each evaluation metric are bold.

According to the results reported in Table 3, the best accuracy (0.835) in bradykinesia detection was obtained employing the combination of CNN with path input and RF classification algorithm, while the best performance in terms of AUC (0.939) was achieved using the approach consisting of a CNN with path input and an MLP block for classification.

Data augmentation methods led to an improvement in performance for all the classification tasks. As for the baseline model, the effect of data augmentation was a slight increase in accuracy and AUC, and a decrease in recall and precision, which is reflected in the F1-score. When data augmentation techniques (SMOTE, signal permutation, and magnitude warping) were applied to DL algorithms, a significant performance improvement was observed. Specifically, accuracy, F1-score, and AUC improved by 15%, 12%, and 25%, respectively. The use of contextual FFT windows did not provide incremental performance regarding the use of raw inertial signals, probably due to the limited amount of data used during training. Finally, using CNNs combined with the RF model, instead of the classic MLP approach, led to an increase of 0.9% in accuracy and 0.8% in Fscore, while AUC decreased from 0.943 to 0.939. However, such small differences can not be considered significant.

On the one hand, the comparison of the results with most of the related literature work is difficult because of the diversity of approaches and validation methodologies. However, when comparing the performance of the best classification methods (CNN with PI) with the best results reported in similar studies (i.e., 0.926 [39]), slightly superior results in terms of AUC (0.939) were achieved. However, in terms of accuracy, higher performance than that achieved in this work (83.5%) was reported (i.e., 90.9% [38]). On the other hand, a direct comparison of the best-proposed method with the (reproduced) baseline shows a significant increment in accuracy (5.2%) and AUC (3%). This presents competitive results for bradykinesia severity rating using a single gyroscope sensor and opens opportunities for the development of unobtrusive solutions for monitoring based on consumer devices.

4.3. Results of the Session-Based Analysis

The results of the session-based analysis were obtained using the best-proposed method (CNN with PI and RF classification). For this task, the window-level predictions of each clinical visit were processed according to the methodology described in Section 3.4. After this process, the session-based results were compared with the reference evaluation obtained from MDS-UPDRS sub-scores during the clinical assessment (see Section 3.1.2).

In addition, to compare the results of the session-based analysis with the window-level evaluation, regression metrics were calculated using the proposed methods. The results of both assessment methods are shown in Table 4.

Table 4. Results of different assessment methods for bradykinesia detection using the proposed CNN with patch input and Random Forest classification. RMSE: root mean square error.

Assessment Method	Accuracy	Precision	Recall	F1-Score	r	RMSE
Window-level	0.835	0.750	0.748	0.746	0.82	0.77
Session-based	0.857	0.733	0.728	0.716	0.94	0.46

According to Table 4, the aggregation of the window-level predictions presents a slight increase in the accuracy (0.857) over the results of window-level detection (accuracy 0.835). However, a decrease in the Precision and Recall is identified in the session-based assessment. On the other hand, an increase in the Pearson r (0.945, $p < 0.001$) and a reduction in the RMSE (0.455) were achieved using the session-based methodology. The results of the session-based analysis indicate that it is feasible to derive a single indicator of the bradykinesia severity using the data aggregated from the three selected MDS-UPDRS exercises. In addition, this indicator shows a high correlation with the clinical assessment of a single clinical visit.

4.4. Summary of the Findings Observed in the Experiments

Overall, the present findings can be summarized as follows.

- The use of the gyroscope sensors presents better results in bradykinesia detection than the use of only the accelerometer data or the combination of accelerometer and gyroscope data.
- DL approaches provided better results than classic ML classifiers fed with extracted features. In addition, the use of the proposed method (CNN with PI and RF classification) presents competitive results for bradykinesia severity detection at the window level.
- Data augmentation methods have a positive effect on classification performance and its effect is more pronounced in the DL-based approaches.
- The use of data transformation methods such as FFT does not provide better results, compared to raw inertial signals.
- Using DL approaches with an RF model as the final classification layer does not significantly improve performance.
- The aggregation of the window-level predictions from a single clinical visit presents a slight increase in the accuracy and increases the correlation between the automatic and the clinical evaluation.

5. Discussion and Conclusions

Bradykinesia is a cardinal symptom for the evaluation of PD. The Objective quantification of this symptom is relevant for diagnosis, treatment adjustment, and a better understanding of disease progression. For this reason, research efforts have been devoted to developing automated systems that seek to diagnose and monitor bradykinesia. However, several limitations and challenges remain to consider.

There is a great heterogeneity of solutions, both in the number and type of sensors and in the methodologies proposed for data analysis. Therefore, the potential use of such technologies in the current clinical practice is limited due to the lack of validation and standardization.

The potential of complementing traditional assessment in medical centers and also extending the diagnosis and monitoring to a home environment suggest that the management of PD could be revolutionized by new wearable systems. However, challenges in scaling up solutions of this type remain, mainly due to the quality and quantity of data recorded to identify PD motor symptoms. This latter varies widely between individuals, and activities, and also evolves over time [41].

The results obtained in this study suggest that it is possible to use commercial smartwatches combined with AI techniques for the detection and evaluation of bradykinesia severity. Moreover, the best performances were obtained by using data recorded by a single tri-axial gyroscope, while combining acceleration and angular velocity data did not provide further improvements. The use of a single sensor embedded in the smartwatch would be beneficial in reducing the computational burden and increasing the battery life, thus enabling continuous monitoring.

The comparison between DL methods and classic ML-based classification approaches revealed the weak performances of the former, due to the limited amount of data. However, the simultaneous application of data augmentation techniques and novel algorithmic approaches led to significantly better performances (AUC 0.939; Accuracy 0.835) than those provided by shallow ML algorithms. Specifically, the use of DL architectures employing CNN patch extraction strategies seems to be a feasible approach to contextualize the raw data from a single sliding window. By employing such a technique, a DNN is capable of extracting specific information from small non-overlapping patches to feed discriminative architectures automatically. In addition, the proposed approach brings opportunities for applications in different tasks involving sequential data such as raw inertial signals. Moreover, the potential of combining this approach with recurrent architectures or transformer-based architectures may allow the development of end-to-end architectures capable of extracting and modeling automatically temporal dependencies. In this line, future studies could evaluate the performance of this approach in tasks where temporal dependencies are relevant, for example in the automatic detection of freezing of gait.

Table 5 reports a comparison of the methods and results of the proposed approach (at window-level) and those reported in the related literature. As can be observed, the classification performance provided by the proposed DL algorithm outperformed the state-of-the-art methods. Specifically, accuracy of 0.84 and AUC of 0.94 were higher than the best results from the related works (accuracy 0.77 and AUC 0.92 [17]). It is worth noticing that an accuracy of 0.91 was obtained in [38]. However, only a binary classification task (presence or absence of bradykinesia) was set in this case, compared to the multi-class classification problem (bradykinesia severity) used in this study.

Table 5. Comparison of different bradykinesia detection methods and results. ML: machine learning; AUC: area under the curve; RMSE: root mean square error; IMU: inertial measurement unit; ADLs: activities of daily living; SVM: support vector machine; kNN: k-nearest neighbors; PCA: principal component analysis; MLP: multi-layer perceptron; RF: random forest; CNN: convolutional neural network; DA: data augmentation; LR: linear regression.

Study	Sensor (Location)	Task	ML Model (Input)	Accuracy	AUC	r	RMSE
[17]	Smartphone (thigh)	Leg agility	MLP (features)	0.77	0.92	0.92	0.42
[31]	Smartwatch, IMU (wrist, hand)	gait, upper limbs exercises	RF (features)	-	0.65	-	-
[33]	IMU (thighs)	Leg agility	kNN (features + PCA)	0.430	-	0.640	-
[35]	IMU (ankles)	Leg agility	SVM (features)	-	-	0.83	0.53
[38]	IMU (forearm)	upper limbs exercises	CNN (raw data)	0.91 (binary)	-	-	-
[39]	IMU (wrist, fingers)	upper limbs exercises	LR (features)	-	0.93	0.85	-
[40]	Accelerometer (wrist)	ADLs	CNN (raw data + DA)	-	-	0.83	-
[41]	IMU (hand)	gait, upper limbs exercises	RF (features)	-	0.73	-	-
Proposed	Smartwatch (wrist)	upper limbs exercises	CNN + RF (raw data)	0.86	0.94	0.94	0.46

As far as the regression results are concerned, the obtained correlation coefficient was found to be larger than that of related literature studies, while slightly smaller errors were obtained in [17]. However, the leg agility task was addressed in this latter study, using sensors on lower limbs. This represents a very specific exercise of the MDS-UPDRS, suitable for in-laboratory examinations but rather uncommon in daily living settings. Moreover, only a single sensor was employed in this study, representing a less invasive solution than those proposed in [31,33,35,39], suitable for passive long-term monitoring of PD patients in home settings. Finally, it is worth noticing that, unlike most related studies, a comprehensive performance evaluation was carried out in this work, providing both classification and regression metrics.

The present work has some limitations. The enrolled subjects' cohort is larger than in [38] and comparable to [31], but it is smaller than in [17,33,35,39–41]. Data augmentation methods were used in this study to increase the data set size and provide more robust results. However, different patients may have different movement patterns, thus a larger population should be investigated to further validate the present findings. Moreover, data were collected during semi-supervised tasks, as was also carried out in most similar literature works [17,31,33,35,38–41]. In order to extend the use of the proposed method in non-supervised settings, a context algorithm for gesture recognition should be developed. Then, the activities recognized by such a model can be analyzed using the computer methods implemented in this work. Alternatively, a new unsupervised data collection procedure may be carried out, as achieved in [40], where the presence or absence of bradykinesia was estimated during unconstrained ADLs.

The proposed solution, further improved and validated on a larger cohort of PD patients, may be used to complement traditional outpatient visits. Specifically, data collected during the sporadic clinical examination can be employed for further training the proposed automatic scoring system. Afterward, the wearable solution can be used in the home setting to passively collect information regarding bradykinesia presence and severity. Finally, the information will be able to be assessed by clinicians for evaluating the evolution of the symptom over time and its fluctuations throughout the day, eventually planning proper therapy adjustments.

The present study is intended to show the potential of consumer wearable technology and DL approaches to detect the severity of bradykinesia by using data recorded during standardized MDS-UPDRS upper limbs' motor tasks. Moreover, different ML and DL methodologies are proposed and compared, further discussing the effect of data augmentation, input type, and architectures.

Future studies will be in the direction of increasing the data set size, by enrolling a larger patient cohort. Then, the development of an automatic scoring system working in non-supervised conditions [40] would pave the way to continuous, long-term, unobtrusive monitoring of PD in home environments.

Author Contributions: L.S.: Conceptualization, Software, Methodology, Validation, Formal analysis, Writing—review and editing. B.D.: Data Curation, Software, Investigation, Methodology, Formal analysis, Writing—Original Draft. L.B.: Conceptualization, Formal analysis, Validation, Writing—review and editing. N.C.: Funding acquisition, Supervision, Writing—review and editing. S.C.: Validation, Formal analysis, Writing—review and editing. P.A.: Resources, Supervision, Writing—review and editing. J.M.L.: Resources, Supervision, Writing—review and editing. G.D.A.: Funding acquisition, Project administration, Writing—review and editing. I.P.: Formal analysis, Project administration, Writing—review and editing. All authors have read and agreed to the published version of the manuscript.

Funding: Part of this research was funded by the project "Tecnologías Capacitadoras para la Asistencia, Seguimiento y Rehabilitación de Pacientes con Enfermedad de Parkinson". Centro Internacional sobre el envejecimiento, CENIE (código 0348_CIE_6_E) Interreg V-A España-Portugal (POCTEP); and (2) FCT—Fundação para a Ciência e Tecnologia within the R&D Units Project Scope: UIDB/00319/2020.

Institutional Review Board Statement: The study was conducted according to the guidelines of the Declaration of Helsinki. In addition, this study was approved by the Institutional Review Board (Ethics Committee) of the Universidad Politécnica de Madrid (date of approval: 18 June 2018) and the Ethics Committee of the University of Minho with the document identification CE.CSH 031/2018 (date of approval: 11 December 2018).

Informed Consent Statement: Informed consent was obtained from all subjects involved in the study.

Data Availability Statement: The data presented in this study are available on request from the corresponding author. The data are not publicly available because they contain protected patient health information.

Acknowledgments: This work has been supported by: (1) Grupo de Investigación en Instrumentación y Acústica Aplicada (I2A2). ETSI Industriales. Universidad Politécnica de Madrid; and (2) ALGORITMI Research Centre, University of Minho (Portugal).

Conflicts of Interest: The authors declare no conflict of interest.

Abbreviations

The following abbreviations are used in this manuscript:

AI	Artificial intelligence
ADAM	Adaptive moment estimation
ADLs	Activities of daily living
AUC	Area under the curve
CNN	Convolutional neural network
CV	Cross-validation
DA	Data augmentation
DL	Deep learning
DNN	Deep neural network
FFT	Fast Fourier Transform
GAP	Global average pooling
kNN	k-nearest neighbors
IMU	Inertial measurement unit
LR	Linear regression
ML	Machine learning
MLP	Multi-layer perceptron
PCA	Principal component analysis
PD	Parkinson's disease
QoL	Quality of life
ReLU	Rectified linear unit
RF	Random forest
ROC	Receiver operating characteristic
RMSE	Root mean square error
SMOTE	Synthetic minority over-sampling technique

References

1. Pringsheim, T.; Jette, N.; Frolkis, A.; Steeves, T.D. The prevalence of Parkinson's disease: A systematic review and meta-analysis. *Mov. Disord.* **2014**, *29*, 1583–1590. [CrossRef] [PubMed]
2. Jankovic, J. Parkinson's disease: Clinical features and diagnosis. *J. Neurol. Neurosurg. Psychiatry* **2008**, *79*, 368–376. [CrossRef] [PubMed]
3. Kalia, L.; Lang, A. Parkinson's disease. *Lancet* **2015**, *386*, 896–912. [CrossRef]
4. Hou, J.; Lai, E. Non-motor Symptoms of Parkinson's Disease. *Int. J. Gerontol.* **2007**, *1*, 53–64. [CrossRef]
5. Williams, D.; Watt, H.; Lees, A. Predictors of falls and fractures in bradykinetic rigid syndromes: A retrospective study. *J. Neurol. Neurosurg. Psychiatry* **2006**, *77*, 468–473. [CrossRef] [PubMed]
6. Bronte-Stewart, H. Postural instability in idiopathic Parkinson's disease: The role of medication and unilateral pallidotomy. *Brain* **2002**, *125*, 2100–2114. [CrossRef] [PubMed]
7. Patel, S.; Lorincz, K.; Hughes, R.; Huggins, N.; Growdon, J.; Standaert, D.; Akay, M.; Dy, J.; Welsh, M.; Bonato, P. Monitoring motor fluctuations in patients with Parkinson's disease using wearable sensors. *IEEE Trans. Inf. Technol. Biomed.* **2009**, *13*, 864–873. [CrossRef] [PubMed]
8. Davie, C. A review of Parkinson's disease. *Br. Med. Bull.* **2008**, *86*, 109–127. [CrossRef]
9. Fahn, S. Parkinson disease, the effect of levodopa, and the ELLDOPA trial. *Arch. Neurol.* **1999**, *56*, 529–535. [CrossRef]
10. Olanow, C.W.; Stern, M.B.; Sethi, K. The scientific and clinical basis for the treatment of Parkinson's disease. *Neurology* **2009**, *72*, S1–S136. [CrossRef] [PubMed]
11. Parkinson Study Group. Evaluation of dyskinesias in a pilot, randomized, placebo-controlled trial of remacemide in advanced Parkinson disease. *Arch. Neurol.* **2001**, *58*, 1660–1668. [CrossRef] [PubMed]
12. Hughes, A.J.; Daniel, S.E.; Blankson, S.; Lees, A.J. A clinicopathologic study of 100 cases of Parkinson's disease. *Arch. Neurol.* **1993**, *50*, 140–148. [CrossRef]
13. Berardelli, A. Pathophysiology of bradykinesia in Parkinson's disease. *Brain* **2001**, *124*, 2131–2146. [CrossRef] [PubMed]
14. Vingerhoets, F.; Schulzer, M.; Calne, D.; Snow, B. Which clinical sign of Parkinson's disease best reflects the nigrostriatal lesion? *Ann. Neurol.* **1997**, *41*, 58–64. [CrossRef] [PubMed]
15. Monje, M.H.G.; Foffani, G.; Obeso, J.; Sànchez-Ferro, Á. New Sensor and Wearable Technologies to Aid in the Diagnosis and Treatment Monitoring of Parkinson's Disease. *Annu. Rev. Biomed. Eng.* **2019**, *21*, 111–143. [CrossRef] [PubMed]
16. Goetz, C.G.; Tilley, B.C.; Shaftman, S.R.; Stebbins, G.T.; Fahn, S.; Martinez-Martin, P.; Poewe, W.; Sampaio, C.; Stern, M.B.; Dodel, R.; et al. Movement Disorder Society-sponsored revision of the Unified Parkinson's Disease Rating Scale (MDS-UPDRS): Scale presentation and clinimetric testing results. *Mov. Disord. Off. J. Mov. Disord. Soc.* **2008**, *23*, 2129–2170. [CrossRef]

17. Borzì, L.; Varrecchia, M.; Sibille, S.; Olmo, G.; Artusi, C.A.; Fabbri, M.; Rizzone, M.G.; Romagnolo, A.; Zibetti, M.; Lopiano, L. Smartphone-Based Estimation of Item 3.8 of the MDS-UPDRS-III for Assessing Leg Agility in People with Parkinson's Disease. *IEEE Open J. Eng. Med. Biol.* **2020**, *1*, 140–147. [CrossRef]
18. Espay, A.J.; Hausdorff, J.M.; Sánchez-Ferro, Á.; Klucken, J.; Merola, A.; Bonato, P.; Paul, S.S.; Horak, F.B.; Vizcarra, J.A.; Mestre, T.A.; et al. A Roadmap for Implementation of Patient-Centered Digital Outcome Measures in Parkinson's disease Obtained Using Mobile Health Technologies. *Mov. Disord.* **2019**, *34*, 657–663. [CrossRef]
19. Luis-Martínez, R.; Monje, M.H.; Antonini, A.; Sánchez-Ferro, Á.; Mestre, T.A. Technology-enabled care: Integrating multidisciplinary care in Parkinson's disease through digital technology. *Front. Neurol.* **2020**, *11*, 575975. [CrossRef]
20. Virginia Anikwe, C.; Friday Nweke, H.; Chukwu Ikegwu, A.; Adolphus Egwuonwu, C.; Uchenna Onu, F.; Rita Alo, U.; Wah Teh, Y. Mobile and wearable sensors for data-driven health monitoring system: State-of-the-art and future prospect. *Expert Syst. Appl.* **2022**, *202*, 117362. [CrossRef]
21. Tauţan, A.M.; Ionescu, B.; Santarnecchi, E. Artificial intelligence in neurodegenerative diseases: A review of available tools with a focus on machine learning techniques. *Artif. Intell. Med.* **2021**, *117*, 102081. [CrossRef] [PubMed]
22. Rovini, E.; Maremmani, C.; Cavallo, F. How Wearable Sensors Can Support Parkinson's Disease Diagnosis and Treatment: A Systematic Review. *Front Neurosci.* **2017**, *11*, 555. [CrossRef] [PubMed]
23. Hubble, R.; Naughton, G.; Silburn, P.; Cole, M.H. Wearable Sensor Use for Assessing Standing Balance and Walking Stability in People with Parkinson's Disease: A Systematic Review. *PLoS ONE* **2015**, *10*, e0123705. [CrossRef]
24. Borzì, L.; Olmo, G.; Artusi, C.A.; Fabbri, M.; Rizzone, M.G.; Romagnolo, A.; Zibetti, M.; Lopiano, L. A new index to assess turning quality and postural stability in patients with Parkinson's disease. *Biomed. Signal Process. Control* **2020**, *62*, 102059. [CrossRef]
25. Mei, J.; Desrosiers, C.; Frasnelli, J. Machine Learning for the Diagnosis of Parkinson's Disease: A Review of Literature. *Front. Aging Neurosci.* **2021**, *13*, 633752. [CrossRef]
26. Cubo, E.; Mir Rivera, P.; Sánchez Ferro, Á. *Manual SEN de Nuevas Tecnologías en Trastornos del Movimiento*; Ediciones SEN: Madrid, Spain, 2021.
27. Heldman, D.A.; Urrea-Mendoza, E.; Lovera, L.C.; Schmerler, D.A.; Garcia, X.; Mohammad, M.E.; McFarlane, M.C.U.; Giuffrida, J.P.; Espay, A.J.; Fernandez, H.H. App-based bradykinesia tasks for clinic and home assessment in Parkinson's disease: Reliability and responsiveness. *J. Park. Dis.* **2017**, *7*, 741–747. [CrossRef] [PubMed]
28. Griffiths, R.I.; Kotschet, K.; Arfon, S.; Xu, Z.M.; Johnson, W.; Drago, J.; Evans, A.; Kempster, P.; Raghav, S.; Horne, M.K. Automated assessment of bradykinesia and dyskinesia in Parkinson's disease. *J. Park. Dis.* **2012**, *2*, 47–55. [CrossRef]
29. Dunnewold, R.J.; Hoff, J.I.; van Pelt, H.C.; Fredrikze, P.Q.; Wagemans, E.A.; van Hilten, B.J. Ambulatory quantitative assessment of body position, bradykinesia, and hypokinesia in Parkinson's disease. *J. Clin. Neurophysiol.* **1998**, *15*, 235–242. [CrossRef] [PubMed]
30. Salarian, A. *Ambulatory Monitoring of Motor Functions in Patients with Parkinson's Disease Using Kinematic Sensors*; Technical Report; EPFL: Lausanne, Switzerland, 2006.
31. Shawen, N.; O'Brien, M.K.; Venkatesan, S.; Lonini, L.; Simuni, T.; Hamilton, J.L.; Ghaffari, R.; Rogers, J.A.; Jayaraman, A. Role of data measurement characteristics in the accurate detection of Parkinson's disease symptoms using wearable sensors. *J. Neuroeng. Rehabil.* **2020**, *17*, 52. [CrossRef] [PubMed]
32. Mahadevan, N.; Demanuele, C.; Zhang, H.; Volfson, D.; Ho, B.; Erb, M.K.; Patel, S. Development of digital biomarkers for resting tremor and bradykinesia using a wrist-worn wearable device. *NPJ Digit. Med.* **2020**, *3*, 5. [CrossRef] [PubMed]
33. Parisi, F.; Ferrari, G.; Giuberti, M.; Contin, L.; Cimolin, V.; Azzaro, C.; Albani, G.; Mauro, A. Body-Sensor-Network-Based Kinematic Characterization and Comparative Outlook of UPDRS Scoring in Leg Agility, Sit-to-Stand, and Gait Tasks in Parkinson's Disease. *IEEE J. Biomed. Health Inform.* **2015**, *19*, 1777–1793. [CrossRef]
34. Giuberti, M.; Ferrari, G.; Contin, L.; Cimolin, V.; Azzaro, C.; Albani, G.; Mauro, A. Automatic UPDRS Evaluation in the Sit-to-Stand Task of Parkinsonians: Kinematic Analysis and Comparative Outlook on the Leg Agility Task. *IEEE J. Biomed. Health Inform.* **2015**, *19*, 803–814. [CrossRef] [PubMed]
35. Aghanavesi, S.; Bergquist, F.; Nyholm, D.; Senek, M.; Memedi, M. Motion sensor-based assessment of Parkinson's disease motor symptoms during leg agility tests: Results from levodopa challenge. *IEEE J. Biomed. Health Inform.* **2019**, *24*, 111–119. [CrossRef] [PubMed]
36. Bengio, Y. Deep learning of representations: Looking forward. In *Statistical Language and Speech Processing, Proceedings of the First International Conference, SLSP 2013, Tarragona, Spain, 29–31 July 2013*; Springer: Berlin/Heidelberg, Germany, 2013; pp. 1–37.
37. Glorot, X.; Bengio, Y. Understanding the difficulty of training deep feedforward neural networks. In Proceedings of the Thirteenth International Conference on Artificial Intelligence and Statistics, Sardinia, Italy, 13–15 May 2010; Teh, Y.W., Titterington, M., Eds.; PMLR: Sardinia, Italy, 2010; Volume 9, pp. 249–256.
38. Eskofier, B.M.; Lee, S.I.; Daneault, J.F.; Golabchi, F.N.; Ferreira-Carvalho, G.; Vergara-Diaz, G.; Sapienza, S.; Costante, G.; Klucken, J.; Kautz, T.; et al. Recent machine learning advancements in sensor-based mobility analysis: Deep learning for Parkinson's disease assessment. In Proceedings of the 2016 38th Annual International Conference of the IEEE Engineering in Medicine and Biology Society (EMBC), Orlando, FL, USA, 16–20 August 2016; pp. 655–658.
39. Park, D.J.; Lee, J.W.; Lee, M.J.; Ahn, S.J.; Kim, J.; Kim, G.L.; Ra, Y.J.; Cho, Y.N.; Jeong, W.B. Evaluation for Parkinsonian Bradykinesia by deep learning modeling of kinematic parameters. *J. Neural Transm.* **2021**, *128*, 181–189. [CrossRef]

40. Pfister, F.M.; Um, T.T.; Pichler, D.C.; Goschenhofer, J.; Abedinpour, K.; Lang, M.; Endo, S.; Ceballos-Baumann, A.O.; Hirche, S.; Bischl, B.; et al. High-Resolution Motor State Detection in Parkinson's Disease Using Convolutional Neural Networks. *Sci. Rep.* **2020**, *10*, 5860. [CrossRef]

41. Lonini, L.; Dai, A.; Shawen, N.; Simuni, T.; Poon, C.; Shimanovich, L.; Daeschler, M.; Ghaffari, R.; Rogers, J.A.; Jayaraman, A. Wearable sensors for Parkinson's disease: Which data are worth collecting for training symptom detection models. *Npj Digit. Med.* **2018**, *1*, 64. [CrossRef]

42. Sigcha, L.; Pavón, I.; Costa, N.; Costa, S.; Gago, M.; Arezes, P.; López, J.M.; De Arcas, G. Automatic Resting Tremor Assessment in Parkinson's Disease Using Smartwatches and Multitask Convolutional Neural Networks. *Sensors* **2021**, *21*, 291. [CrossRef]

43. Hoehn, M.M.; Yahr, M.D. Parkinsonism: Onset, progression, and mortality. *Neurology* **1998**, *50*, 318. [CrossRef]

44. Bao, L.; Intille, S.S. Activity recognition from user-annotated acceleration data. In *Pervasive Computing, Proceedings of theSecond International Conference, PERVASIVE 2004, Vienna, Austria, 21–23 April 2004*; Springer: Berlin/Heidelberg, Germany, 2004; pp. 1–17.

45. Bonato, P.; Sherrill, D.M.; Standaert, D.G.; Salles, S.S.; Akay, M. Data mining techniques to detect motor fluctuations in Parkinson's disease. In Proceedings of the 26th Annual International Conference of the IEEE Engineering in Medicine and Biology Society, Chicago, IL, USA, 26–30 August 2004; Volume 2, pp. 4766–4769.

46. Channa, A.; Ifrim, R.C.; Popescu, D.; Popescu, N. A-WEAR bracelet for detection of hand tremor and bradykinesia in Parkinson's patients. *Sensors* **2021**, *21*, 981. [CrossRef]

47. San-Segundo, R.; Navarro-Hellín, H.; Torres-Sánchez, R.; Hodgins, J.; De la Torre, F. Increasing robustness in the detection of freezing of gait in Parkinson's disease. *Electronics* **2019**, *8*, 119. [CrossRef]

48. Sigcha, L.; Costa, N.; Pavón, I.; Costa, S.; Arezes, P.; López, J.; De Arcas, G. Deep Learning Approaches for Detecting Freezing of Gait in Parkinson's Disease Patients through On-Body Acceleration Sensors. *Sensors* **2020**, *20*, 1895. [CrossRef]

49. Sigcha, L.; Borzì, L.; Pavón, I.; Costa, N.; Costa, S.; Arezes, P.; López, J.M.; De Arcas, G. Improvement of Performance in Freezing of Gait detection in Parkinson's Disease using Transformer networks and a single waist-worn triaxial accelerometer. *Eng. Appl. Artif. Intell.* **2022**, *116*, 105482. [CrossRef]

50. Chawla, N.V.; Bowyer, K.W.; Hall, L.O.; Kegelmeyer, W.P. SMOTE: Synthetic minority over-sampling technique. *J. Artif. Intell. Res.* **2002**, *16*, 321–357. [CrossRef]

51. Um, T.T.; Pfister, F.M.J.; Pichler, D.; Endo, S.; Lang, M.; Hirche, S.; Fietzek, U.; Kulić, D. Data augmentation of wearable sensor data for parkinson's disease monitoring using convolutional neural networks. In Proceedings of the 19th ACM International Conference on Multimodal Interaction, Glasgow, UK, 13–17 November 2017. [CrossRef]

52. Jost, L. Entropy and diversity. *Oikos* **2006**, *113*, 363–375. [CrossRef]

53. Trockman, A.; Kolter, J.Z. Patches are all you need? *arXiv* **2022**, arXiv:2201.09792.

54. Hassani, A.; Walton, S.; Shah, N.; Abuduweili, A.; Li, J.; Shi, H. Escaping the big data paradigm with compact transformers. *arXiv* **2021**, arXiv:2104.05704.

55. Li, L.; Jamieson, K.; DeSalvo, G.; Rostamizadeh, A.; Talwalkar, A. Hyperband: A Novel Bandit-Based Approach to Hyperparameter Optimization. *J. Mach. Learn. Res.* **2017**, *18*, 6765–6816.

56. Kingma, D.P.; Ba, J. Adam: A Method for Stochastic Optimization. *arXiv* **2014**, arXiv:1412.6980.

57. Japkowicz, N.; Shah, M. *Evaluating Learning Algorithms: A Classification Perspective*; Cambridge University Press: Cambridge, UK, 2011.

58. Breiman, L. Random Forests. *Mach. Learn.* **2001**, *45*, 5–32. [CrossRef]

electronics   **MDPI**

Article

Classification of Parkinson's Disease Patients—A Deep Learning Strategy

Helber Andrés Carvajal-Castaño [1,2,*], Paula Andrea Pérez-Toro [1,3] and Juan Rafael Orozco-Arroyave [1,3,*]

[1]   GITA Lab, Electronics Engineering and Telecommunications Department, Faculty of Engineering, Universidad de Antioquia, Medellín 050010, Colombia

[2]   GIBIC Lab, Bioengineering Department, Faculty of Engineering, Universidad de Antioquia, Medellín 050010, Colombia

[3]   Pattern Recognition Lab, Friedrich-Alexander-Universität Erlangen-Nürnberg, 91054 Erlangen, Germany

*   Correspondence: helber.carvajal@udea.edu.co (H.A.C.-C.); rafael.orozco@udea.edu.co (J.R.O.-A.)

Abstract: (1) Background and objectives: Parkinson's disease (PD) is one of the most prevalent neurodegenerative diseases whose typical symptoms include bradykinesia, abnormal gait and posture, shortened strides, and other movement disorders. In this study, we present a novel framework to evaluate PD gait patterns using state of the art deep learning algorithms. A comparative analysis with three different approaches is presented and evaluated upon three groups of subjects: PD patients, Young Healthy Controls (YHC), and Elderly Healthy Controls (EHC). (2) Methods: The three approaches used in the study include: (i) The energy content of the gait signals in the frequency domain is captured with spectrograms that are used to feed a CNN model, (ii) Temporal information is incorporated by creating GRU networks, (iii) Temporal and spectral information is simultaneously captured by creating a new architecture based on CNNs and GRUs. (3) Results: Accuracies of up to 83.7% and 92.7% are found for the classification between PD vs. EHC and PD vs. YHC, respectively. According to our observations, the proposed approach based on the combination of temporal and spectral information, yields better results than others reported in the state of the art. (4) Conclusions: The results obtained in this study suggest that the combination of temporal and spectral information is more accurate than individual approaches used to classify and evaluate gait patterns in PD patients. To the best of our knowledge, this is the first study in gait analysis where temporal and spectral information is combined in an architecture of deep learning.

Keywords: gait analysis; Parkinson's disease; convolutional neural networks; gate recurrent units; deep learning

Citation: Carvajal-Castaño, H.A.; Pérez-Toro, P.A.; Orozco-Arroyave, J.R. Classification of Parkinson's Disease Patients—A Deep Learning Strategy. *Electronics* **2022**, *11*, 2684. https://doi.org/10.3390/electronics11172684

Academic Editor: Byung-Gyu Kim

Received: 26 July 2022
Accepted: 24 August 2022
Published: 27 August 2022

Publisher's Note: MDPI stays neutral with regard to jurisdictional claims in published maps and institutional affiliations.

1. Introduction

Parkinson's Disease (PD) is a neurodegenerative disease that produces movement disorders including tremor, rigidity, postural instability and lack of coordination which affect patients' gait [1–3]. PD patients are characterized by abnormal gait patterns associated with bradykinesia (slowness of movement), less steady walk, reduced stride length and shuffling steps or impaired gait initiation [4–6]. The symptoms of PD may appear about 10 years prior to the clinical manifestations [7], besides, several studies show that PD mainly impacts elderly people [8]. An important fact is that the prevalence of the disease is increasing with age worldwide [9,10]. Neurologists usually use clinical scales to evaluate and quantify the neurological state of the patient. The most used is the Movement Disorder Society—Unified Parkinson's Disease Rating Scale (MDS-UPDRS) [11]. This scale allows neurologists to evaluate the patient's state and it is useful to follow up on therapies. The MDS-UPDRS scale is composed of four sections. The third one is called MDS-UPDRS-III and corresponds to the assessment of routine motor activities including 33 tasks, therefore it ranges from 0 to 132.

Gait patterns allow to obtain information about different movement disorders that are sometimes associated to Parkinson's disease (PD) symptoms. In the literature, the use of Inertial Movement Units (IMU) has increased considerably since they allow to capture gait patterns to study the movement dynamics of patients. This includes the study of kinematic characteristics of PD patients [12,13], nonlinear dynamics [14,15], stability and deep learning approaches [16], among others.

Computer vision methods and force platforms are used in laboratories to evaluate gait disorders [17,18]; however, they are expensive and difficult to access. Conversely, wearable sensors allow for designing low-cost and unobtrusive solutions that enable continuous monitoring of patients [19]. The most common wearable sensors for gait analyses are those based on plantar pressure systems [20–22] and IMU sensors [23,24]

Gait analysis is of great interest for the research community due to its suitability to perform unobtrusive automatic and continuous evaluation of motor symptoms of PD patients.

1.1. State of the Art

The study of gait patterns is related to the human locomotion and includes the study of people move while walking. Models can be created considering different gait features, related to kinematics, such as: stride length, stride velocity, turning angle, swing phase, and others [25,26]. The analysis of abnormal gait patterns have been typically performed considering Inertial Movement Units (IMU). An IMU is an electronic device usually consisting of accelerometer and gyroscope sensors, and in some cases also include magnetometer sensors. In [27] the authors presented a complete study related to the use of IMU sensors. The aim of the authors is to model abnormal gait patterns. According to the authors IMU sensors in gait analysis are used due to their low cost and the potential for designing wearable devices for continuous monitoring. Even though the study was presented about seven years ago, the same claim continues to be valid today. In [28] the authors proposed the use of one IMU sensor in each foot to analyse patients with different neurological conditions. The authors extracted several kinematic gait measures like stride length, stance time, swing time, and cycle time. The proposed method was tested with a dataset comprised of 22 healthy control (HC) subjects recorded with a camera-based system. A clinical discussion using a dataset of 17 subjects with different neurological disorders was also presented. According to the authors, it is possible to obtain relevant information on different neurodegenerative diseases, even outside clinical settings. In [20] the classification of PD patients and HC subjects was performed by using several spatial-temporal measures like stride length, cadence, stance time, and swing time. Different classifiers were tested including Random Forest (RF), Support Vector Machine (SVM) and Kernel Fisher Discriminant. The best result was found with a RF classifier (92.6%). A multi-modal study for the discrimination between PD patients and HC subjects, considering information of three bio-signals: speech, handwriting, and gait, was presented in [16]. To merge the information of each bio-signal a Convolutional Neural Networks (CNN) was implemented. The authors reported the highest accuracy with the combination of the three bio-signals (97.6%). Another approach in gait analysis is based on non-linear dynamics (NLD) measures. In [14] the authors extracted several NLD and Entropy measures. Three classifiers were compared: SVM, RF and k-nearest neighbours (KNN). Accuracies up to 92% were reported In [15] the authors proposed a new strategy considering Poincaré sections. Accuracies up to 89% in the classification of PD vs. HC were reported, besides, the authors' proposal includes experiments with PD patients in three different stages of the disease: mild, moderate, and severe where accuracies up to 67.2% were reported. Recently, in [13] the authors computed three sets of features named kinematics, NLD, and stability, and proposed a clinical interpretation based on the most discriminant feature per subset. The authors reported accuracies of up to 92% when using only three of the features.

In this paper, we use raw gait signals captured using IMU sensors to assess the ability of different deep learning architectures to classify PD patients vs. HC subjects. Three architectures were considered: Convolutional Neural Networks (CNN), Gate Recurrent

Units (GRU) and a new approach that considers energy information at the input of a CNN and temporal information with a GRU. In order to consider the effect of age three groups of subjects are examined: Young Healthy Control (YHC), Elderly Healthy Control (EHC), and PD patients. The EHC group and the PD group are matched in age. Accuracies up to 85.3% were reported in the PD vs. EHC scenario, and accuracies of up to 92.7% were found in PD vs. YHC.

1.2. Contributions of This Study

Three different deep learning architectures, namely CNN, GRU and CNN + GRU were evaluated in this study to classify between PD patients and HC subjects. Models based on CNNs yielded good results but did not consider temporal information, therefore we decided to evaluate an architecture based on GRUs to incorporate relevant information possibly encoded in the evolution of the patterns, i.e., temporal information. The combination of CNNs and GRUs in the same model was introduced to take advantage of incorporating temporal and frequency information in the same model, which potentially enables clinical interpretation. We believe that the CNN+GRU model did not show better results due to the small amount of data available for the present experiments.

In this study two tasks were considered: 2×10 m task corresponds to a 10 m walk performed twice and 4×10 m task corresponds to a 10 m walk which is performed 4 times. In general terms, the 4×10 m task is better than the 2×10 m one. We think that this is because longer tasks allow to collect more information and therefore increase the chances to find abnormal patterns in the gait signal.

2. Materials and Methods

2.1. Methodology

The general methodology proposed in this study is summarized in Figure 1. Gait signals are collected using wearable IMU sensors. Note that the main characteristic of the proposed methodology is that there is no a sophisticated feature extraction stage. The segmentation process is based on sliding windows of fixed length and, in the case of the CNN architecture, we compute the spectrogram that is used as input. Information of each foot and their combination are considered. In the following subsections, the stages of this methodology are explained.

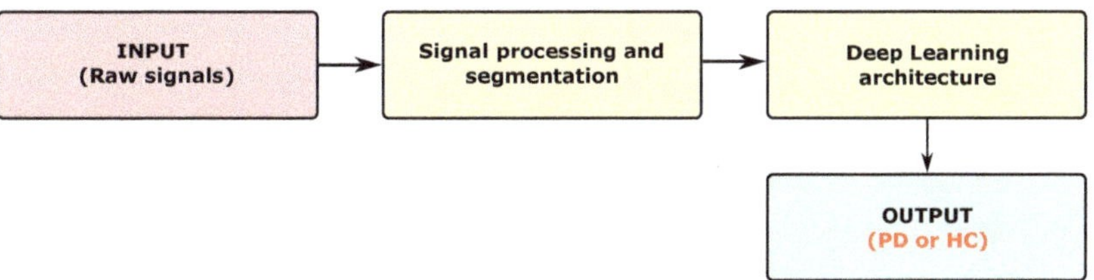

Figure 1. Scheme of the general methodology addressed in the study.

2.2. Data Collection and Participants

The eGaIT system (Embedded Gait analysis using Intelligent Technology), was used to record gait signals. eGaIT consists of a 6 degrees of freedom sensor to capture accelerometer and gyroscope signals. The accelerometer allows to measure the acceleration in a range of ± 6 g and 200 mV/g of sensibility. Gyroscope allows to measure rotational velocities in a range of $\pm 500\ ^\circ$/s and ± 2 mV/g of sensitivity. A representation of the eGait system and the position of the sensor in the shoe is shown in Figure 2. Signals are collected using an Android Application.

The sensor used captures motion patterns at a sampling frequency (F_s) of 102.4 Hz with 12 bits of resolution. Besides, this value enables capturing information with a resolution good enough to model low-frequency patterns such as those related to the patient's gait.

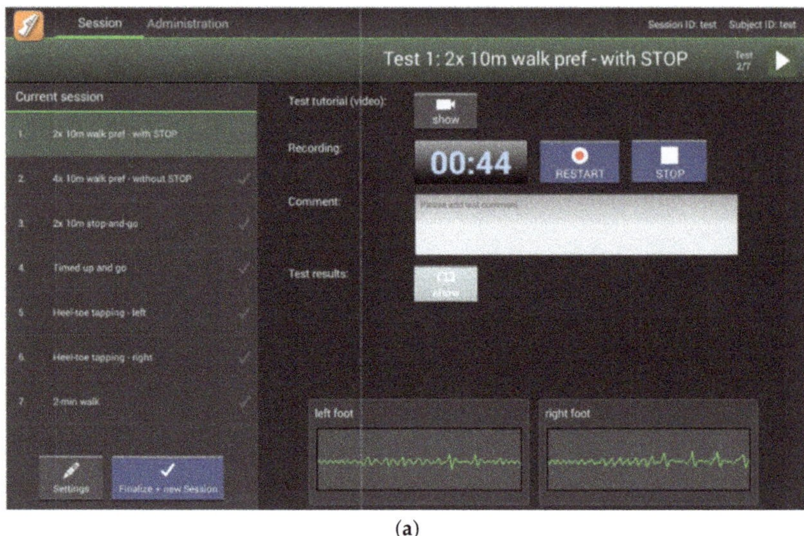

(a)

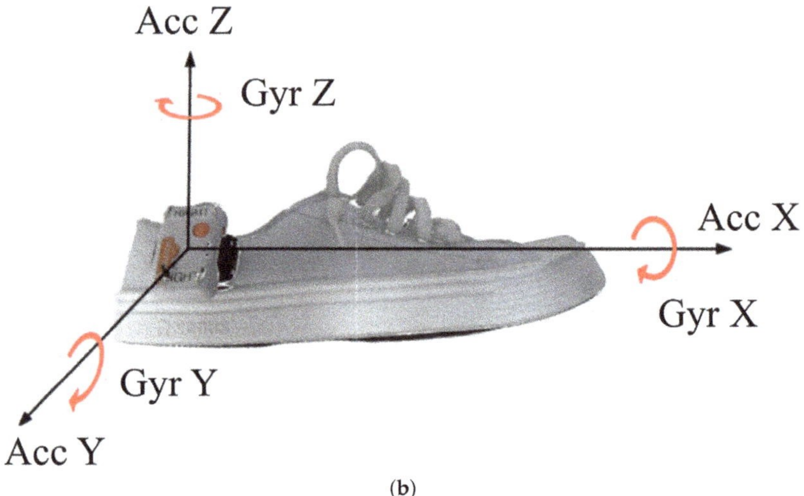

(b)

Figure 2. (**a**) Interface of the eGait software, (**b**) Location of the eGait sensor in the shoe.

In this study, two tasks were considered:

i. 2×10 m task:

- The subject starts standing.
- The subject walks 10 m in a straight line.
- The subject stop.
- The subject turns right and returns to the starting point.

ii. 4×10 m task:

- The subject starts standing
- The subject walks 10 m straight.
- The subject turns right and returns to the starting point.
- The subject turns right walks 10 m.
- Finally, the subject turns right, again, and returns to the starting point.

The dataset used in this study consists of 134 recordings where 45 are PD patients and 89 HC subjects. The HC group is divided into two groups: 44 YHC subjects under 45 years old and 45 EHC subjects of people older than 45 years. In Table 1 the information of the dataset is presented. The age of the EHC group is balanced with respect to the PD participants.

Table 1. Details about the participants.

Group	Gender	# Subjects	Age & Range	MDS-UPDRS-III & Range
PD	Male	17	65.0 ± 10 & [41–82]	37.6 ± 21 & [8–82]
	Female	28	58.9 ± 11 & [29–75]	33.5 ± 21 & [9–106]
EHC	Male	23	63.3 ± 11 & [49–85]	-
	Female	22	58.9 ± 10 & [45–83]	-
YHC	Male	26	25.3 ± 5 & [21–42]	-
	Female	18	22.9 ± 3 & [19–32]	-

Age and MDS-UPDRS-III score are presented in terms of mean $\pm$ standard deviation. There are no significant differences in the age of PD vs EHC (t-student test, p-value $\ll 0.05$). The last column includes the MDS-UPDRS-III values associated to PD patients.

2.3. Convolutional Neural Network (CNN)

A CNN is a deep learning architecture typically used for image analysis where convolution and pooling layers are used with the aim to obtain relevant information of the input [29]. The main advantage of a CNN is that it requires minimal or sometimes no pre-processing for the input to implement the architecture. Let's define the input of a CNN as a tensor as follows:

$$\mathbf{X} \in \mathbb{R}^{p \times q \times r} \tag{1}$$

where p, q and r correspond to the number of vertical pixels, horizontal pixels and channels of the image, respectively. The convolution process is performed between the input tensor $\mathbf{X}$ and a convolutional filter, named kernel, represented as follow:

$$\mathbf{W} \in \mathbb{R}^{n \times n \times d} \tag{2}$$

where n is the size of the kernel and d is the number of kernels in the convolutional layer. The result of the convolution between $\mathbf{X}$ and $\mathbf{W}$ per channel produces a hidden representation $\mathbf{H}$ as follows:

$$\mathbf{H} = \mathbf{X} * \mathbf{W} \tag{3}$$

where:

$$\mathbf{H} \in \mathbb{R}^{(p-n+1) \times (q-n+1) \times d} \tag{4}$$

Note that tensor $\mathbf{H}$ represents the extracted features obtained from the input $\mathbf{X}$. A pooling layer is implemented after each convolution step. The pooling layer reduces the size of the hidden representation $\mathbf{H}$. One of the aims of the pooling layer is to reduce the computational cost required to process the information, in addition, it is useful to remove some invariant features [29]. Finally, a fully connected layer with h hidden units followed by an activation function is implemented to obtain the final decision of the classification process.

It is important to note that different CNN architectures can be created depending on the problem. Figure 3 presents an illustration of a CNN architecture with two convolutional hidden layers.

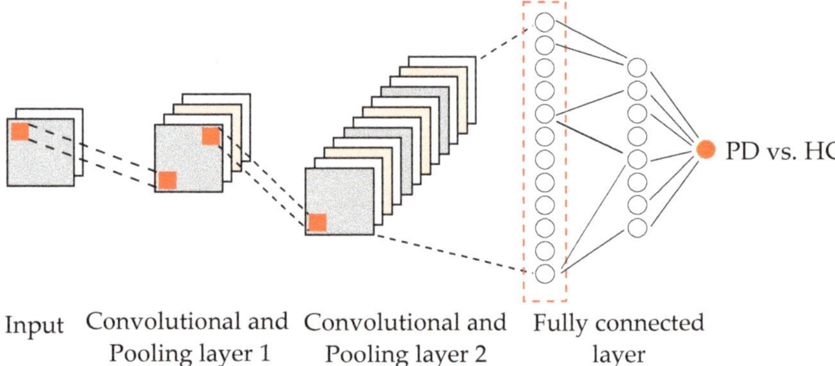

Figure 3. Illustration of a CNN with two convolutional layers.

For the case of gait signals, the CNN architecture corresponds to a two-dimensional (2D) CNN. The input to the CNN consists of $r = 12$ channels when the two feet are considered. The channels have information of the accelerometer and gyroscope signals in the x, y, and z-axes.

With the aim to guarantee at leas 3 quasi-periods in the gait signal, segments of 3 s are considered. The Short Time Fourier Transform (STFT) is computed to create the input to the CNN. Figure 4 shows four examples of STFT computed upon two PD patients (a and b), one EHC subject (c), and one YHC subject (d). In the four cases images are extracted from gyroscope signals (z-axis) of the left foot during the 2×10 task.

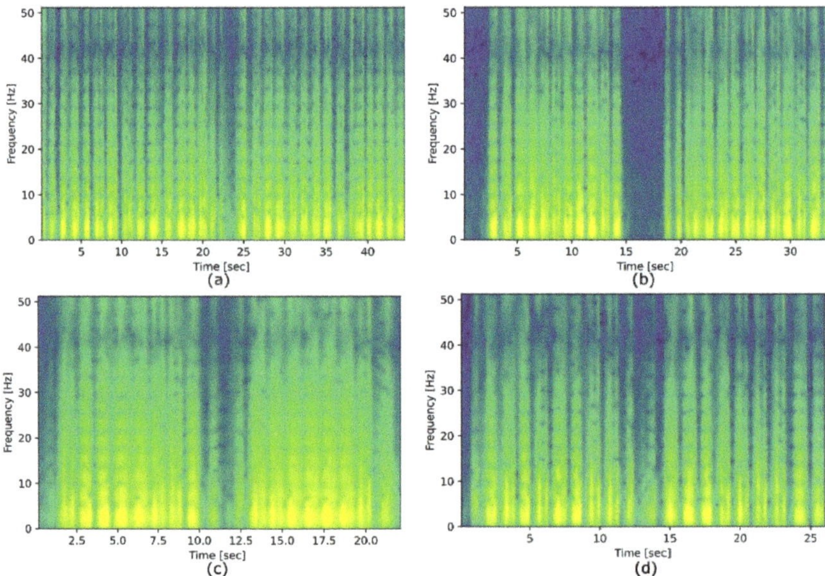

Figure 4. Resulting STFT computed to: (**a**) PD female patient, Lower limps score: 50, Age: 75; (**b**) PD female patient, Lower limps score: 10, Age: 65; (**c**) EHC female patient, Age: 50; (**d**) YHC female patient, Age: 20.

The CNN was trained using the stochastic gradient descent (SGD) algorithm. The loss function is the cross-entropy between the label of the training data y and the prediction $\hat{y}$. An Exponential Linear Unit (elu) is used as activation function for the convolutional layer. Dropout is included to avoid over-fitting in the training process. The architecture of the CNN for this study includes two convolutional layers with max-pooling, dropout for regularization, and five fully connected hidden layers. A sigmoid activation function is used at the output. Figure 5 summarizes the details of the architecture.

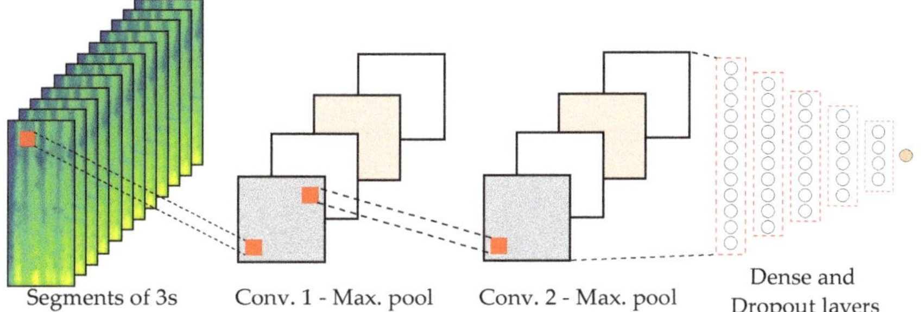

Figure 5. Architecture of the CNN implemented in this study.

2.4. Gate Recurrent Network, GRU

The paradigm of GRU was proposed in [30,31] as a variation to recurrent neural networks (RNN). A GRU is composed of two gates: update and reset, whose objective is to only pass relevant information through the network to improve the predictions. Among the advantages of the GRU over other recurrent networks are the fact that they require less memory, therefore their training process is faster. Figure 6 illustrates a single GRU unit.

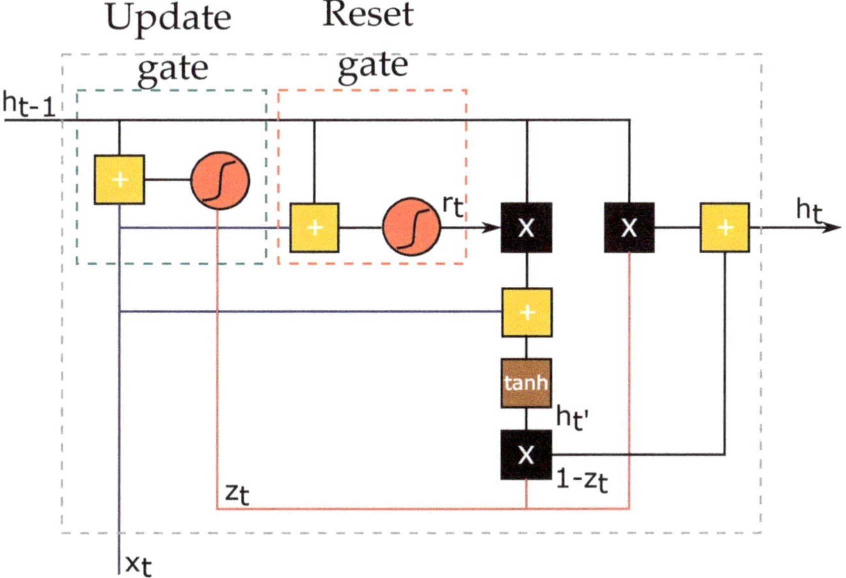

Figure 6. Single GRU unit.

The computation of a GRU starts with the calculation in the step time t for the update gate z_t, as follow:

$$z_t = \sigma(W^{(z)}x_t + U^{(z)}h_{t-1}) \tag{5}$$

x_t is multiplied by $W(z)$, which is its own weight. The same process is performed with h_{t-1}, which has information of the previous step time $t-1$ and is multiplied by its own weight $U(z)$. A sigmoid activation function is applied to the sum of both products. The aim of the update gate is to define the information to be considered in the future.

The reset gate intends to find the information to be forgotten, in this case it is called r_t, defined as follows:

$$r_t = \sigma(W^{(r)}x_t + U^{(r)}h_{t-1}) \tag{6}$$

Which is similar to the equation of the update gate except for the weights. The current memory content h'_t is calculated as follow:

$$h'_t = \tanh(Wx_t + r_t \odot Uh_{t-1}) \tag{7}$$

where $\odot$ is the Hadamard matrix product. The final memory at the time step t is:

$$h_t = z_t \odot h_{t-1} + (1 - z_t) \odot h_t \tag{8}$$

A GRU architecture is able to process information of time series such as the one existing in raw gait signals. In this work, the input to the GRU consists of 12 raw signals captured with the IMU sensor. Figure 7 shows four examples of signals collected from two PD patients (a and b), one EHC subject (c), and one YHC subject (d). The GRU architecture implemented in this study is presented in Figure 8.

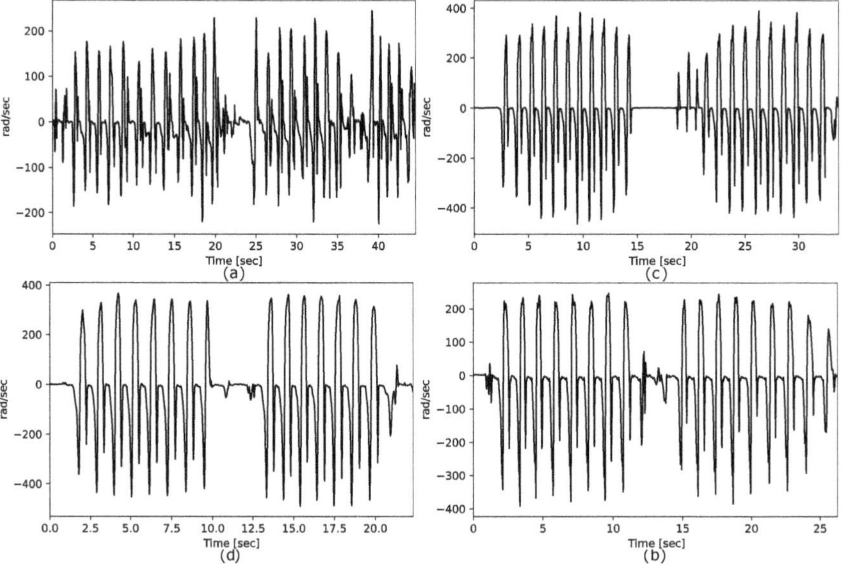

Figure 7. Comparison between the raw time series of: (**a**) PD female patient, Lower limps score: 50, Age: 75; (**b**) PD female patient, Lower limps score: 10, Age: 65; (**c**) EHC female patient, Age: 50; (**d**) YHC female patient, Age: 20.

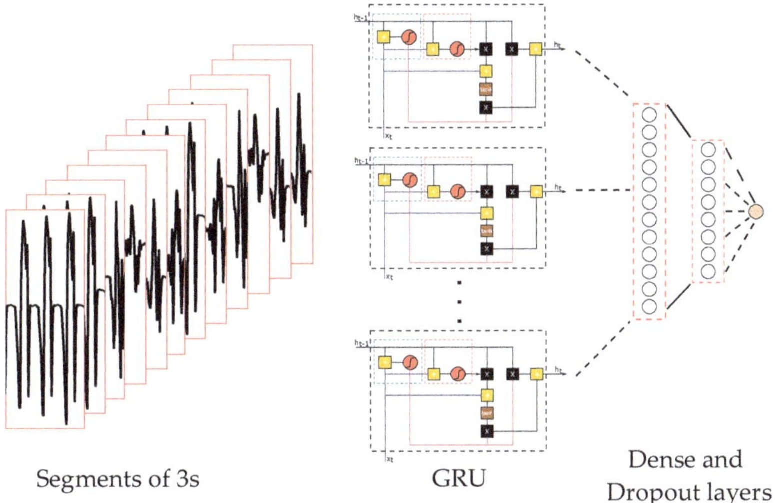

Figure 8. Architecture of the GRU implemented in this study.

2.5. Training Process and Classification

A 5-fold cross-validation strategy was used to evaluate the proposed approach along the experiments. Four folds were used for training and one-fold for testing. Each experiment was repeated ten times and the reported results correspond to the average over those repetitions. Adam optimizer [32] and binary cross-entropy were used in the classification stage of all experiments.

The acquisition of the acceleration data performed in this work is by-default normalized between -6 g and $+6$ g and the gyroscope signals between $+500\,°/s$ and $-500\,°/s$. Therefore, it is not necessary to perform any additional normalization. Besides, the architecture of the Neural Network by itself performs an "internal batch" normalization according to the patterns that it is observing during the training process. Further details of the batch normalization can be found in [33].

3. Experiments and Results

Three different experiments were considered: only with the CNN, only with the GRU, and with the combination of both architectures. Each experiment considers two scenarios: PD vs. EHC and PD vs. YHC. The two gait tasks were considered independently. Results are reported in terms of accuracy (Acc), sensitivity (Sen), specificity (Spe), and Area Under ROC Curve (AUC) [34]. Two different accuracy values are reported in each experiment, accuracy in development refers to the result obtained within the 4 folds considered during the training process and accuracy in test refers to the result obtained in the external fold that did not participate in the optimization process. Standard deviation values appear because the experiments were repeated ten times independently to perform a fair evaluation of the proposed approach.

3.1. Classification with CNN

The general scheme of the proposed CNN architecture is presented in Figure 5. This approach includes two convolutional layers and five fully connected hidden layers, besides Max-pooling and dropout layers are included to avoid overfitting. Details of the implemented architecture are presented in Appendix A, Table A1. Figure 9 shows details of the pre-processing stages applied in this experiment before feeding the CNN architecture. Notice that the raw input contains 12 channels, therefore there is the same number of spectrograms before the segmentation step. The STFT is computed upon segments of 3s per channel with an overlap of 80%.

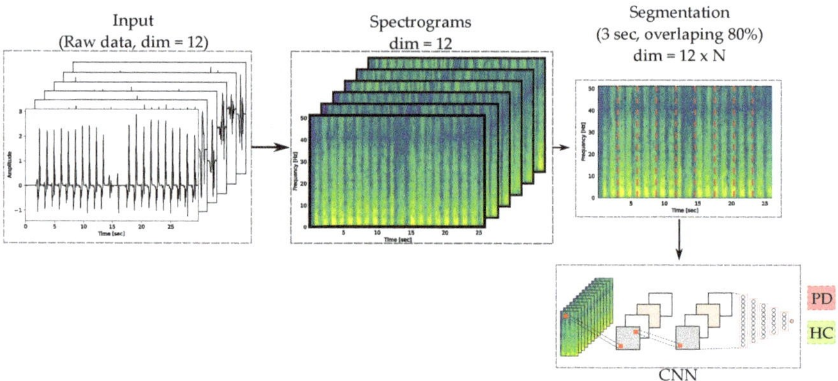

Figure 9. Methodology for the classification based on CNNs.

Note that when both feet are considered the dimension of the input is $r = 12$ which corresponds to three accelerometer signals and three gyroscope signals per foot. Table 2 shows the results obtained in the classification with the 2×4 task. The highest accuracy in test is 82.4% for the PD vs. EHC scenario, while 87.5% for PD vs. YHC. Table 3 presents the results obtained with the 4×10 task. Notice that in this case the results are higher compared to those obtained in the previous task. In the PD vs. EHC scenario, the highest accuracy in test is 82.7% while in PD vs. YHC it is 92.1%. This improvement could be associated to the fact that this task is longer than the previous one, therefore there are more chances to observe abnormal patterns in the gait signals. Also, a longer task likely produces more fatigue in the participants, especially the patients.

Table 2. Results using CNN and 2×10 m task.

Scenario	Foot	Acc. Dev. [%]	Acc. Test [%]	Sen [%]	Spe [%]	AUC
	Left	79.1 ± 3	81.3 ± 4	74.3 ± 4	84.3 ± 4	0.87
PD vs. EHC	Right	79.5 ± 4	80.1 ± 4	79.3 ± 4	80.3 ± 4	0.88
	Both	80.2 ± 3	82.4 ± 3	76.2 ± 3	87.3 ± 4	0.88
	Left	83.9 ± 5	85.4 ± 5	87.2 ± 3	88.0 ± 4	0.90
PD vs. YHC	Right	81.7 ± 4	83.8 ± 4	88.1 ± 3	85.0 ± 4	0.90
	Both	84.3 ± 5	87.5 ± 4	88.7 ± 3	87.1 ± 4	0.91

Acc. Test: accuracy in test, Acc. Dev.: accuracy in development, Sen: Sensitivity, Spe: Specificity, AUC: Area under the ROC curve.

Table 3. Results using CNN and 4×10 m task.

Scenario	Foot	Acc. Dev. [%]	Acc. Test [%]	Sen [%]	Spe [%]	AUC
	Left	83.2 ± 5	85.3 ± 4	83.0 ± 6	91.3 ± 3	0.92
PD vs. EHC	Right	83.3 ± 4	79.5 ± 6	83.3 ± 5	82.9 ± 6	0.91
	Both	83.5 ± 6	82.7 ± 4	85.6 ± 2	87.9 ± 2	0.92
	Left	87.4 ± 4	88.5 ± 4	88.3 ± 6	90.5 ± 5	0.95
PD vs. YHC	Right	85.3 ± 6	86.5 ± 4	91.4 ± 4	87.9 ± 5	0.94
	Both	88.5 ± 5	92.1 ± 5	91.2 ± 3	88.4 ± 5	0.95

Acc. Test: accuracy in test, Acc. Dev.: accuracy in development, Sen: Sensitivity, Spe: Specificity, AUC: Area under the ROC curve.

3.2. Classification with GRU

The general scheme of the GRU architecture used in this work is presented in Figure 8. In this case, the raw input with 12 channels is first segmented into windows of 3 s with 80%

overlap. Each window is segmented into N number of steps. Details of the implemented architecture are presented in Appendix A, Table A2. Notice that since every person can produce a different number of steps during the time window, the number of steps needs to be variable in order to make it the method robust and flexible. This segmentation procedure is shown in Figure 10. Table 4 shows results of the two classification scenarios: PD vs. EHC and PD vs. YHC when the 2×10 task is considered. Similarly, Table 5 includes results obtained with the 4×10 m task. Note that the GRU architecture yields better results in most of the experiments, compared to those obtained with the CNN. Similar to what we observed in the previous experiment, the 4×10 task yields better results. In the case of the classification between PD patients and EHC subjects the highest accuracy was 82.7% and in the case of PD vs. YHC the best result was 92.5%. Signals of both feet provided the best results in both scenarios, as it was also observed in the experiment with CNN.

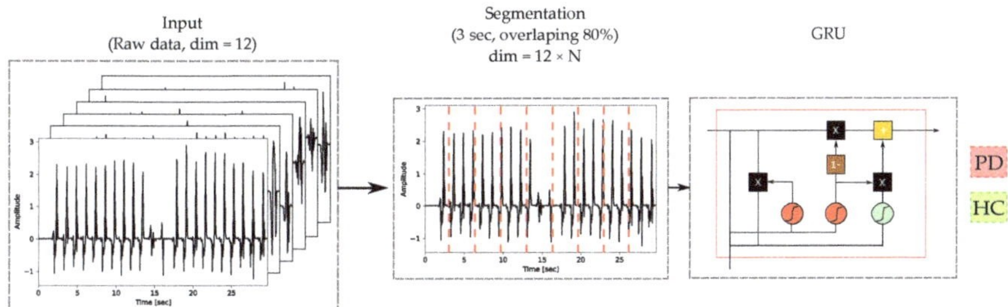

Figure 10. Methodology used for the classification with a GRU architecture and both feet.

Table 4. Results using GRU and 2×10 m task.

Scenario	Foot	Acc. Dev. [%]	Acc. Test [%]	Sen [%]	Spe [%]	AUC
	Left	75.7 ± 4	75.2 ± 4	70.6 ± 6	74.3 ± 4	0.82
PD vs. EHC	Right	74.9 ± 4	73.3 ± 5	75.7 ± 4	78.8 ± 6	0.81
	Both	78.5 ± 6	78.7 ± 5	72.8 ± 3	83.3 ± 5	0.84
	Left	84.7 ± 4	83.4 ± 5	88.4 ± 4	87.2 ± 5	0.89
PD vs. YHC	Right	81.3 ± 6	82.6 ± 4	87.7 ± 5	84.4 ± 4	0.88
	Both	86.7 ± 5	84.5 ± 4	89.1 ± 3	87.9 ± 5	0.90

Acc. Test: accuracy in test, Acc. Dev.: accuracy in development, Sen: Sensitivity, Spe: Specificity, AUC: Area under the ROC curve

Table 5. Results using GRU and 4×10 m task.

Scenario	Foot	Acc. Dev. [%]	Acc. Test [%]	Sen [%]	Spe [%]	AUC
	Left	77.5 ± 4	76.6 ± 3	75.1 ± 2	80.8 ± 3	0.88
PD vs. EHC	Right	75.6 ± 4	74.5 ± 3	77.3 ± 5	82.1 ± 5	0.87
	Both	84.1 ± 4	82.7 ± 6	83.8 ± 4	86.3 ± 6	0.92
	Left	91.8 ± 6	90.3 ± 4	89.3 ± 6	90.1 ± 6	0.94
PD vs. YHC	Right	89.5 ± 5	88.6 ± 6	90.4 ± 4	92.4 ± 5	0.94
	Both	93.7 ± 4	92.5 ± 5	92.6 ± 3	94.1 ± 6	0.96

Acc. Test: accuracy in test, Acc. Dev.: accuracy in development, Sen: Sensitivity, Spe: Specificity, AUC: Area under the ROC curve.

3.3. Classification with CNN + GRU

To consider temporal and frequency information of the gait signals simultaneously, a novel strategy is proposed in this work. Details of the implemented architecture are presented in Appendix A, Table A3. The input to the proposed architecture are the spectrograms and also the raw signals. Figure 11 shows how the two approaches can be considered simultaneously to perform the final decision of whether a subject belongs to the PD or HC group.

Results presented in Tables 6 and 7 show that this methodology yields results slightly better than those obtained with the GRU model. When observing the 4×10 m task, the highest accuracy in the PD vs. EHC scenario was 83.7%, while in PD vs. YHC the result was 92.7%.

Although the results of the CNN + GRU model are not much higher than those obtained with the GRU architecture, we believe that this is due to the small amount of data considered in this work. We are currently working on the collection of more data to validate whether these kinds of architectures yield results significantly better than others where only temporal or frequency information is considered separately.

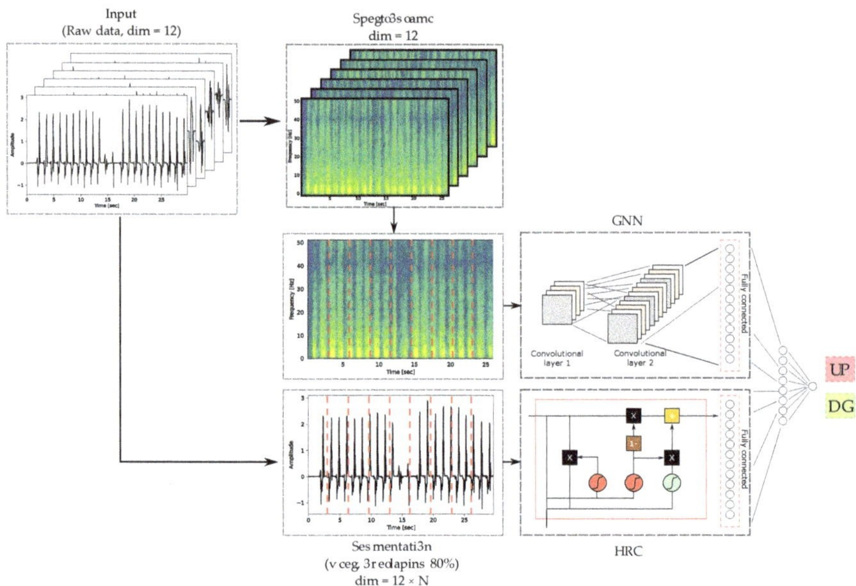

Figure 11. Proposed methodology considering a CNN + GRU architecture

Table 6. Results using CNN + GRU and 2×10 m task.

Scenario	Foot	Acc. Dev. [%]	Acc. Test [%]	Sen [%]	Spe [%]	AUC
	Left	76.7 ± 5	74.6 ± 6	71.3 ± 5	75.2 ± 4	0.83
PD vs. EHC	Right	75.2 ± 3	73.6 ± 5	74.9 ± 6	79.3 ± 4	0.81
	Both	80.4 ± 5	78.7 ± 4	73.3 ± 5	84.1 ± 5	0.85
	Left	85.4 ± 5	83.7 ± 4	89.2 ± 5	86.9 ± 4	0.88
PD vs. YHC	Right	83.5 ± 6	82.0 ± 6	88.4 ± 4	85.2 ± 6	0.87
	Both	88.5 ± 4	85.7 ± 5	90.2 ± 3	86.5 ± 5	0.89

Acc. Test: accuracy in test, Acc. Dev.: accuracy in development, Sen: Sensitivity, Spe: Specificity, AUC: Area under the ROC curve.

Table 7. Results using CNN + GRU and 4×10 m task.

Scenario	Foot	Acc. Dev. [%]	Acc. Test [%]	Sen [%]	Spe [%]	AUC
PD vs. EHC	Left	76.7 ± 6	74.7 ± 4	75.2 ± 5	77.8 ± 5	0.84
	Right	74.2 ± 7	72.4 ± 6	74.2 ± 4	78.5 ± 5	0.82
	Both	85.7 ± 3	83.7 ± 4	84.7 ± 5	85.9 ± 4	0.93
PD vs. YHC	Left	90.9 ± 6	91.1 ± 3	89.5 ± 6	91.1 ± 5	0.94
	Right	90.3 ± 5	87.9 ± 5	90.6 ± 3	92.2 ± 6	0.95
	Both	93.3 ± 4	92.7 ± 4	91.9 ± 3	94.4 ± 5	0.96

Acc. Test: accuracy in test, Acc. Dev.: accuracy in development, Sen: Sensitivity, Spe: Specificity, AUC: Area under the ROC curve.

Figure 12 presents the best results of the 4×10 m task in the two scenarios and the three experiments. The distribution of the scores/posteriors obtained in the classification stage of each scenario (PD vs. EHC and PD vs. YHC) are included in Figure 12a,c. Although both scenarios are clearly separable due to the robustness of the proposed approach based on a GRU+CNN architecture, it can be observed that there is more overlap in the first scenario. Regarding Figure 12b,d, they include the ROC curves resulting from the three experiments (CNN, GRU, and CNN + GRU) in each scenario. Notice that in both cases the CNN + GRU architecture yields the highest AUC values, which confirms its superiority compared to other approaches.

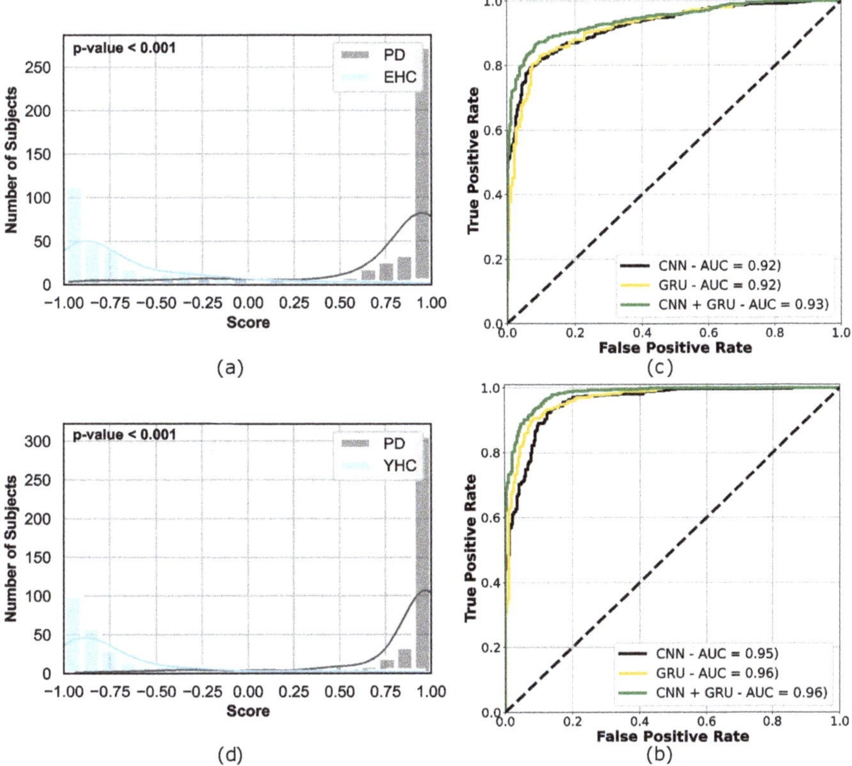

Figure 12. Comparison of the best results considering the 4×10 m task and both feet. (**a**) Distribution of the scores in PD vs. EHC using the GRU + CNN architecture. (**b**) Comparison of the ROC curves in the PD vs. EHC scenario. (**c**) Distribution of the scores in PD vs. YHC using the GRU + CNN architecture. (**d**) Comparison of the ROC curves in the PD vs. YHC scenario.

Table 8 summarizes the results obtained with the different architectures with the 4×10 task.

Table 8. Summary of the accuracy in test for the best results considering the 4×10 m task.

Scenario	Foot	CNN	GRU	CNN + GRU
	Left	85.3 ± 4	76.6 ± 3	74.7 ± 4
PD vs. EHC	Right	79.5 ± 6	74.5 ± 3	72.4 ± 6
	Both	82.7 ± 4	82.7 ± 6	83.7 ± 4
	Left	88.5 ± 4	90.3 ± 4	91.1 ± 3
PD vs. YHC	Right	86.5 ± 4	88.6 ± 6	87.9 ± 5
	Both	92.1 ± 5	92.5 ± 5	92.7 ± 4

Data correspond to the accuracy in test by each architecture.

4. Discussion

Three different deep learning architectures were considered for the classification of PD vs. HC subjects. Two subgroups of healthy healthy subjects were included, elderly (EHC) and young (YHC). The architectures evaluated in this work correspond to the state of the art in gait analysis and are based on CNNs and GRUs. Previous works suggest that CNN architectures are a suitable approach when considering the STFT of gait signals [16,35]. In [16] the use of a CNN is introduced to classify gait signals and the authors reported an accuracy of 88%. We found comparable results in our present study, where the best accuracy obtained with the CNN is 82.7% considering the YHC group and the best result with the EHC group is 82.4%. The algorithm performed better always when the 4×10 m task was considered. We believe that this is because a longer task allows to capture more information about possible abnormal gait patterns which provides better classification results and also improves the generalization of the algorithms. The GRU model presented here allowed modelling information of gait signals without any pre-processing. In [36] the authors explored the use of RNNs to predict gait phases. They showed that these kinds of architectures are promising for the analysis of gait signals. In our experiments, we could observe that the GRU architecture improved the accuracy in most of the experiments. Similar to the CNN case, better results were obtained with the 4×10 m task. The classification of PD vs. EHC yields an accuracy of 82.7%, while with the YHC group, the accuracy is 92.5%. Besides CNN and GRU architectures evaluated individually, in this paper we proposed a model where CNN and GRU architectures are considered together. We hypothesized that results could improve when temporal and frequency information were combined in a single model. We found that, in general, results were similar to those obtained with the GRU architecture, with accuracies of 83.7% and 92.7% in the PD vs. EHC and PD vs. YHC scenarios, respectively. We believe that the results of the CNN + GRU model were not higher due to the small amount of data that we could consider here. Further research with a larger group of participants is required to validate whether this could lead to better results. We consider that this work is a step forward in the development of deep learning models for the automatic classification of PD patients.

In addition, it is necessary to consider other deep learning architectures seeking to improve the results, such as transfer learning, data argumentation and combinations of classifiers. Perhaps the most realistic approach would be to do transfer learning based on existing datasets, however, it is necessary to perform the experiments to raise strong conclusions.

5. Conclusions

GRU architectures clearly yielded better results than the CNN ones and this is likely due to the fact that temporal information is incorporated when the first approach is considered. Besides, we validated that the combination of CNN and GRU methods are suitable and provide similar results to those observed with GRUs only. Although we expected to find higher accuracies with the combination of methods, it was not possible to prove our hypothesis. We believe that it was due to the small amount of data considered in this study.

Regarding the comparison of gait tasks, we could validate that the 4×10 m task is more suitable and systematically yields better results than the 2×10 m task. Very likely this is because longer tasks allow us to collect more information and also give more chances to observe abnormal patterns in the gait signals.

Besides the evaluation of state-of-the-art deep learning architectures, the results obtained in this paper are comparable to others reported in the literature, so we believe that this study is a contribution to the topic gait analysis in PD patients.

We are aware that one of the limitations of this study is the small amount of data. We expect to perform experiments with more participants in the near future to make it possible to validate further hypotheses.

Author Contributions: Conceptualization, H.A.C.-C. and J.R.O.-A.; methodology, H.A.C.-C. and J.R.O.-A.; software, H.A.C.-C. and P.A.P.-T.; validation, H.A.C.-C. and J.R.O.-A.; formal analysis, H.A.C.-C. and J.R.O.-A.; investigation, H.A.C.-C. and J.R.O.-A.; resources, J.R.O.-A.; data curation, H.A.C.-C., P.A.P.-T. and J.R.O.-A.; writing—original draft preparation, H.A.C.-C., P.A.P.-T. and J.R.O.-A.; visualization, H.A.C.-C.; supervision, J.R.O.-A.; project administration, J.R.O.-A.; funding acquisition, J.R.O.-A. All authors have read and agreed to the published version of the manuscript.

Funding: H.A.C.-C. is under grants of the Colombian Ministry of Science "Programa de Becas de Excelencia Doctoral del Bicentenario—Cohorte 1" grant # 20230017-19-20. This work was also funded by CODI at UdeA grant # PRG2020-34068.

Institutional Review Board Statement: This study was approved by the Ethical Research Committee of the University of Antioquia and according to the Helsinki declaration (1964) and its later amendments.

Informed Consent Statement: Informed consent was obtained from all the participants of the study.

Data Availability Statement: Not applicable.

Appendix A

Details of the architectures used in the models.

Table A1. Details of the CNN architecture considering both feet (12 channels).

Layer	Output Shape	Number of Parameters
Input	(12, 257, 12)	0
Convolution 2D	(10, 255, 8)	872
Convolution 2D	(8, 253, 16)	1168
Max pooling	(4, 126, 16)	0
Flatten	2016	0
Dense 1	256	2,064,640
Dropout	256	0
Dense 2	128	32,896
Dropout	128	0
Dense 3	64	8256
Dropout	64	0
Dense 4	32	2080
Dropout	32	0
Dense 5	16	528
Dropout	16	0
Output	1	17

Table A2. Details of the GRU architecture considering both feet (12 channels).

Layer	Output Shape	Number of Parameters
Input	(12, 306)	0
GRU	128	148,608
Dense 1	256	8256
Dropout	256	0
Dense 2	128	1040
Dropout	128	0
Output	1	17

For the CNN + GRU architecture we combined the dense 5th layer of the CNN architecture with the dense 2nd layer of the GRU architecture as follow:

Table A3. Details of the CNN + GRU architecture considering both feet (12 channels).

Layer	Output Shape	Number of Parameters
Input	Dense 5th of CNN and dense 2nd of GRU	0
Dense	4	36
Output	1	5

References

1. Cacabelos, R. Parkinson's disease: From pathogenesis to pharmacogenomics. *Int. J. Mol. Sci.* **2017**, *18*, 551. [CrossRef]
2. Pahwa, R.; Lyons, K.E. *Handbook of Parkinson's Disease*, 5th ed.; CRC Press: Boca Raton, FL, USA, 2013.
3. Jankovic, J.; Tolosa, E. *Parkinson's Disease & Movement Disorders*, 6th ed.; Wolters Kluwer: Alphen aan den Rijn, The Netherlands, 2015.
4. Kressig, R.W.; Gregor, R.J.; Oliver, A.; Waddell, D.; Smith, W.; O'Grady, M.; Curns, A.T.; Kutner, M.; Wolf, S.L. Temporal and spatial features of gait in older adults transitioning to frailty. *Gait Posture* **2004**, *20*, 30–35. [CrossRef] [PubMed]
5. Ellis, T.; Cavanaugh, J.T.; Earhart, G.M.; Ford, M.P.; Foreman, K.B.; Dibble, L.E. Which measures of physical function and motor impairment best predict quality of life in Parkinson's disease? *Park. Relat. Disord.* **2011**, *17*, 693–697. [CrossRef]
6. Hannink, J.; Kautz, T.; Pasluosta, C.F.; Gasmann, K.G.; Klucken, J.; Eskofier, B.M. Sensor-Based Gait Parameter Extraction with Deep Convolutional Neural Networks. *IEEE J. Biomed. Health Inform.* **2017**, *21*, 85–93. [CrossRef]
7. Sheinerman, K.S.; Umansky, S.R. Early detection of neurodegenerative diseases: Circulating brain-enriched microRNA. *Cell Cycle* **2013**, *12*, 1–2. [CrossRef]
8. De Rijk, M.C.; Launer, L.J.; Berger, K.; Breteler, M.M.; Dartigues, J.F.; Baldereschi, M.; Fratiglioni, L.; Lobo, A.; Martinez-Lage, J.; Trenkwalder, C.; et al. Prevalence of Parkinson's disease in Europe: A collaborative study of population-based cohorts. *Neurology* **2000**, *54*, 21–23. [CrossRef]
9. Pringsheim, T.; Jette, N.; Frolkis, A.; Steeves, T. The prevalence of Parkinson's disease: A systematic review and meta-analysis. *Mov. Disord.* **2014**, *29*, 1583–1590. [CrossRef]
10. Ou, Z.; Pan, J.; Tang, S.; Duan, D.; Yu, D.; Nong, H.; Wang, Z. Global trends in the incidence, prevalence, and years lived with disability of parkinson's disease in 204 countries/territories from 1990 to 2019. *Front. Public Health* **2021**, *9*, 776847.
11. Goetz, C.G.; Tilley, B.C.; Shaftman, S.R.; Stebbins, G.T.; Fahn, S.; Martinez-Martin, P.; Poewe, W.; Sampaio, C.; Stern, M.B.; Dodel, R.; et al. Movement Disorder Society-Sponsored Revision of the Unified Parkinson's Disease Rating Scale (MDS-UPDRS): Scale presentation and clinimetric testing results. *Mov. Disord.* **2008**, *23*, 2129–2170. [CrossRef]
12. Roth, N.; Küderle, A.; Ullrich, M.; Gladow, T.; Marxreiter, F.; Klucken, J.; Eskofier, B.; Kluge, F. Hidden Markov Model based Stride Segmentation on Unsupervised Free-living Gait Data in Parkinson's Disease Patients. *J. Neuroeng. Rehabil.* **2021**, *18*, 1–15 [CrossRef]
13. Carvajal-Castaño, H.A.; Lemos-Duque, J.D.; Orozco-Arroyave, J.R. Effective detection of abnormal gait patterns in Parkinson's disease patients using kinematics, nonlinear, and stability gait features. *Hum. Mov. Sci.* **2022**, *81*, 102891. [CrossRef] [PubMed]
14. Pérez-Toro, P.A.; Vásquez-Correa, J.C.; Arias-Vergara, T.; Garcia-Ospina, N.; Orozco-Arroyave, J.R.; Nöth, E. A Non-linear Dynamics Approach to Classify Gait Signals of Patients with Parkinson's Disease. *Commun. Comput. Inf. Sci.* **2018**, *916*, 268–278. [CrossRef]
15. Pérez-Toro, P.A.; Vásquez-Correa, J.C.; Arias-Vergara, T.; Nöth, E.; Orozco-Arroyave, J.R. Nonlinear dynamics and Poincaré sections to model gait impairments in different stages of Parkinson's disease. *Nonlinear Dyn.* **2020**, *100*, 3253–3276. [CrossRef]

16. Vásquez-Correa, J.C.; Arias-Vergara, T.; Orozco-Arroyave, J.R.; Eskofier, B.; Klucken, J.; Nöth, E. Multimodal Assessment of Parkinson's Disease: A Deep Learning Approach. *IEEE J. Biomed. Health Inform.* **2019**, *23*, 1618–1630. [CrossRef]
17. González, I.; López-Nava, I.H.; Fontecha, J.; Muñoz-Meléndez, A.; Pérez-SanPablo, A.I.; Quiñones-Urióstegui, I. Comparison between passive vision-based system and a wearable inertial-based system for estimating temporal gait parameters related to the GAITRite electronic walkway. *J. Biomed. Inform.* **2016**, *62*, 210–223. [CrossRef] [PubMed]
18. Taborri, J.; Palermo, E.; Rossi, S.; Cappa, P. Gait partitioning methods: A systematic review. *Sensors* **2016**, *16*, 40–42. [CrossRef]
19. Channa, A.; Popescu, N.; Ciobanu, V. Wearable solutions for patients with parkinson's disease and neurocognitive disorder: A systematic review. *Sensors* **2020**, *20*, 2713. [CrossRef]
20. Wahid, F.; Begg, R.K.; Hass, C.J.; Halgamuge, S.; Ackland, D.C. Classification of Parkinson's disease gait using spatial-temporal gait features. *IEEE J. Biomed. Health Inform.* **2015**, *19*, 1794–1802. [CrossRef]
21. Ramirez-Bautista, J.A.; Huerta-Ruelas, J.A.; Chaparro-Cárdenas, S.L.; Hernández-Zavala, A. A Review in Detection and Monitoring Gait Disorders Using In-Shoe Plantar Measurement Systems. *IEEE Rev. Biomed. Eng.* **2017**, *10*, 299–309. [CrossRef]
22. Wang, F.; Yan, L.; Xiao, J. A new hidden markov model algorithm to detect human gait phase based on information fusion combining inertial with plantar pressure. *Sens. Mater.* **2019**, *31*, 2637–2655. [CrossRef]
23. Ghassemi, N.H.; Hannink, J.; Martindale, C.F.; Gaßner, H.; Müller, M.; Klucken, J.; Eskofier, B.M. Segmentation of gait sequences in sensor-based movement analysis: A comparison of methods in Parkinson's disease. *Sensors* **2018**, *18*, 145. [CrossRef]
24. Renggli, D.; Graf, C.; Tachatos, N.; Singh, N.; Meboldt, M.; Taylor, W.R.; Stieglitz, L.; Schmid Daners, M. Wearable Inertial Measurement Units for Assessing Gait in Real-World Environments. *Front. Physiol.* **2020**, *11*, 90. [CrossRef] [PubMed]
25. Hubble, R.P.; Naughton, G.A.; Silburn, P.A.; Cole, M.H. Wearable sensor use for assessing standing balance and walking stability in people with Parkinson's disease: A systematic review. *PLoS ONE* **2015**, *10*, e0123705.
26. Vienne, A.; Barrois, R.P.; Buffat, S.; Ricard, D.; Vidal, P.P. Inertial sensors to assess gait quality in patients with neurological disorders: A systematic review of technical and analytical challenges. *Front. Psychol.* **2017**, *8*, 817. [CrossRef]
27. Sprager, S.; Juric, M.B. Inertial sensor-based gait recognition: A review. *Sensors* **2015**, *15*, 22089–22127. [CrossRef] [PubMed]
28. Tunca, C.; Pehlivan, N.; Ak, N.; Arnrich, B.; Salur, G.; Ersoy, C. Inertial sensor-based robust gait analysis in non-hospital settings for neurological disorders. *Sensors* **2017**, *17*, 825. [CrossRef]
29. Goodfellow, I.; Bengio, Y.; Courville, A. *Deep Learning*; MIT Press: Cambridge, MA, USA, 2016. Available online: http://www.deeplearningbook.org (accessed on 30 June 2022).
30. Cho, K.; van Merriënboer, B.; Bahdanau, D.; Bengio, Y. On the properties of neural machine translation: Encoder–decoder approaches. In Proceedings of the SSST-8, Eighth Workshop on Syntax, Semantics and Structure in Statistical Translation, Doha, Qatar, 25–29 October 2014; Association for Computational Linguistics: Doha, Qatar, 2014; pp. 103–111. [CrossRef]
31. Cho, K.; van Merriënboer, B.; Gulcehre, C.; Bahdanau, D.; Bougares, F.; Schwenk, H.; Bengio, Y. Learning phrase representations using RNN encoder–decoder for statistical machine translation. In Proceedings of the 2014 Conference on Empirical Methods in Natural Language Processing (EMNLP), Doha, Qatar, 25–29 October 2014; Association for Computational Linguistics: Doha, Qatar, 2014; pp. 1724–1734. [CrossRef]
32. Kingma, D.P.; Ba, J. Adam: A Method for Stochastic Optimization. *arXiv* **2014**, arXiv:1412.6980.
33. Han, J.; Pei, J.; Tong, H. *Data Mining*, 4th ed.; Morgan Kaufmann Publishers: Burlington, MA, USA, 2017. [CrossRef]
34. Orozco, J.R.; Vasquez, J.; Noeth, E. Current methods and new trends in signal processing and pattern recognition for the automatic assessment of motor impairments: The case of Parkinson's disease. *Neurol. Disord. Imaging Phys.* **2020**, *5*, 8-1–8-57. [CrossRef]
35. Ashfaque Mostafa, T.; Soltaninejad, S.; McIsaac, T.L.; Cheng, I. A Comparative Study of Time Frequency Representation Techniques for Freeze of Gait Detection and Prediction. *Sensors* **2021**, *21*, 6446. [CrossRef]
36. Sarshar, M.; Polturi, S.; Schega, L. Gait phase estimation by using lstm in imu-based gait analysis—Proof of concept. *Sensors* **2021**, *21*, 5749. [CrossRef]

Article

Age-Associated Changes on Gait Smoothness in the Third and the Fourth Age

Massimiliano Pau [1], Giuseppina Bernardelli [2,3], Bruno Leban [1], Micaela Porta [1], Valeria Putzu [4], Daniela Viale [4,5], Gesuina Asoni [4,5], Daniela Riccio [4,5], Serena Cerfoglio [6,7], Manuela Galli [6] and Veronica Cimolin [6,7,*]

[1] Department of Mechanical, Chemical and Materials Engineering, Piazza d'Armi, 09123 Cagliari, Italy
[2] Department of Clinical Sciences and Community Health (DISCCO), University of Milan, 20122 Milano, Italy
[3] Exercise Medicine Unit, IRCCS Istituto Auxologico Italiano, Via Pier Lombardo 22, 20135 Milano, Italy
[4] Center for Cognitive Disorders and Dementia, ASL Cagliari District, 09121 Cagliari, Italy
[5] Geriatric Unit, SS. Trinità Hospital, 09121 Cagliari, Italy
[6] Department of Electronics, Information and Bioengineering, Politecnico di Milano, Piazza Leonardo da Vinci 32, 20133 Milano, Italy
[7] Istituto Auxologico Italiano, IRCCS, S. Giuseppe Hospital, Piancavallo, 28824 Oggebbio, Italy
* Correspondence: veronica.cimolin@polimi.it; Tel.: +39-02-2399-3359

Abstract: Although gait disorders represent a highly prevalent condition in older adults, the alterations associated with physiologic aging are often not easily differentiable from those originated by concurrent neurologic or orthopedic conditions. Thus, the detailed quantitative assessment of gait patterns represents a crucial issue. In this context, the study of trunk accelerations may represent an effective proxy of locomotion skills in terms of symmetry. This can be carried out by calculating the Harmonic Ratio (HR), a parameter obtained through the processing of trunk accelerations in the frequency domain. In this study, trunk accelerations during level walking of 449 healthy older adults (of age > 65) who were stratified into three groups (Group 1: 65–74 years, $n = 175$; Group 2: 75–85 years, $n = 227$; Group 3: >85 years, $n = 47$) were acquired by means of a miniaturized Inertial Measurement Unit located in the low back and processed to obtain spatio-temporal parameters of gait and HR, in antero-posterior (AP), medio-lateral (ML) and vertical (V) directions. The results show that Group 3 exhibited a 16% reduction in gait speed and a 10% reduction in stride length when compared with Group 1 ($p < 0.001$ in both cases). Regarding the cadence, Group 3 was characterized by a 5% reduction with respect to Groups 1 and 2 ($p < 0.001$ in both cases). The analysis of HR revealed a general trend of linear decrease with age in the three groups. In particular, Group 3 was characterized by HR values significantly lower (-17%) than those of Group 1 in all three directions and significantly lower than Group 2 in ML and V directions (-10%). Taken together, such results suggest that HR may represent a valid measure to quantitatively characterize the progressive deterioration of locomotor abilities associated with aging, which seems to occur until the late stages of life.

Keywords: gait; harmonic ratio (HR); smoothness; symmetry; older adults; inertial sensor

Citation: Pau, M.; Bernardelli, G.; Leban, B.; Porta, M.; Putzu, V.; Viale, D.; Asoni, G.; Riccio, D.; Cerfoglio, S.; Galli, M.; et al. Age-Associated Changes on Gait Smoothness in the Third and the Fourth Age. *Electronics* **2023**, *12*, 637. https://doi.org/10.3390/electronics12030637

Academic Editors: Gabriella Olmo and Florenc Demrozi

Received: 22 December 2022
Revised: 18 January 2023
Accepted: 22 January 2023
Published: 27 January 2023

1. Introduction

Gait is a fundamental physical activity of daily life and represents an important factor for independent living. However, gait efficiency undergoes significant changes with age [1–3]. In fact, the physiologic decline of the musculoskeletal system and cognitive performance associated with aging [4] leads to reduced movement smoothness and cognitive reserve, thus impairing several aspects of mobility associated with daily life tasks. This primarily affects walking, which results in altered automaticity and skill [5,6] but also affects other movements, such as turning and sitting to standing and vice versa [7]. Moreover, it should be recalled that, in older adults, most falls occur while walking, which emphasizes the importance of having a stable gait as a preventive countermeasure against such hazardous events [8].

In this context, the detailed analysis of the gait characteristics appears crucial to define the current status of the individual and, where necessary, plan specific interventions able to ensure that a sufficient degree of mobility is preserved during the late stages of life. Gait is usually investigated in both spatial and temporal domains [9], and its main features can be classified into relatively independent domains, with pace, rhythm, variability, symmetry, and postural control being probably the most important ones [10]. The parameters belonging to each of these domains can be assessed using several kinds of systems, such as motion-capture systems, electronic walkways, and, more recently, wearable Inertial Measurement Units (IMUs). IMUs employed for human movement analysis are basically stand-alone microelectromechanical systems that integrate multiaxial inertial sensors. A typical configuration, which includes a 3-axis accelerometer, a 3-axis gyroscope, and a 3-axis magnetometer, allow measuring acceleration, angular speed, and magnetic vector field of a moving object in a three-dimensional space, providing up to six degrees of freedom [11,12]. Since modern IMUs are designed to be small, lightweight, economic, and unobtrusive, their use quickly gained popularity among researchers involved in human movement analysis. To date, they are considered a reliable and affordable solution to assess gait in a variety of environments, as they do not require dedicated spaces or complex laboratory settings [13,14]. In particular, contrary to the equipment present in the traditional movement analysis laboratories, they allow individuals to be tested while wearing their usual clothes and shoes, thus ensuring good ecological validity [15]. IMUs can provide a new dimension of granularity for gait analysis and are increasingly used in research studies [16,17]. Although the number and placement of sensors can be variable, the simple setup which makes use of a single sensor (usually located in the low back) is widely employed as it ensures a minimum encumbrance for tested individuals, thus allowing gait to be performed freely under habitual conditions and type of terrains [18]. Several metrics derived from trunk accelerations during gait have been associated with specific features such as pattern regularity (through Recurrence Quantification Analysis [19]), motor complexity (through Multiscale Entropy Analysis [20]), gait stability (using short Lyapunov exponents [20]), and step-to-step symmetry or rhythmicity/smoothness (through calculation of the Harmonic Ratio, HR, [21]).

Particularly the HR, which is obtained by processing trunk accelerations in the frequency domain for antero-posterior (AP), vertical (V), and medio-lateral (ML) directions, has been demonstrated as a valid and robust metric useful to quantify step-to-step symmetry and to describe the overall smoothness/rhythmicity of gait. As higher values of HRs are associated with greater smoothness/symmetry, this parameter can be considered a good indicator of whole-body balance during gait [22,23], and, to date, some evidence supports the pivotal role of HRs in discriminating gait variations consequent to neurologic [24–26] and orthopaedic [27] conditions. Moreover, HRs are sensitive to subtle changes in gait smoothness which may occur even in the presence of normal spatio-temporal parameters [26,28].

Among other applications, HR has been employed to characterize age-associated changes in the smoothness of gait as its value has been found to increase when passing from childhood to adolescence and maturity (where a maximum is reached), while it tends to decrease during aging [23,29,30]. In this context, such parameters would potentially be useful to discriminate physiologic gait alterations from those associated with specific pathologic conditions, including cognitive deficits, in older adults. However, it is noteworthy that there are few applications of this approach to investigate the role of aging in terms of smoothness modifications [21,23,29,31–34]. Brach et al. [29] aimed to validate the discriminative power of HR by testing groups of young and old participants across different walking conditions (i.e., straight and curved path, dual task). They found that older adults had lower HR in the AP direction, indicating a less smooth strategy in the direction of motion. Lowry et al. [23] examined age-related differences in HRs across a range of self-selected overground walking speeds, finding that young and older adults exhibited similar HRs in all directions of motion across speeds, while old-old adults exhibited lower HR in AP and V directions. However, no differences were observed in HRs calculated for natural and faster speeds, with the exception of reduced HR in the V direction in the very fast condition for the older groups. The HR in the ML direction was not different between groups and varied less across speeds. Lowry et al. [31] investigated age-related differences in locomotor strategies during an adaptive walking task (i.e., walking with narrow and wide step widths). They demonstrated that, compared to young adults, older adults generally had greater reductions in the variables used to describe forward progression (HR in AP direction) in both narrow and wide step width. In contrast, the pattern of results for ML control was similar between young and older adults. In the study by Misu et al. [32], HR was employed to assess possible changes associated with nutritional status in a group of community-dwelling older adults. They found significantly reduced HR in the ML direction in those characterized by a poor nutritional status and hypothesized that this aspect could affect lateral trunk control. Asai et al. [33] used HR to assess whether fall history and the fear of falling contribute to the smoothness of lower trunk oscillation during walking in older adults living in the local community. Row Lazzarini et al. [34] examined the effects of speed and treadmill walking (TW) on the smoothness and rhythmicity of 40 men and women aged 70–96 years. They concluded that the use of treadmills for gait smoothness and rhythmicity studies in older adults is problematic as some participants were not able to achieve overground speed during TW; walking at the overground speed on a treadmill improves rhythmicity and ML smoothness, and walking at the slower preferred treadmill walking speed worsens vertical and AP gait smoothness. At last, Pau et al. [35] reported that, in older adults, the existence of a cognitive deficit is associated with a significant reduction of HR in AP and V directions with respect to cognitively intact individuals and that HR values in all the three directions resulted moderately correlated with the cognitive performance assessed using either Mini Mental State Examination (MMSE) or Addenbrooke's Cognitive Examination Revised (ACE-R).

The existing literature seems to support the hypothesis that HR may represent a suitable measure to describe the changes in gait smoothness associated with aging. However, studies on this topic are quite limited and often carried out in small groups and/or approximately around the age of 70–75 years. Moreover, only one study [23] included the presence of a small sample (13 participants) of the oldest-old adults (i.e., those aged 85 and over). As the effects of aging on gait become significantly stronger approximately around the age of 80 years [36], it could be interesting to specifically investigate the reductions of gait smoothness in such individuals.

Based on the aforementioned considerations, in this study, we aim to provide reference values of HR during gait useful to characterize the changes occurring during aging in a large cohort of healthy individuals aged 65 and over, including the oldest-old participants. Our hypothesis is that aging is associated, other than with changes in spatio-temporal parameters previously recognized [1,37], also by modifications of gait smoothness that may indicate a progressive deterioration of locomotor abilities.

2. Materials and Methods

2.1. Participants

During the period November 2019 to June 2022, 863 older adults were screened for eligibility at the Center for Cognitive Disorders and Dementia (in collaboration with the Geriatric Unit of "SS. Trinità" General Hospital, Cagliari, Italy) and the University of Milan (Milan, Italy). Eligibility criteria included: (1) age over 65 years; (2) ability to walk independently (i.e., without an assistive device or the assistance of another person); (3) being free from either neuromuscular disorders impairing movement (including but not limited to Parkinson's disease, stroke, and multiple sclerosis) or spinal disorder affecting accelerometer placement; (4) being cognitively intact (i.e., MMSE score > 26); and (5) being free from depressive symptoms (i.e., score on 30-item Geriatric Depression Scale > 10).

Four hundred forty-nine individuals matched the inclusion criteria and were enrolled in the study and stratified into three groups as follows:

- Group 1 (age 65–74 years, n = 175);
- Group 2 (age 75–85 years, n = 227);
- Group 3 (age >85 years, n = 47).

The selection process is shown in Figure 1.

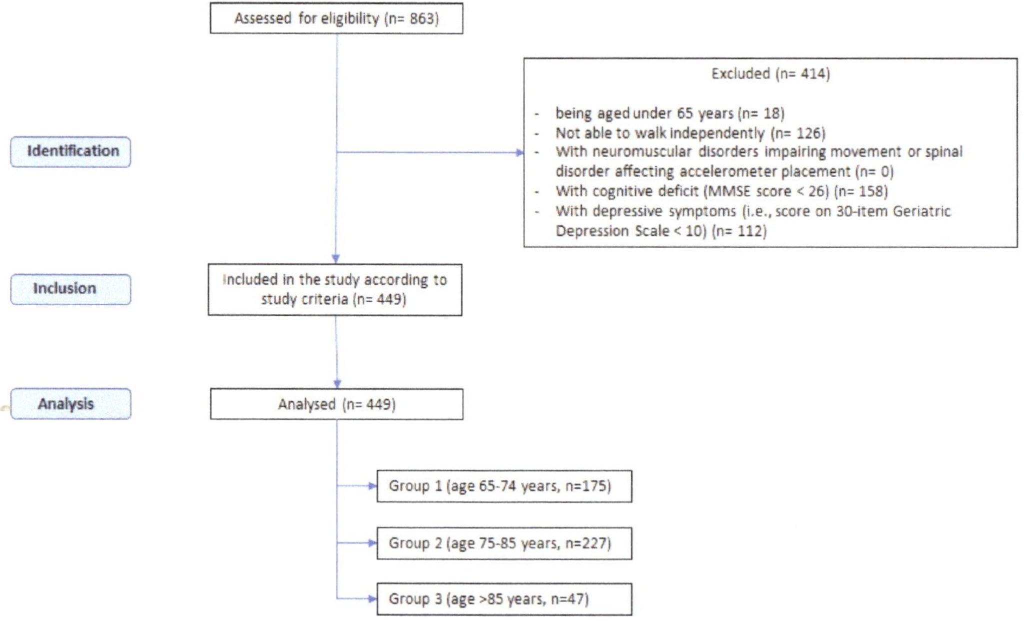

Figure 1. Process of participants' selection.

The anthropometric features of the participants are reported in Table 1.

Table 1. Participant's characteristics. Values are expressed as mean ± SD.

	Group 1 (65–74 Years)	Group 2 (75–85 Years)	Group 3 (>85 Years)
Participants # (F, M)	175 (103 F, 72 M)	228 (128 F, 100 M)	47 (28 F, 19 M)
Participants percentage (F, M)	F 59%, M 41%	F 56%, M 44%	F 60%, M 40%
Age (years)	70.4 ± 2.5	79.1 ± 2.8 [a]	86.5 ± 1.7 [a,b]
Body Mass (kg)	66.8 ± 12.4	65.6 ± 11.4	61.3 ± 13.6 [a,b]
Height (cm)	162.0 ± 8.4	160.0 ± 8.7 [a]	158.6 ± 8.5 [a]

The symbol [a] indicates a significant difference for Group 1; the symbol [b] indicates a significant difference for Group 2.

The study, which was conducted in accordance with the Declaration of Helsinki of 1964 and its latest amendments, was approved by the ethical committees of the University of Milan (authorization number 12_2019) and ATS Sardegna, Italy (authorization number 300/2021/CE). Written informed consent was obtained from all participants.

2.2. Data Acquisition

A small, lightweight inertial sensor (G-Sensor®, BTS Bioengineering, Italy), previously validated for the assessment of gait spatio-temporal parameters in healthy individuals [38] and previously used to assess gait in older adults [33,39,40], was attached to participants' trunk (at the L4-L5 vertebrae level) using a dedicated semi-elastic belt (see Figure 2). After a short familiarization period, participants were required to walk, at a self-selected speed and in the most natural manner, along a 30 m hallway following a straight trajectory. The device acquired the linear accelerations in the three directions (AP, ML, and V) at 100 Hz frequency, then transmitted in real-time via Bluetooth to a personal computer to be stored as ASCII files. Subsequently, data were processed by means of a custom Matlab® routine to calculate the gait parameters of interest. In the first 5 s of the acquisition, the participant is required to stand without moving; this period was employed to confirm the sensor orientation and to adjust the acceleration vector data during the data collection. The most relevant spatio-temporal parameters (gait speed, cadence, stride length, stance, and double support phase duration) were computed starting from the raw acceleration data, according to the peak-detection algorithm formulated by Zijlstra et al. [41].

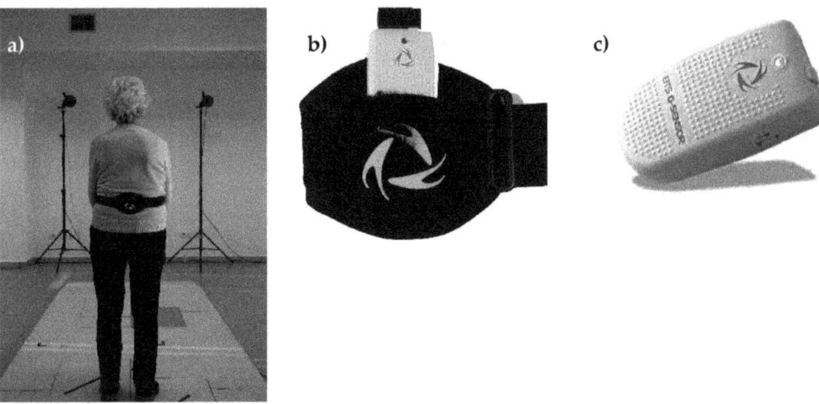

Figure 2. From left to right: (**a**) participant equipped with the IMU for the gait trials; (**b**) detail of the IMU positioning inside the semi-elastic belt; (**c**) the BTS G-Sensor IMU used for the experimental tests.

Instead, HRs were calculated using the approach proposed by Menz, Lord, and Fitzpatrick in 2003 [19]. In short, the accelerations of the trunk collected by the IMU in the three orthogonal directions are handled in the frequency domain via a finite Fourier series. Then, the HRs for the AP and V directions are calculated using Eq. 1 as the ratio between the sum of the amplitudes (A) of the first ten even harmonics (associated with the in-phase components of the signal) and the sum of the amplitudes of the first ten odd harmonics (which contain the out-of-phase components), the latter being minimized as gait smoothness improves. In the case of ML direction, the calculation is slightly different (see Eq. (2)). In fact, since the acceleration pattern is characterized by one peak per stride (thus resulting in the dominance of the first harmonic and subsequent odd harmonics), in this case, HR ML is obtained by dividing the sum of the amplitudes of the odd harmonics divided by the sum of the amplitudes of the even harmonics.

$$HR_{AP-V} = \frac{\sum A_{even\ harmonics}}{\sum A_{odd\ harmonics}} \tag{1}$$

$$HR_{ML} = \frac{\sum A_{odd\ harmonics}}{\sum A_{even\ harmonics}} \tag{2}$$

HR values are quite simple to interpret, being lower values indicative of a less smooth/symmetrical gait. Previous studies reported that healthy older adults are characterized by values of HR approximately from 3–4 in the AP and V directions and from 2.1–2.6 in the ML direction [21,23,25,34].

2.3. Statistical Analysis

A two-way multivariate analysis of variance (MANOVA) was used to verify the presence of differences among the three groups in terms of spatio-temporal parameters and HRs. In particular, regarding HRs, previous studies indicated its sensitivity to gait speed (i.e., higher speed originates higher HR values [23]), and its value is expected to differ across different age groups. Thus, it appears necessary to include it in the analysis as a covariate.

The independent variables were the participant's age group, while the dependent variables were, in one case, the six spatio-temporal parameters and, in the other, the three HRs. The statistical significance level was set at $p < 0.05$, and the effect sizes were evaluated via the eta-squared (η^2) coefficient. Univariate analysis of variance (ANOVA) was used as a post hoc test through a reduction of the significance level according to the Bonferroni correction for multiple comparisons ($p = 0.008, 0.05/6$) for spatio-temporal parameters and $p = 0.016, 0.05/3$) for HRs). Data were analyzed using the IBM SPSS Statistics v.23 software (IBM, Armonk, NY, USA).

3. Results

Table 2 reports the experimental test results regarding the spatio-temporal parameters of gait and HRs for each age group. A significant main effect of the group [$F(12, 878) = 4.38$, $p < 0.001$, Wilks $\lambda = 0.89$, $\eta^2 = 0.06$] was found by the MANOVA on spatio-temporal parameters of gait. In particular, the oldest participants (Group 3) were characterized by a 16% reduction in gait speed and by a 10% reduction in stride length ($p < 0.001$) when compared with the performance of the Group 1 ($p < 0.001$ in both cases), resulting after the post hoc analysis. Slightly smaller (yet statistically significant) reductions in speed and stride length (approximately 9% in both cases) were also observed between individuals of Group 1 and Group 2. Regarding the cadence, the statistical analysis revealed that Group 3 was characterized by a 5% reduction with respect to Groups 1 and 2 ($p < 0.001$ in both cases). In contrast, no significant differences were observed for the duration of the stance, swing, and double support phases.

Table 2. Mean and standard deviation values of the spatio-temporal and HR (Harmonic Ratio) parameters in the three considered groups.

		Group 1 (65–74 Years)	Group 2 (75–85 Years)	Group 3 (>85 Years)
Spatial-temporal parameters of gait	Gait speed (m s^{-1})	1.08 ± 0.24	0.98 ± 0.24 [a]	0.91 ± 0.26 [a]
	Stride length (m)	1.16 ± 0.21	1.05 ± 0.22 [a]	1.04 ± 0.27 [a]
	Cadence (steps min^{-1})	111.20 ± 9.46	111.61 ± 10.75	105.20 ± 13.00 [a,b]
	Stance phase (% of the GC)	60.46 ± 2.55	60.82 ± 1.97	61.28 ± 2.74
	Swing phase (% of the GC)	39.42 ± 3.00	39.06 ± 2.73	39.04 ± 3.32
	Double support phase (% of the GC)	10.56 ± 2.08	10.80 ± 1.98	11.21 ± 2.71
Harmonic Ratio *	AP direction *	3.63 ± 1.03	3.15 ± 0.97 [a]	3.01 ± 0.89 [a]
	ML direction *	2.60 ± 0.80	2.42 ± 0.69 [a]	2.17 ± 0.58 [a,b]
	V direction *	3.57 ± 0.97	3.33 ± 0.86 [a]	2.96 ± 0.88 [a,b]

The symbol [a] indicates a significant difference for Group 1 after Bonferroni correction ($p = 0.016$), the symbol [b] indicates a significant difference for Group 2 after Bonferroni correction ($p = 0.016$); * controlled for gait speed; GC: Gait Cycle.

The trend of HR for AP, ML, and V directions across the analysis groups is shown in Figure 3. Including gait speed as the covariate, MANCOVA detected a significant main effect of age on HR values [$F(6,882) = 3.10$, $p = 0.005$, Wilks $\lambda = 0.96$, $\eta^2 = 0.02$]. The post hoc analysis showed that older participants (i.e., Group 3) exhibited a quite uniform reduction of HR with respect to those of Group 1 for all three directions of approximately 17% ($p < 0.001$), while differences vs. Group 2 involved only the ML and V directions and were smaller (-10%). Finally, significant differences were also found between Group 2 and

Group 1 for HR in all three directions, as those aged 75–85 exhibited reduced HR values (approximately between 6 and 13%).

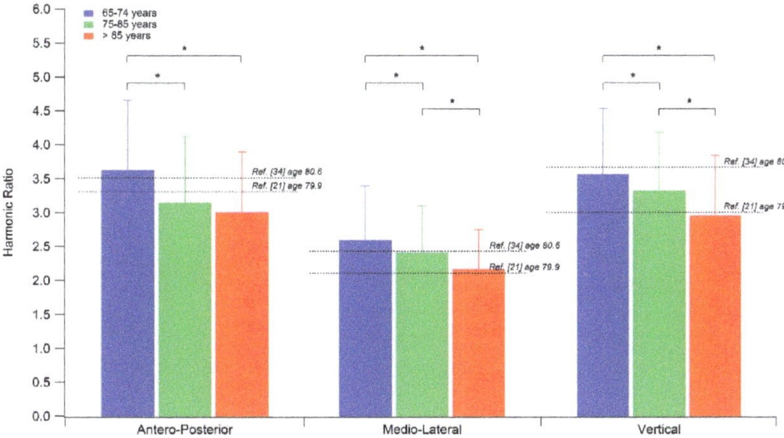

Figure 3. Harmonic Ratio values in AP, ML, and V directions for the three groups of tested older adults. Data from the present study are compared with those previously reported for similar age ranges (Refs. [21,34]).

4. Discussion

In this study, we aimed to quantitatively investigate the existence of age-related alterations of gait patterns in a large cohort of healthy older adults aged 65 and over, including the oldest-old participants, assessed in a clinical environment via a wearable inertial sensor. The hypothesis was explored by analyzing HRs computed from trunk acceleration and the most common spatio-temporal parameters. In particular, HRs, representative of gait smoothness, can be considered an effective indicator of whole-body balance during gait, and it was already demonstrated to be a measure suitable to describe the changes in ambulation associated with aging. In this regard, although there is previous evidence of age-related reductions of HRs during walking [29,33,42], little data is available as regards the oldest-old adults. We attempted here to overcome such a limitation by testing a large cohort of healthy individuals, which also included a group of 47 participants aged 85–90.

Our data confirm that aging is associated with significant changes in spatio-temporal parameters. In particular, as for gait speed and stride length, the youngest participants (Group 1) exhibited higher values when compared with the other two groups, while as regards the cadence, significantly lower values were found in Group 3 compared to Group 1 and Group 2. In contrast, no significant differences among groups were found in terms of stance, swing, and double support phase duration, although a consistent trend of variation with age (i.e., stance and double support phases increase, swing phase duration decreases) was observed. It is noteworthy that the observed speed changes are consistent with those reported in previous studies, which showed a continuous reduction in gait speed in older adults, especially from the seventh decade [43,44]. These data align with Hollman et al., 2011 [36], which presents the normative spatio-temporal gait parameters in older adults, and a more recent study [45] which reports the reference values for usual gait speed in community-dwelling older adults living in Western Europe. Moreover, since such a reduction is accompanied by a correspondent step/stride shortening (particularly when passing from 65 to 75 years), our data confirm that older adults adopt a cautious strategy in order to achieve better stability in locomotion and, consequently, decrease the risk of falling [46–49].

As regards HRs, the values calculated in the present study are quite consistent with those previously reported for individuals of similar age ranges [21,23,33–35]. For example, in Lowry et al. [21], the adults aged 80–86 years exhibited lower HR in the AP and V directions compared to those aged 60–69 years, and no differences in the ML direction were detected between the Groups. In Brach et al. (2011) [29], older adults (mean age: 77.5 years) had lower HR in the AP direction than young adults (mean age = 24.4 years).

This indicates that, despite the possible differences involving equipment, measurement protocol, and data processing, the HR parameter could represent a sensitive approach. The main finding which emerges from the trunk acceleration analysis is that aging is associated with a substantially linear decrease of HRs in all directions, with the oldest-old participants characterized by the lowest values in all three directions. As previously mentioned, it is known that older adults are usually characterized by lower HRs compared to young adults [23,29], and among them, the co-existence of cognitive decline enhances this phenomenon [33]. In this regard, our results demonstrate that smoothness tends to further worsen in those aged 85 and over (Group 3), thus suggesting that changes in motor control abilities continue to occur until the late stages of life. However, it is difficult to perform a comparison as the age ranges are not similar among studies, and also equipment and measurement protocols are not uniform [23,36]. Alterations in limb dynamics and a different distribution of joint torques and powers on lower limb joints towards proximal segments (i.e., older adults tend to perform more work at the hip and less at the ankle) have been identified as possible factors able to increase irregularity of trunk accelerations and thus consequently able originate reduced smoothness [23]. However, it should also be recalled that the control of ML motion during the stance phase of gait represents a main risk factor for falls in older adults [50,51]. It is, thus, possible that reduced HRs (in particular those of ML directions) are one of the expressions of the cautious strategy adopted to keep the center of mass safely between their feet and thus preserve balance. In fact, an optimal balance during walking requires continuous integrative control, particularly in a lateral direction, due to inherent instability associated with single limb support [52]. There is now considerable evidence for the effects of age on ML motion [53–56], some of which have been associated with increased fall risk [57].

As HR is a parameter that is very sensitive to even subtle changes in gait smoothness, it could be used as an outcome measure of rehabilitation/training programs aimed at improving gait in older people in combination with the conventional spatio-temporal parameters. In particular, the literature suggests that structured exercise programs, prescribed and designed according to individual clinical conditions, age, and goal/s to reach, can now be considered a strategy to maintain and improve physical function in older people [58–60]. In particular, it was demonstrated that exercise programs reduce the rate of falls and the number of people experiencing falls in older people living in the community [61] and that physical exercises, including functional mobility training, especially walking, have better results than physical programs with only static, resistance, and flexibility training, especially in those with cognitive deficit [62]. It is also important to stimulate them to regularly perform physical activity to improve general well-being and cognition, also considering home-based exercise programs.

Some limitations of the study should be acknowledged. Firstly, the proposed stratification resulted in a Group (i.e., those aged 85 and over) that was markedly smaller than the others. This reflects, to some extent, the need to include participants free from significant mobility restrictions and cognitively intact, which is not easy to achieve considering that such issues are quite common during the late stage of life. Secondly, since this study focused solely on spatio-temporal parameters and HRs, we could only speculate about the mechanisms underlying the pattern of results. Future developments of the study should aim to combine kinematic and kinetic features of gait with HRs and other trunk acceleration-derived measures, to have a detailed and exhaustive picture of the control of body motion during walking and even to understand which measures are most sensitive to age-related changes in gait.

5. Conclusions

In the present study, the possible changes in spatio-temporal parameters and smoothness of gait associated with aging were explored in a cohort of youngest-, middle-, and oldest-old, using parameters derived from trunk acceleration collected in a clinical context using a simple setup composed by a single miniaturized IMU. Our data showed the presence of significant alterations in gait according to aging, reporting a reduced speed and stride length and a reduction of HR in the three directions. The latter changes were similar in magnitude across the three groups and suggested that smoothness similarly worsens in all directions until the late stage of life. Considering the sensitivity of HR to the presence of physical and cognitive conditions which interfere with mobility, the analysis of smoothness of gait may be considered a useful and valid tool for the early detection of subtle changes in gait in older adults, which that the spatio-temporal parameters alone could fail to highlight. This parameter could also be used in clinical practice by a physician and a physical therapist.

Author Contributions: Conceptualization, M.P. (Massimiliano Pau); G.B.; M.G. and V.C.; methodology, M.P. (Massimiliano Pau) and V.C.; software, B.L.; M.P. (Micaela Porta) and S.C.; formal analysis, M.P. (Massimiliano Pau); Clinical Assessment of the Participants, G.B.; V.P.; D.V.; D.R. and G.A.; data curation, M.P. (Massimiliano Pau) and V.C.; writing—original draft preparation, M.P. (Massimiliano Pau); B.L.; M.P. (Micaela Porta); S.C. and V.C.; writing—review and editing, M.P. (Massimiliano Pau), G.B. and V.C.; supervision, M.P. (Massimiliano Pau) and V.C. All authors have read and agreed to the published version of the manuscript.

Funding: This research received no external funding.

Institutional Review Board Statement: The study was conducted in accordance with the Declaration of Helsinki, and approved by the ethical committees of the University of Milan (authorization number 12_2019) and ATS Sardegna, Italy (authorization number 300/2021/CE).

Informed Consent Statement: Informed consent was obtained from all subjects involved in the study.

Data Availability Statement: Data available on request due to restrictions, e.g., privacy or ethical.

Conflicts of Interest: The authors declare no conflict of interest.

References

1. DeVita, P.; Hortobagyi, T. Age Causes a Redistribution of Joint Torques and Powers during Gait. *J. Appl. Physiol.* **2000**, *88*, 1804–1811. [CrossRef] [PubMed]
2. McGibbon, C.A.; Krebs, D.E. Age-Related Changes in Lower Trunk Coordination and Energy Transfer during Gait. *J. Neurophysiol.* **2001**, *85*, 1923–1931. [CrossRef] [PubMed]
3. Ko, S.; Stenholm, S.; Metter, E.J.; Ferrucci, L. Age-Associated Gait Patterns and the Role of Lower Extremity Strength - Results from the Baltimore Longitudinal Study of Aging. *Arch. Gerontol. Geriatr.* **2012**, *55*, 474–479. [CrossRef]
4. Harada, C.N.; Natelson Love, M.C.; Triebel, K.L. Normal Cognitive Aging. *Clin. Geriatr. Med.* **2013**, *29*, 737–752. [CrossRef]
5. Lajoie, Y.; Teasdale, N.; Bard, C.; Fleury, M. Upright Standing and Gait: Are There Changes in Attentional Requirements Related to Normal Aging? *Exp. Aging Res.* **1996**, *22*, 185–198. [CrossRef]
6. Woollacott, M.; Shumway-Cook, A. Attention and the Control of Posture and Gait: A Review of an Emerging Area of Research. *Gait Posture* **2002**, *16*, 1–14. [CrossRef] [PubMed]
7. Sunderaraman, P.; Maidan, I.; Kozlovski, T.; Apa, Z.; Mirelman, A.; Hausdorff, J.M.; Stern, Y. Differential Associations between Distinct Components of Cognitive Function and Mobility: Implications for Understanding Aging, Turning and Dual-Task Walking. *Front. Aging Neurosci.* **2019**, *11*, 1–13. [CrossRef] [PubMed]
8. Boyé, N.D.A.; Mattace-Raso, F.U.S.; Van der Velde, N.; Van Lieshout, E.M.M.; De Vries, O.J.; Hartholt, K.A.; Kerver, A.J.H.; Bruijninckx, M.M.M.; Van der Cammen, T.J.M.; Patka, P.; et al. Circumstances Leading to Injurious Falls in Older Men and Women in the Netherlands. *Injury* **2014**, *45*, 1224–1230. [CrossRef]
9. Zijlstra, W.; Hof, A.L. Assessment of Spatio-Temporal Gait Parameters from Trunk Accelerations during Human Walking. *Gait Posture* **2003**, *18*, 1–10. [CrossRef]
10. Lord, S.; Galna, B.; Verghese, J.; Coleman, S.; Burn, D.; Rochester, L. Independent Domains of Gait in Older Adults and Associated Motor and Nonmotor Attributes: Validation of a Factor Analysis Approach. *J. Gerontol Ser. A Biol. Sci. Med. Sci.* **2013**, *68*, 820–827. [CrossRef]
11. Roetenberg, D.; Luinge, H.; Slycke, P. Xsens MVN: Full 6DOF Human Motion Tracking Using Miniature Inertial Sensors. *Xsens Motion Technol. BV* **2009**, 1–7.

12. Seel, T.; Raisch, J.; Schauer, T. IMU-Based Joint Angle Measurement for Gait Analysis. *Sensors* **2014**, *14*, 6891–6909. [CrossRef] [PubMed]

13. Shrivastava, S.; Trung, T.Q.; Lee, N.E. Recent Progress, Challenges, and Prospects of Fully Integrated Mobile and Wearable Point-of-Care Testing Systems for Self-Testing. *Chem. Soc. Rev.* **2020**, *49*, 1812–1866. [CrossRef] [PubMed]

14. Ji, X.; Rao, Z.; Zhang, W.; Liu, C.; Wang, Z.; Zhang, S.; Zhang, B.; Hu, M.; Servati, P.; Xiao, X. Airline Point-of-Care System on Seat Belt for Hybrid Physiological Signal Monitoring. *Micromachines* **2022**, *13*, 1880. [CrossRef]

15. Cho, Y.-S.; Jang, S.-H.; Cho, J.-S.; Kim, M.-J.; Lee, H.D.; Lee, S.Y.; Moon, S.-B. Evaluation of Validity and Reliability of Inertial Measurement Unit-Based Gait Analysis Systems. *Ann. Rehabil. Med.* **2018**, *42*, 872–883. [CrossRef] [PubMed]

16. Horak, F.; King, L.; Mancini, M. Role of Body-Worn Movement Monitor Technology for Balance and Gait Rehabilitation. *Phys. Ther.* **2015**, *95*, 461–470. [CrossRef]

17. Maetzler, W.; Rochester, L. Body-Worn Sensors–the Brave New World of Clinical Measurement? *Mov. Disord. Off. J. Mov. Disord. Soc.* **2015**, *30*, 1203–1205. [CrossRef]

18. Del Din, S.; Godfrey, A.; Galna, B.; Lord, S.; Rochester, L. Free-Living Gait Characteristics in Ageing and Parkinson's Disease: Impact of Environment and Ambulatory Bout Length. *J. Neuroeng. Rehabil.* **2016**, *13*, 1–12. [CrossRef]

19. Labini, F.S.; Meli, A.; Ivanenko, Y.P.; Tufarelli, D. Recurrence Quantification Analysis of Gait in Normal and Hypovestibular Subjects. *Gait Posture* **2012**, *35*, 48–55. [CrossRef]

20. Bisi, M.C.; Riva, F.; Stagni, R. Measures of Gait Stability: Performance on Adults and Toddlers at the Beginning of Independent Walking. *J. Neuroeng. Rehabil.* **2014**, *11*, 131. [CrossRef]

21. Menz, H.B.; Lord, S.R.; Fitzpatrick, R.C. Acceleration Patterns of the Head and Pelvis When Walking Are Associated with Risk of Falling in Community-Dwelling Older People. *J. Gerontol. Ser. A Biol. Sci. Med. Sci.* **2003**, *58*, M446–M452. [CrossRef]

22. Bellanca, J.L.; Lowry, K.A.; VanSwearingen, J.M.; Brach, J.S.; Redfern, M.S. Harmonic Ratios: A Quantification of Step to Step Symmetry. *J. Biomech.* **2013**, *46*, 828–831. [CrossRef] [PubMed]

23. Lowry, K.A.; Lokenvitz, N.; Smiley-Oyen, A.L. Age- and Speed-Related Differences in Harmonic Ratios during Walking. *Gait Posture* **2012**, *35*, 272–276. [CrossRef] [PubMed]

24. Menz, H.B.; Lord, S.R.; St George, R.; Fitzpatrick, R.C. Walking Stability and Sensorimotor Function in Older People with Diabetic Peripheral Neuropathy. *Arch. Phys. Med. Rehabil.* **2004**, *85*, 245–252. [CrossRef]

25. Lowry, K.A.; Smiley-Oyen, A.L.; Carrel, A.J.; Kerr, J.P. Walking Stability Using Harmonic Ratios in Parkinson's Disease. *Mov. Disord. Off. J. Mov. Disord. Soc.* **2009**, *24*, 261–267. [CrossRef]

26. Pau, M.; Mandaresu, S.; Pilloni, G.; Porta, M.; Coghe, G.; Marrosu, M.G.; Cocco, E. Smoothness of Gait Detects Early Alterations of Walking in Persons with Multiple Sclerosis without Disability. *Gait Posture* **2017**, *58*, 307–309. [CrossRef] [PubMed]

27. Iosa, M.; Paradisi, F.; Brunelli, S.; Delussu, A.S.; Pellegrini, R.; Zenardi, D.; Paolucci, S.; Traballesi, M. Assessment of Gait Stability, Harmony, and Symmetry in Subjects with Lower-Limb Amputation Evaluated by Trunk Accelerations. *J. Rehabil. Res. Dev.* **2014**, *51*, 623–634. [CrossRef]

28. Cimolin, V.; Cau, N.; Sartorio, A.; Capodaglio, P.; Galli, M.; Tringali, G.; Leban, B.; Porta, M.; Pau, M. Symmetry of Gait in Underweight, Normal and Overweight Children and Adolescents. *Sensors* **2019**, *19*, 2054. [CrossRef]

29. Brach, J.S.; McGurl, D.; Wert, D.; Vanswearingen, J.M.; Perera, S.; Cham, R.; Studenski, S. Validation of a Measure of Smoothness of Walking. *J. Gerontol. Ser. A Biol. Sci. Med. Sci.* **2011**, *66*, 136–141. [CrossRef]

30. Leban, B.; Cimolin, V.; Porta, M.; Arippa, F.; Pilloni, G.; Galli, M.; Pau, M. Age-Related Changes in Smoothness of Gait of Healthy Children and Early Adolescents. *J. Mot. Behav.* **2020**, *52*, 694–702. [CrossRef] [PubMed]

31. Lowry, K.A.; Sebastian, K.; Perera, S.; Van Swearingen, J.; Smiley-Oyen, A.L. Age-Related Differences in Locomotor Strategies During Adaptive Walking. *J. Mot. Behav.* **2017**, *49*, 435–440. [CrossRef] [PubMed]

32. Misu, S.; Asai, T.; Doi, T.; Sawa, R.; Ueda, Y.; Saito, T.; Nakamura, R.; Murata, S.; Sugimoto, T.; Yamada, M.; et al. Association between Gait Abnormality and Malnutrition in a Community-Dwelling Elderly Population. *Geriatr. Gerontol. Int.* **2017**, *17*, 1155–1160. [CrossRef]

33. Asai, T.; Misu, S.; Sawa, R.; Doi, T.; Yamada, M. The Association between Fear of Falling and Smoothness of Lower Trunk Oscillation in Gait Varies According to Gait Speed in Community-Dwelling Older Adults. *J. Neuroeng. Rehabil.* **2017**, *14*, 1–9. [CrossRef]

34. Row Lazzarini, B.S.; Kataras, T.J. Treadmill Walking Is Not Equivalent to Overground Walking for the Study of Walking Smoothness and Rhythmicity in Older Adults. *Gait Posture* **2016**, *46*, 42–46. [CrossRef] [PubMed]

35. Pau, M.; Mulas, I.; Putzu, V.; Asoni, G.; Viale, D.; Mameli, I.; Leban, B.; Allali, G. Smoothness of Gait in Healthy and Cognitively Impaired Individuals: A Study on Italian Elderly Using Wearable Inertial Sensor. *Sensors* **2020**, *20*, 3577. [CrossRef]

36. Hollman, J.H.; McDade, E.M.; Petersen, R.C. Normative Spatiotemporal Gait Parameters in Older Adults. *Gait Posture* **2011**, *34*, 111–118. [CrossRef] [PubMed]

37. Brach, J.S.; Perera, S.; Studenski, S.; Katz, M.; Hall, C.; Verghese, J. Meaningful Change in Measures of Gait Variability in Older Adults. *Gait Posture* **2010**, *31*, 175–179. [CrossRef]

38. Bugané, F.; Benedetti, M.G.; Casadio, G.; Attala, S.; Biagi, F.; Manca, M.; Leardini, A. Estimation of Spatial-Temporal Gait Parameters in Level Walking Based on a Single Accelerometer: Validation on Normal Subjects by Standard Gait Analysis. *Comput. Methods Programs Biomed.* **2012**, *108*, 129–137. [CrossRef] [PubMed]

39. Ilaria Viscione; Francesca D'Elia; Rodolfo Vastola; Maurizio Sibilio the Correlation Between Technologies and Rating Scales in Gait Analysis. *J. Sport. Sci.* **2016**, *4*, 119–123. [CrossRef]
40. Pau, M.; Porta, M.; Pilloni, G.; Corona, F.; Fastame, M.C.; Hitchcott, P.K.; Penna, M.P. Texting While Walking Induces Gait Pattern Alterations in Healthy Older Adults. *Proc. Hum. Factors Ergon. Soc. Annu. Meet.* **2018**, *62*, 1908–1912. [CrossRef]
41. Zijlstra, W. Assessment of Spatio-Temporal Parameters during Unconstrained Walking. *Eur. J. Appl. Physiol.* **2004**, *92*, 39–44. [CrossRef] [PubMed]
42. Mazzà, C.; Iosa, M.; Pecoraro, F.; Cappozzo, A. Control of the Upper Body Accelerations in Young and Elderly Women during Level Walking. *J. Neuroeng. Rehabil.* **2008**, *5*, 1–10. [CrossRef] [PubMed]
43. Abellan van Kan, G.; Rolland, Y.; Andrieu, S.; Bauer, J.; Beauchet, O.; Bonnefoy, M.; Cesari, M.; Donini, L.M.; Gillette Guyonnet, S.; Inzitari, M.; et al. Gait Speed at Usual Pace as a Predictor of Adverse Outcomes in Community-Dwelling Older People an International Academy on Nutrition and Aging (IANA) Task Force. *J. Nutr. Health Aging* **2009**, *13*, 881–889. [CrossRef]
44. Studenski, S.; Perera, S.; Patel, K.; Rosano, C.; Faulkner, K.; Inzitari, M.; Brach, J.; Chandler, J.; Cawthon, P.; Connor, E.B.; et al. Gait Speed and Survival in Older Adults. *JAMA* **2011**, *305*, 50–58. [CrossRef] [PubMed]
45. Dommershuijsen, L.J.; Ragunathan, J.; Ruiter, T.R.; Groothof, D.; Mattace-Raso, F.U.S.; Ikram, M.A.; Polinder-Bos, H.A. Gait Speed Reference Values in Community-Dwelling Older Adults – Cross-Sectional Analysis from the Rotterdam Study. *Exp. Gerontol.* **2022**, *158*, 111646. [CrossRef] [PubMed]
46. Aboutorabi, A.; Arazpour, M.; Bahramizadeh, M.; Hutchins, S.W.; Fadayevatan, R. The Effect of Aging on Gait Parameters in Able-Bodied Older Subjects: A Literature Review. *Aging Clin. Exp. Res.* **2016**, *28*, 393–405. [CrossRef] [PubMed]
47. Brach, J.S.; Studenski, S.A.; Perera, S.; VanSwearingen, J.M.; Newman, A.B. Gait Variability and the Risk of Incident Mobility Disability in Community-Dwelling Older Adults. *J. Gerontol. Ser. A* **2007**, *62*, 983–988. [CrossRef]
48. Brach, J.S.; Studenski, S.; Perera, S.; VanSwearingen, J.M.; Newman, A.B. Stance Time and Step Width Variability Have Unique Contributing Impairments in Older Persons. *Gait Posture* **2008**, *27*, 431–439. [CrossRef]
49. Jones, L.M.; Waters, D.L.; Legge, M. Walking Speed at Self-Selected Exercise Pace Is Lower but Energy Cost Higher in Older versus Younger Women. *J. Phys. Act. Health* **2009**, *6*, 327–332. [CrossRef]
50. Kavanagh, J.J.; Barrett, R.S.; Morrison, S. Age-Related Differences in Head and Trunk Coordination during Walking. *Hum. Mov. Sci.* **2005**, *24*, 574–587. [CrossRef]
51. Mesure, S.; Azulay, J.P.; Pouget, J.; Amblard, B. Strategies of Segmental Stabilization during Gait in Parkinson's Disease. *Exp. Brain Res.* **1999**, *129*, 573–581. [CrossRef] [PubMed]
52. O'Connor, S.M.; Kuo, A.D. Direction-Dependent Control of Balance during Walking and Standing. *J. Neurophysiol.* **2009**, *102*, 1411–1419. [CrossRef]
53. Brodie, M.A.D.; Menz, H.B.; Lord, S.R. Age-Associated Changes in Head Jerk While Walking Reveal Altered Dynamic Stability in Older People. *Exp. Brain Res.* **2014**, *232*, 51–60. [CrossRef] [PubMed]
54. Chou, L.-S.; Kaufman, K.R.; Hahn, M.E.; Brey, R.H. Medio-Lateral Motion of the Center of Mass during Obstacle Crossing Distinguishes Elderly Individuals with Imbalance. *Gait Posture* **2003**, *18*, 125–133. [CrossRef] [PubMed]
55. Cofré Lizama, L.E.; Pijnappels, M.; Reeves, N.P.; Verschueren, S.M.; van Dieën, J.H. Centre of Pressure or Centre of Mass Feedback in Mediolateral Balance Assessment. *J. Biomech.* **2015**, *48*, 539–543. [CrossRef]
56. Singer, J.C.; McIlroy, W.E.; Prentice, S.D. Kinetic Measures of Restabilisation during Volitional Stepping Reveal Age-Related Alterations in the Control of Mediolateral Dynamic Stability. *J. Biomech.* **2014**, *47*, 3539–3545. [CrossRef]
57. Maki, B.E. Gait Changes in Older Adults: Predictors of Falls or Indicators of Fear. *J. Am. Geriatr. Soc.* **1997**, *45*, 313–320. [CrossRef]
58. Casas-Herrero, A.; Anton-Rodrigo, I.; Zambom-Ferraresi, F.; Sáez de Asteasu, M.L.; Martinez-Velilla, N.; Elexpuru-Estomba, J.; Marin-Epelde, I.; Ramon-Espinoza, F.; Petidier-Torregrosa, R.; Sanchez-Sanchez, J.L.; et al. Effect of a Multicomponent Exercise Programme (VIVIFRAIL) on Functional Capacity in Frail Community Elders with Cognitive Decline: Study Protocol for a Randomized Multicentre Control Trial. *Trials* **2019**, *20*, 362. [CrossRef]
59. Amorese, A.J.; Ryan, A.S. Home-Based Tele-Exercise in Musculoskeletal Conditions and Chronic Disease: A Literature Review. *Front. Rehabil. Sci.* **2022**, *3*. [CrossRef]
60. Baez, M.; Khaghani Far, I.; Ibarra, F.; Ferron, M.; Didino, D.; Casati, F. Effects of Online Group Exercises for Older Adults on Physical, Psychological and Social Wellbeing: A Randomized Pilot Trial. *PeerJ* **2017**, *5*, e3150. [CrossRef]
61. Sherrington, C.; Fairhall, N.J.; Wallbank, G.K.; Tiedemann, A.; Michaleff, Z.A.; Howard, K.; Clemson, L.; Hopewell, S.; Lamb, S.E. Exercise for Preventing Falls in Older People Living in the Community. *Cochrane Database Syst. Rev.* **2019**, *1*, CD012424. [CrossRef] [PubMed]
62. Zhang, W.; Low, L.-F.; Gwynn, J.D.; Clemson, L. Interventions to Improve Gait in Older Adults with Cognitive Impairment: A Systematic Review. *J. Am. Geriatr. Soc.* **2019**, *67*, 381–391. [CrossRef] [PubMed]

 *electronics*

Article

Harmonic Distortion Aspects in Upper Limb Swings during Gait in Parkinson's Disease

Luca Pietrosanti [1], Alexandre Calado [1], Cristiano Maria Verrelli [1], Antonio Pisani [2,3], Antonio Suppa [4,5], Francesco Fattapposta [4], Alessandro Zampogna [4], Martina Patera [4], Viviana Rosati [6], Franco Giannini [1] and Giovanni Saggio [1,*]

[1] Department of Electronic Engineering, University of Rome Tor Vergata, 00133 Rome, Italy
[2] Department of Brain and Behavioral Sciences, University of Pavia, 27100 Pavia, Italy
[3] IRCCS Mondino Foundation, 27100 Pavia, Italy
[4] Department of Human Neurosciences, Sapienza University of Rome, 00185 Rome, Italy
[5] IRCCS Neuromed, 86077 Pozzilli, Italy
[6] A.O.U. Policlinico Umberto I, 00161 Rome, Italy
* Correspondence: saggio@uniroma2.it

Abstract: Parkinson's disease (PD) is responsible for a broad spectrum of signs and symptoms, including relevant motor impairments generally rated by clinical experts. In recent years, motor measurements gathered by technology-based systems have been used more and more to provide objective data. In particular, wearable devices have been adopted to evidence differences in the gait capabilities between PD patients and healthy people. Within this frame, despite the key role that the upper limbs' swing plays during walking, no studies have been focused on their harmonic content, to which this work is devoted. To this end, we measured, by means of IMU sensors, the walking capabilities of groups of PD patients (both de novo and under-chronic-dopaminergic-treatment patients when in an off-therapy state) and their healthy counterparts. The collected data were FFT transformed, and the frequency content was analyzed. According to the results obtained, PD determines upper limb rigidity objectively evidenced and correlated to lower harmonic contents.

Keywords: Parkinson's disease; neurological disorders; wearable sensors; frequency harmonics; gait analysis; gait impairments

Citation: Pietrosanti, L.; Calado, A.; Verrelli, C.M.; Pisani, A.; Suppa, A.; Fattapposta, F.; Zampogna, A.; Patera, M.; Rosati, V.; Giannini, F.; et al. Harmonic Distortion Aspects in Upper Limb Swings during Gait in Parkinson's Disease. *Electronics* **2023**, *12*, 625. https://doi.org/10.3390/electronics12030625

Academic Editor: Enzo Pasquale Scilingo

Received: 24 December 2022
Revised: 23 January 2023
Accepted: 24 January 2023
Published: 27 January 2023

1. Introduction

Parkinson's disease (PD) is one of the most common neurodegenerative disorders worldwide, with an increasing incidence over the last 30 years [1] and prevalence in people aged >65 years [2]. PD is associated with a progressive loss of dopaminergic neurons in a specific brainstem area called *Substantia Nigra pars compacta*, which is responsible for the occurrence of cardinal motor signs, including rest tremor (i.e., a 4–6 Hz tremor in fully resting limbs), bradykinesia (i.e., slowness of movements) and rigidity (i.e., muscles become stiff or inflexible). PD patients may also suffer from postural instability [3] and falls, especially in the advanced stages of the disease [4], and voice disorders [5,6]. Moreover, a broad spectrum of non-motor symptoms such as anxiety, depression and urogenital dysfunction also frequently arise, negatively impacting patients' quality of life [7].

The clinical evaluation of parkinsonian signs and symptoms is generally carried out by using rating scales, the most adopted one being the standardized Movement Disorder Society-Unified Parkinson's Disease Rating Scale (MDS-UPDRS) [8], divided into four parts. The third (III) part concerns motor assessment, with a set of motor tasks subjectively ranked by expert clinical staff with discrete values (0–4). The UPDRS protocol demonstrates great validity but also limits, such as moderate intra- and inter-rater reliability [9,10], some observer inconsistencies [11] and a coarse assessment due to the limited scale with only four discrete values. Consequently, researchers have been more and more open to the adoption of objective technology-based systems [12,13] aimed at fine quantitative assessments of PD motor signs, such as optical apparatuses (e.g., RGB cameras) [14,15], smartphones [16], EMG tools [17,18], flex [19,20] and force sensors [21] and wearable devices (e.g., inertial measurement units, hereafter IMUs) [22–24]. In particular, the latter have been demonstrated to be effective in furnishing reliable data to be used for feature extraction and data classification purposes with high correlations to standards. For example, Bobić et al. [25] assessed bradykinesia by means of gyroscopes sensors placed on the thumb and index fingernails during finger tapping, di Biase et al. [26] determined rigidity with IMUs measuring passive oscillation of arms, and Dai et al. [27] adopted IMUs to objectively evidence tremors.

Apart from single aspects of motor deficiencies, IMUs have been adopted to reveal an ensemble of motor signs, too. Ricci et al. [28] adopted a set of IMUs to assess rigidity, bradykinesia, postural instability and gait abnormalities, Zampieri et al. [29] adopted IMUs to objectively evidence reduced arm swing, rigidity and bradykinesia during gait tests and Lewek et al. [30] evidenced asymmetry in the upper limb swing in early PD subjects on the basis of measurements gathered by IMUs.

Despite the detailed motor investigations of the published papers, as far as we know, no work has focused on analyzing the frequency content of forearm swings during gait tests. However, since forearm movements represent a key aspect of human walking [31], here we highlight their contribution through IMU-based measurements during a walking task performed by groups of PD patients and healthy subjects to evidence differences in their spectral components.

This work is arranged to present the subjects involved in the study (Section 2.1), the adopted technologies (Section 2.2), the procedures (Sections 2.3 and 2.4) and the results (Section 3) to determine the impact of key relevant walking motor features (Sections 4 and 5).

2. Materials and Methods

2.1. Subjects

A total of 89 subjects (Table 1), comprising 58 PD patients and 31 age-matched healthy people (i.e., the control group, HC hereafter), were recruited from the Movement Disorders Outpatient Service of both Tor Vergata university hospital and Sapienza University of Rome (Italy). In particular, the PD patients were 44 drug-free de novo (i.e., newly diagnosed and not yet in therapy, PD-DN hereafter), and 14 PD patients under chronic dopaminergic treatment, examined when in the off state of therapy (i.e., after drug withdrawal for at least 12 h; PD-OFF hereafter) [27,28]. All patients were evaluated by medical experts and rated according to the MDS-UPDRS part III, Hoen and Yahr (H&Y) and Mini-Mental State Examination (MMSE) scales.

Table 1. Clinical and demographic features of the participants.

	PD-DN	PD-OFF	HC
Age [years]	62 ± 9.7	70 ± 9	69 ± 11.2
Gender	29 M, 15 F	9 M, 5 F	10 M, 21 F
Height [cm]	170.7 ± 8.6	169.4 ± 11.5	163.1 ± 10.7
Weight [kg]	73.7 ± 12.7	70.8 ± 16.8	67.5 ± 12.4
MDS-UPDRS III	22.2 ± 8.8	29.8 ± 10.7	
H&Y	1.75 ± 0.5	1.87 ± 0.3	

Inclusion criteria for PD subjects were diagnosis of idiopathic PD, ability to walk independently, absence of comorbidities possibly affecting gait and an MMSE score of >24.

All subjects voluntarily agreed to participate in this study, furnishing informed consent according to the local ethics committees following the Declaration of Helsinki.

2.2. Wearable Electronic Devices

We adopted three lightweight (<20 g each), unobtrusive (4 cm $\times$ 3 cm $\times$ 1.5 cm), not-hindering-movements (placed by Velcro strips) wearable devices (wearables hereafter), each termed Movit G1 (by Captiks Srl, Rome, Italy). Each Movit G1 hosts IMUs with a triaxial accelerometer and a triaxial gyroscope, already validated with respect to gold standard system [32], to collect kinematic data (i.e., acceleration, angular velocity and orientation) from different anatomic segments (i.e., forearms and upper back, Figure 1a,b) of each participant. We configured the accelerometer to ±8 g with 16.384 LSB/g sensitivity and the gyroscope to $\pm2000°$/s with 32.8 LSB/$°$/s sensitivity. Signals were acquired at a sampling rate of 50 Hz and sent to a receiver connected to a personal computer that runs a dedicated application named Captiks Motion Studio (by Captiks Srl, Rome, Italy).

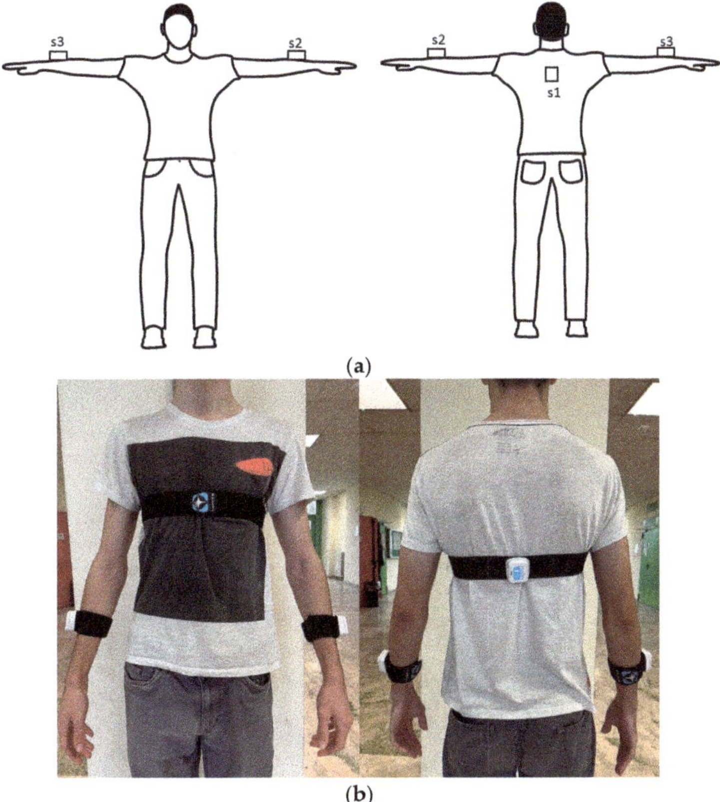

Figure 1. Wearable sensors as placed on the subject's body ((**a**) scheme, (**b**) real), s1 on the upper shoulder, s2 and s3 on the forearms.

2.3. Testing Procedure

Due to our walking-focused work, we adopted the Timed Up and Go (TUG) test as a standard procedure for gait assessment.

The TUG is a sequence of subjects' common movements, as explained below:

- Sit to stand: from seated on a chair with arms crossed over the chest to standing up;
- First walking: a walk for 6 m at a comfortable speed;
- First turning: a first 180° rotation;
- Second walking: a walk from the turning point back to the chair;
- Second turning: a second 180° rotation;

- Stand to sit: from standing to sitting down.

A schematic representation of TUG test phases is reported in Figure 2.

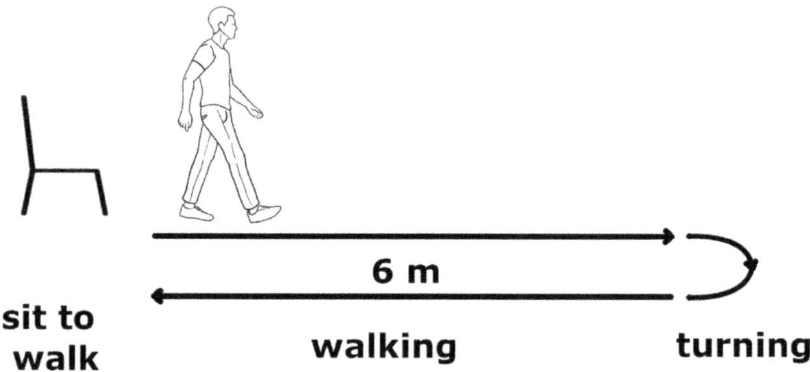

Figure 2. Graphical representation of different phases of Timed Up and Go test.

The test was fully performed by all participants. Gait was clinically assessed through discrete scores assigned according to the MDS-UPDRS part III. The frequency contents were later extrapolated from the walking part only.

2.4. Data Analysis

The TUG test was segmented to isolate the walking phase. This segmentation was empirically obtained by evidencing the start of the first walk (when the subject releases the arms from the crossed position), the end of the first walk (when the trunk rotates more than 10 degrees) and similarly for the second walking phase.

Occasionally, some sample data were not correctly transmitted from the IMUs to the receiving unit; when this occurred, we interpolated the data streaming (even if empirically, we could consider this issue not relevant). Signals were also windowed with the Tukey window function and zero-padded to guarantee a minimum of 1024 samples.

Data were gathered to determine the frequency content (by means of the Fast Fourier Transform algorithm) of the forearm swings (for both left and right upper limbs) and, in particular, those motor features of the harmonics related to PD, and seven features were determined accordingly, as reported in Table 2. Since PD patients, especially those in the first stages of the disease, behave asymmetrically in arm swings, we considered the differences between the two upper limbs empirically differentiated into the "most affected" and "least affected" sides according to their range of motion (ROM) during the walking phase ("least affected" was for the higher ROM).

A two-sample *t*-test (p-value = 0.05) was performed to determine if the features' distributions showed significant differences between PD and HC populations. The comparison was undertaken distinctly for PD-DN and PD-OFF patients and separately for the "most affected" and "least affected" sides. Moreover, Spearman's rank correlation coefficient was computed between the motor features and MDS-UPDRS III scores to relate the features and PD signs by considering the MDS-UPDRS items, namely no. 3.3 (rigidity), no. 3.10 (gait), no. 3.14 (body bradykinesia and hypokinesia) and no. 3.16 (action tremor), as provided by clinicians. We chose these items because they have already demonstrated a good correlation with the features obtained from gait [28].

The described data analysis was homemade by means of an ad hoc algorithm written in MATLAB 2022b (by Mathworks Inc., Natick, MA, USA).

Table 2. Motor features of the harmonics in upper limb swings during gait.

Features	Description
Hamp1	Maximum amplitude of the fundamental frequency
Hamp2	Maximum amplitude of the second harmonic
Hamp3	Maximum amplitude of the third harmonic
freq	Fundamental frequency
HD2	Second-harmonic distortion, computed as ratio of Hamp2 and Hamp1
HD3	Third-harmonic distortion, computed as ratio of Hamp3 and Hamp1
Asym	Difference in angular velocities of arms, calculated as the difference between Hamp1 of left and right arm, divided by the highest one
THD	Total harmonic distortion, computed considering the first seven harmonics

3. Results

Figure 3 shows the Hamp1 and Hamp2 distributions for HC (Figure 3a,b), PD-DN (Figure 3c,d) and PD-OFF (Figure 3e,f) populations.

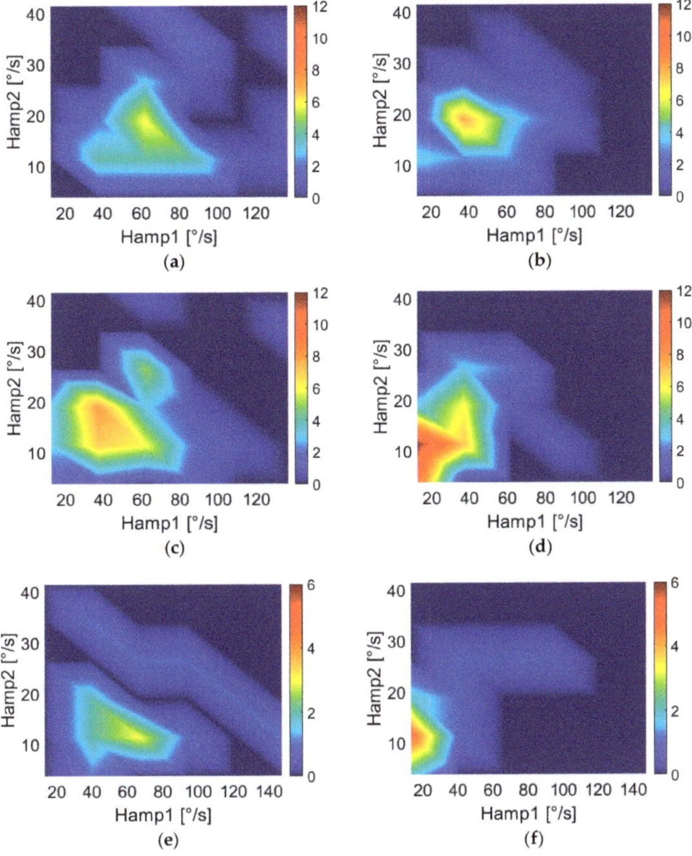

Figure 3. Two-dimensional features' distributions for HC subjects ((**a**) "least affected" side, (**b**) "most affected" side), PD-DN patients ((**c**) "least affected" side, (**d**) "most affected" side) and PD-OFF patients ((**e**) "least affected" side, (**f**) "most affected" side). The colors are related to the number of subjects. Color bar limits of PD-OFF plots are lower with respect to the other plots because of the smaller population.

Table 3 reports the mean and standard deviations of each feature for PD-DN, PD-OFF and HC, respectively, and the *p*-values obtained from *t*-tests separately for the "most affected" and "least affected" sides.

Table 3. Values of the motor features related to PD-DN, PD-OFF and HC.

Side	Feature	PD-DN (Mean ± Std)	PD-OFF (Mean ± Std)	HC (Mean ± Std)	*p*-Value * PD-DN	*p*-Value * PD-OFF
Least affected	Hamp1	50.94 ± 26.61	70.59 ± 38.00	71.79 ± 29.77	**0.002**	0.909
	Hamp2	15.74 ± 7.42	16.43 ± 8.07	18.41 ± 8.86	0.161	0.480
	Hamp3	7.13 ± 4.30	8.12 ± 4.79	9.53 ± 5.93	**0.046**	0.441
	freq	0.87 ± 0.07	0.96 ± 0.10	0.87 ± 0.06	0.571	**0.002**
	THD	-8.92 ± 4.21	-10.89 ± 5.13	-10.31 ± 3.99	0.154	0.686
	HD2	0.38 ± 0.26	0.28 ± 0.17	0.28 ± 0.13	**0.050**	0.943
	HD3	0.16 ± 0.09	0.13 ± 0.08	0.14 ± 0.09	0.432	0.703
Most affected	Hamp1	26.20 ± 19.25	30.72 ± 21.79	45.30 ± 19.62	**<0.001**	**0.031**
	Hamp2	13.21 ± 6.75	14.60 ± 7.62	17.07 ± 6.38	**0.014**	0.263
	Hamp3	5.75 ± 3.60	6.18 ± 3.95	8.92 ± 4.42	**0.001**	0.053
	freq	0.89 ± 0.10	0.96 ± 0.11	0.87 ± 0.07	0.525	**0.003**
	THD	-6.23 ± 4.40	-5.61 ± 2.96	-7.96 ± 4.56	0.103	0.086
	HD2	0.66 ± 0.35	0.58 ± 0.38	0.46 ± 0.29	**0.010**	0.225
	HD3	0.32 ± 0.27	0.32 ± 0.41	0.22 ± 0.11	**0.046**	0.228
	Asym	0.48 ± 0.24	0.52 ± 0.25	0.35 ± 0.21	**0.017**	**0.021**

* Statistically relevant *p*-Values are highlighted in bold.

Table 4 shows the correlation coefficients and related *p*-values as results of Spearman's rank correlation between features and gait and body bradykinesia scores of the MDS-UPDRS III for PD-DN subjects.

Table 4. Spearman's rank correlation coefficients for the "most affected" and "least affected" sides in de novo parkinsonian patients.

Side	Feature	MDS-UPDRS III r	MDS-UPDRS III *p*-Val *	Gait (Item 3.10) r	Gait (Item 3.10) *p*-Val *	Body Bradykinesia (Item 3.14) r	Body Bradykinesia (Item 3.14) *p*-Val *
Least affected	Hamp1	**−0.58**	**<0.001**	**−0.35**	**0.030**	**−0.43**	**0.006**
	Hamp2	−0.27	0.093	−0.07	0.651	−0.13	0.413
	Hamp3	**−0.49**	**0.001**	−0.15	0.350	−0.28	0.080
	freq	−0.25	0.116	−0.20	0.210	−0.19	0.238
	THD	0.14	0.373	**0.32**	**0.046**	0.16	0.335
	HD2	**0.43**	**0.005**	**0.33**	**0.040**	**0.35**	**0.029**
	HD3	0.22	0.175	0.28	0.078	0.20	0.224
Most affected	Hamp1	**−0.65**	**<0.001**	−0.27	0.099	**−0.60**	**<0.001**
	Hamp2	**−0.55**	**<0.001**	−0.21	0.188	**−0.45**	**0.003**
	Hamp3	**−0.42**	**0.007**	−0.22	0.170	**−0.40**	**0.011**
	freq	−0.04	0.794	−0.31	0.053	−0.05	0.723
	THD	−0.17	0.294	−0.13	0.408	−0.13	0.410
	HD2	**0.40**	**0.012**	0.11	0.506	**0.40**	**0.011**
	HD3	**0.41**	**0.009**	0.20	0.228	**0.37**	**0.019**
	Asym	0.27	0.089	−0.11	0.506	**0.49**	**0.001**

* Statistically relevant correlation coefficients and relative *p*-Values are highlighted in bold.

Table 5 shows the correlations between motor features and specific MDS-UPDRS III items for the upper limbs (i.e., action tremor and rigidity) for PD-DN subjects.

Table 5. Spearman's rank correlation coefficients in de novo parkinsonian patients.

Feature	Action Tremor (Item 3.16)		Rigidity (Item 3.3)	
	r	*p*-Val *	r	*p*-Val *
Hamp1	−0.20	0.066	**−0.54**	**<0.001**
Hamp2	−0.09	0.406	**−0.36**	**<0.001**
Hamp3	0.23	0.038	**0.51**	**<0.001**
freq	0.08	0.444	−0.06	0.542
THD	0.16	0.141	0.01	0.881
HD2	0.16	0.150	**0.37**	**<0.001**
HD3	<0.01	0.992	0.17	0.111

* Statistically relevant correlation coefficients and relative *p*-Values are highlighted in bold.

Table 6 shows the correlation coefficients and related *p*-values as results of Spearman's correlation between features and items related to gait and body bradykinesia of the MDS-UPDRS III relative to PD-OFF patients.

Table 6. Spearman's rank correlation coefficients for the "most affected" and "least affected" sides in patients with Parkinson's disease when in off state of therapy.

Side	Feature	UPDRS III		Gait (Item 3.10)		Body Bradykinesia (Item 3.14)	
		r	*p*-Val *	r	*p*-Val *	r	*p*-Val *
Least affected	Hamp1	−0.13	0.657	−0.10	0.718	−0.44	0.113
	Hamp2	**−0.57**	**0.030**	−0.23	0.413	**−0.59**	**0.024**
	Hamp3	−0.33	0.234	−0.02	0.923	−0.23	0.414
	freq	−0.45	0.103	−0.34	0.223	**−0.56**	**0.034**
	THD	−0.22	0.439	0.08	0.765	0.06	0.827
	HD2	−0.19	0.506	0.05	0.847	0.03	0.906
	HD3	−0.18	0.516	0.05	0.847	0.20	0.487
Most affected	Hamp1	−0.41	0.145	**−0.64**	**0.013**	−0.34	0.226
	Hamp2	−0.40	0.155	−0.27	0.348	**−0.67**	**0.008**
	Hamp3	−0.11	0.685	0.05	0.860	0.01	0.959
	freq	−0.38	0.180	−0.28	0.328	**−0.54**	**0.042**
	THD	−0.04	0.886	0.22	0.440	−0.31	0.275
	HD2	−0.05	0.851	0.44	0.114	−0.30	0.287
	HD3	0.11	0.701	0.43	0.114	0.33	0.244
	Asym	0.22	0.448	**0.58**	**0.027**	−0.01	0.946

* Statistically relevant correlation coefficients and relative *p*-Values are highlighted in bold.

Table 7 reports the results of the correlations between motor features and specific MDS-UPDRS III items for the upper limbs (i.e., action tremor and rigidity).

Table 7. Spearman's rank correlation coefficients in patients with Parkinson's disease when in off state of therapy.

Feature	Action Tremor (Item 3.16)		Rigidity (Item 3.3)	
	r	*p*-Val *	r	*p*-Val *
Hamp1	−0.29	0.123	**−0.53**	**0.003**
Hamp2	−0.20	0.292	**−0.37**	**0.048**
Hamp3	−013	0.484	−0.19	0.33
freq	−0.20	0.305	−0.28	0.14
THD	−0.06	0.760	−0.02	0.91
HD2	0.164	0.402	0.31	0.099
HD3	0.10	0.596	0.28	0.136

* Statistically relevant correlation coefficients and relative *p*-Values are highlighted in bold.

4. Discussion

In this study, we investigated if the frequency analysis of forearm swing during walking can be a discriminating tool for distinguishing PD patients (both at early, called PD-DN, or chronically treated, called PD-OFF, stages) from HC and if it allows quantifying different motor signs.

Our measurements evidenced the (expected) "compound pendulum" behavior of the upper limb swings, which is characterized by multiple higher-order harmonics. We focused on the relationship between these spectral components and PD signs. A set of seven features was used to quantify the nonlinear behavior of the arm swing.

As highlighted in Figure 3, walking arm-swing characteristics of subjects are significantly different for PD-DN and HC, demonstrating lower values of Hamp1 (i.e., the maximum amplitude of the fundamental frequency) and Hamp2 (i.e., the maximum amplitude of the second harmonic) for both arms in PD-DN than in HC. This evidence was confirmed by the results of t-tests reported in Table 3, where, in particular, the p-values demonstrated significant differences in PD-DN vs. HC in the swings of both arms. In particular, it was observed that the features of the "most affected" side were more significant than those of the "least affected" side, especially for features Hamp2 and HD2 (i.e., second-harmonic distortion), which showed no difference between HC and PD-DN for the "least affected" arm. This result agrees with what has been reported in the literature, namely that the early stages of this disease are characterized by a greater asymmetry of motor signs. This asymmetry in the arm swing has been quantified by means of the named Asym feature.

Table 3 also shows the t-test results for PD-OFF patients. In this case, differences from HC are evidenced for the "most affected" arm but not for the "least affected" one. Indeed, residual effects of the dopaminergic therapy, due to the known long-duration response to L-Dopa, may have smoothed possible differences also involving the "least affected" side despite a drug withdrawal of at least 12 h [33,34]. In particular, only features Hamp2 and freq (fundamental frequency) of the "most affected" side show significant p-values, although the other features of the "most affected" side have mean and standard deviation values close to those reported for PD-DN.

According to the results evidenced in Table 4, for PD-DN subjects, there is a meaningful correlation between the features of the "least affected" and "most affected" arms. In particular, motor features of the "most affected" arm are better correlated to the overall MDS-UPDRS III scores with respect to the "least affected" one, and in a special way with item no. 3.14 (bradykinesia).

Results shown in Table 5 evidence a significant correlation between rigidity- and amplitude-related features to the first three spectral components and the second-order harmonic distortion, while no correlation of the features is found with item no. 3.16 (action tremor).

As Table 6 shows, the motor feature Hamp2 is particularly useful to evidence bradykinesia issues in PD-OFF subjects. This is interesting in relation to the significance shown in Figure 3, which underlines lower values of Hamp2 in PD patients.

Table 7 reports how item no. 3.16 (action tremor) is uncorrelated to any features, while the Hamp1 and Hamp2 features present moderate to good correlation with item no. 3.3.

From the previous outcomes, it is possible to assert that the information obtained from the analysis of the first three spectral components of the arm swing can be useful for discriminating PD patients from HC subjects, and they also present a good level of correlation with MDS-UPDRS scores for bradykinesia and rigidity.

Figure 4 shows the distribution of Hamp1 and Hamp2 for different scores of bradykinesia assigned by clinicians for both PD-DN and PD-OFF patients.

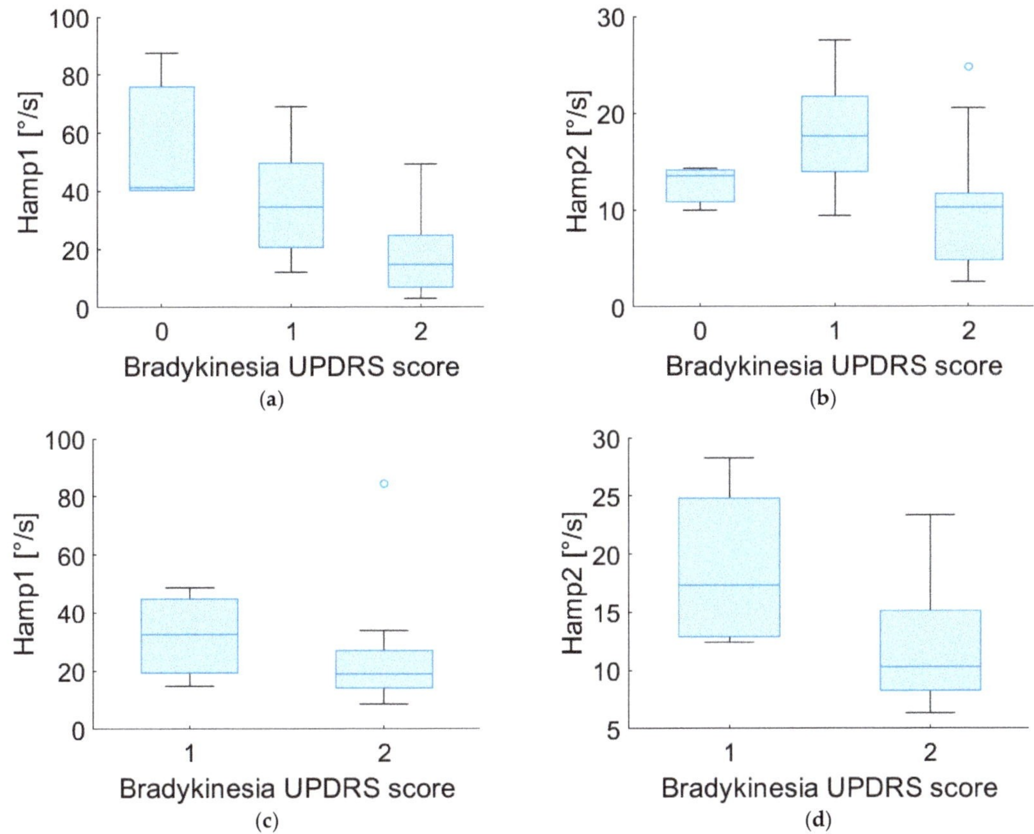

Figure 4. Boxplot of Hamp1 and Hamp2 of "most affected" side for PD-DN (**a**,**b**) and PD-OFF (**c**,**d**) subjects for different scores of UPDRS item no. 3.14 (bradykinesia).

Of particular interest is the effect of the second harmonic on the arm swing, as it is possible to provide a physical interpretation of the phenomenon.

In fact, second- and third-harmonic distortion is a well-known phenomenon in nonlinear systems such as power amplifiers [35,36]. These effects are due to the nonlinear effects of the system's physical constraints. In nonlinear systems, the presence of even and odd harmonics is expected, but in the human arm, the constraint of the elbow forces the forearm into an asymmetric flexion–extension movement with a consequent swing, such as the one reported in Figure 5, where three different distortions are shown. This "asymmetry" generates a prevalence of even harmonics. Accordingly, our results highlight the presence of the second harmonic in HC subjects, while PD patients show a decrease in the second-harmonic component, meaning a reduced flexion–extension of the elbow. This finding may reflect the presence of rigidity in patients' limbs. Indeed, as demonstrated in [37,38], PD subjects show abnormal EMG patterns in the biceps brachii and triceps surae muscles during passive movements. Since the considered muscles are responsible for elbow movements, an alteration in the EMG activity can alter the flexion–extension of the joint with a consequent variation of the second harmonic. These remarks suggest a connection between higher-order harmonics and parkinsonian rigidity. As further support to this hypothesis, we found significant correlations between motor features (i.e., Hamp2, Hamp3, DH2 and DH3) and rigidity, as clinically assessed through MDS-UPDRS III scores.

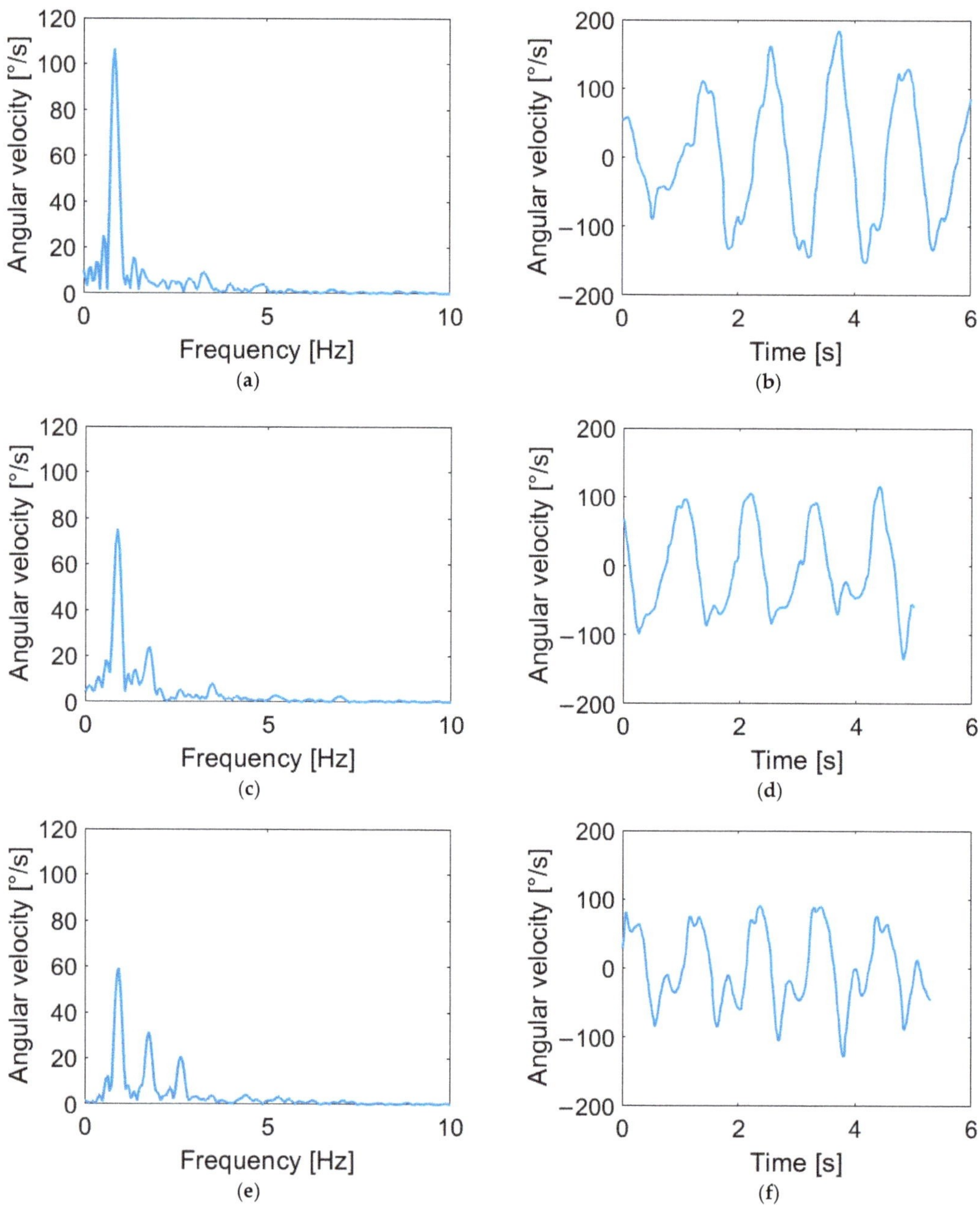

Figure 5. Arm-swing signals in frequency and time domains for different harmonic content. Signals reported show negligible distortion (**a,b**), second-harmonic distortion (**c,d**) and third-harmonic distortion (**e,f**). As shown in the figure, the second harmonic corresponds to a flattening of negative peaks in the time domain, while the third harmonic causes a flattening of both positive and negative peaks.

Figure 6 shows the distribution of Hamp1 and Hamp2 for different rigidity scores in PD-DN patients. Rigidity scores 0 and 2 present distinct distributions. Thus, it can be asserted that the study of second harmonics can provide interesting information on the rigidity severity of upper limbs in parkinsonian subjects.

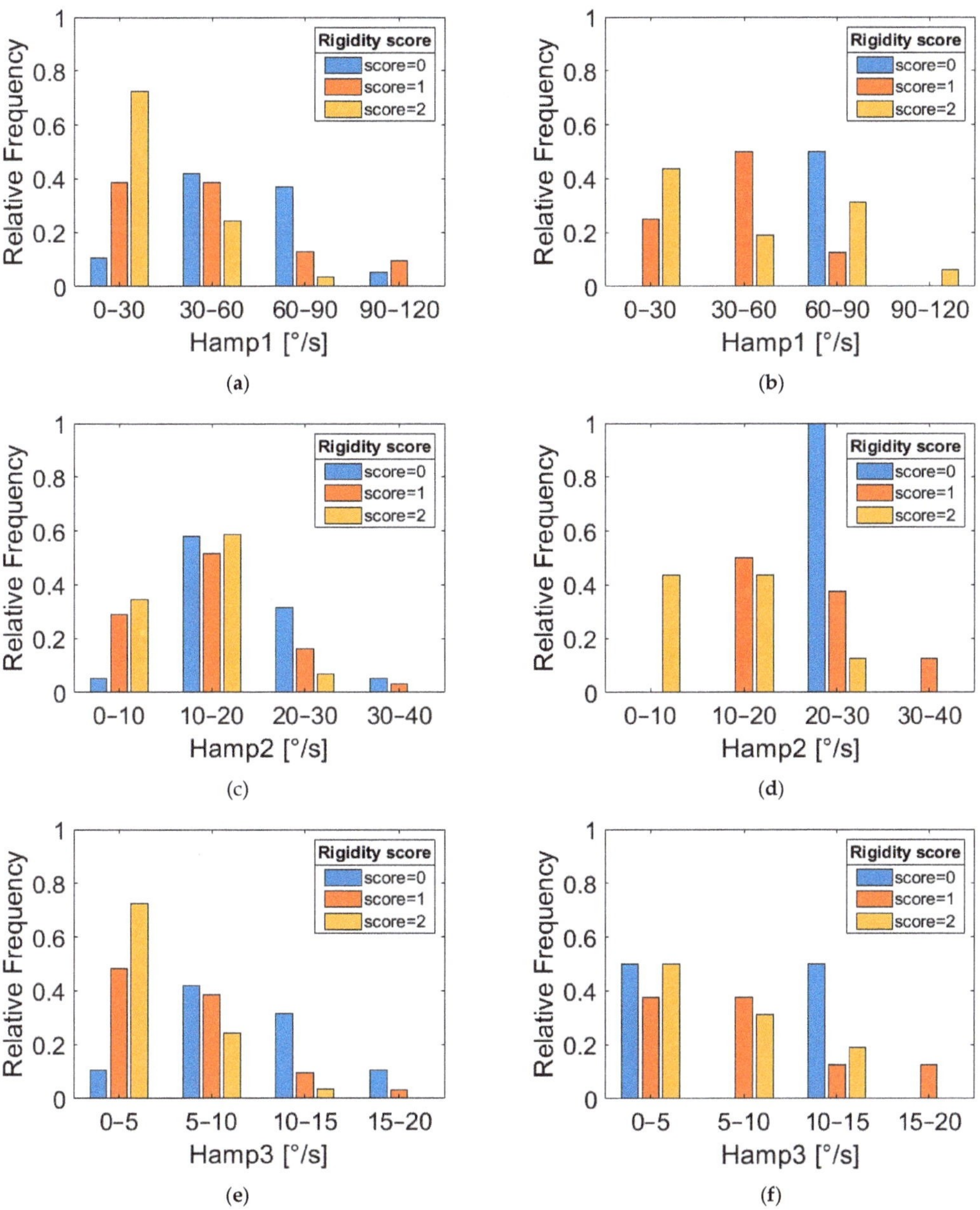

Figure 6. Relative frequency distributions of features Hamp1, Hamp2 and Hamp3 for PD-DN (**a,c,e**) and PD-OFF (**b,d,f**) subjects grouped according to three different UPDRS rigidity scores (item 3.3).

5. Conclusions

We propose an approach for PD assessment based on the evaluation of harmonic distribution and the distortion of arm swings during walking tasks. This assessment was applied to two different PD groups (de novo subjects, called PD-DN, and PD patients under chronic dopaminergic treatment when in off state of therapy, called PD-OFF) and their healthy group (called HC) counterparts.

Findings in harmonic distribution and distortion highlight the nonlinear behaviors more evident in the HC group.

In particular, features related to the first three spectral frequencies are here demonstrated to be statistically significant for the discrimination of PD-DN patients, while it is mainly the fundamental frequency that is relevant for PD-OFF subjects. However, for both PD-DN and PD-OFF, asymmetry is a key feature for objectively discerning the HC group.

Another key result is the correlation of bradykinesia and limb rigidity with harmonic distortion. Although studies have been conducted on this topic, rigidity assessment through wearable systems is still under examination and, as far as we know, without validated methods.

Although this work presents encouraging results, the authors are aware that the number of investigated subjects must be enlarged for improved statistical validity.

Author Contributions: Conceptualization, L.P., F.G. and G.S.; methodology, L.P., A.S., A.P., C.M.V. and G.S.; software, L.P.; validation, G.S.; formal analysis, L.P. and G.S.; investigation, L.P., A.S., A.P., G.S., A.Z., M.P. and V.R.; resources, A.S., A.P., G.S. and F.F.; data curation, L.P.; writing—original draft preparation, L.P.; writing—review and editing, G.S. and A.C.; visualization, L.P.; supervision, G.S.; project administration, G.S.; funding acquisition, G.S. All authors have read and agreed to the published version of the manuscript.

Funding: This research received no external funding.

Data Availability Statement: Data are available upon reasonable request.

Acknowledgments: We want to thank Captiks Srl for providing the wearable sensors.

Conflicts of Interest: The authors declare no conflict of interest.

References

1. Ou, Z.; Pan, J.; Tang, S.; Duan, D.; Yu, D.; Nong, H.; Wang, Z. Global Trends in the Incidence, Prevalence, and Years Lived with Disability of Parkinson's Disease in 204 Countries/Territories from 1990 to 2019. *Front. Public Health* **2021**, *9*, 776847. [CrossRef] [PubMed]
2. Simon, D.K.; Tanner, C.M.; Brundin, P. Parkinson Disease Epidemiology, Pathology, Genetics, and Pathophysiology. *Clin. Geriatr. Med.* **2020**, *36*, 1–12. [CrossRef] [PubMed]
3. Verrelli, C.M.; Iosa, M.; Roselli, P.; Pisani, A.; Giannini, F.; Saggio, G. Generalized Finite-Length Fibonacci Sequences in Healthy and Pathological Human Walking: Comprehensively Assessing Recursivity, Asymmetry, Consistency, Self-Similarity, and Variability of Gaits. *Front. Hum. Neurosci.* **2021**, *15*, 649533. [CrossRef] [PubMed]
4. Zampogna, A.; Cavallieri, F.; Bove, F.; Suppa, A.; Castrioto, A.; Meoni, S.; Pélissier, P.; Schmitt, E.; Bichon, A.; Lhommée, E.; et al. Axial Impairment and Falls in Parkinson's Disease: 15 Years of Subthalamic Deep Brain Stimulation. *NPJ Park. Dis.* **2022**, *8*, 121. [CrossRef] [PubMed]
5. Suppa, A.; Asci, F.; Saggio, G.; Marsili, L.; Casali, D.; Zarezadeh, Z.; Ruoppolo, G.; Berardelli, A.; Costantini, G. Voice Analysis in Adductor Spasmodic Dysphonia: Objective Diagnosis and Response to Botulinum Toxin. *Park. Relat. Disord.* **2020**, *73*, 23–30. [CrossRef]
6. Saggio, G.; Costantini, G. Worldwide Healthy Adult Voice Baseline Parameters: A Comprehensive Review. *J. Voice* **2020**, *36*, 637–649. [CrossRef]
7. Poewe, W. Non-Motor Symptoms in Parkinson's Disease. *Eur. J. Neurol.* **2008**, *15*, 14–20. [CrossRef]
8. Goetz, C.G.; Tilley, B.C.; Shaftman, S.R.; Stebbins, G.T.; Fahn, S.; Martinez-Martin, P.; Poewe, W.; Sampaio, C.; Stern, M.B.; Dodel, R.; et al. Movement Disorder Society-Sponsored Revision of the Unified Parkinson's Disease Rating Scale (MDS-UPDRS): Scale Presentation and Clinimetric Testing Results. *Mov. Disord.* **2008**, *23*, 2129–2170. [CrossRef] [PubMed]
9. Fahn, S.; Marsden, C.; Goldstein, M.; Calne, D. State of the Art Review—The Unified Parkinson's Disease Rating Scale (UPDRS): Status and Recommendations. *Mov. Disord.* **2003**, *18*, 738–750.

10. Siderowf, A.; McDermott, M.; Kieburtz, K.; Blindauer, K.; Plumb, S.; Shoulson, I. Test-Retest Reliability of the Unified Parkinson's Disease Rating Scale in Patients with Early Parkinson's Disease: Results from a Multicenter Clinical Trial. *Mov. Disord.* **2002**, *17*, 758–763. [CrossRef]

11. Oung, Q.W.; Muthusamy, H.; Lee, H.L.; Basah, S.N.; Yaacob, S.; Sarillee, M.; Lee, C.H. Technologies for Assessment of Motor Disorders in Parkinson's Disease: A Review. *Sensors* **2015**, *15*, 21710–21745. [CrossRef]

12. Di Lazzaro, G.; Ricci, M.; Al-Wardat, M.; Schirinzi, T.; Scalise, S.; Giannini, F.; Mercuri, N.B.; Saggio, G.; Pisani, A. Technology-Based Objective Measures Detect Subclinical Axial Signs in Untreated, de Novo Parkinson's Disease. *J. Park. Dis.* **2020**, *10*, 113–122. [CrossRef]

13. Zampogna, A.; Mileti, I.; Palermo, E.; Celletti, C.; Paoloni, M.; Manoni, A.; Mazzetta, I.; Costa, G.D.; Pérez-López, C.; Camerota, F.; et al. Fifteen Years of Wireless Sensors for Balance Assessment in Neurological Disorders. *Sensors* **2020**, *20*, 3247. [CrossRef]

14. Rocha, A.P.; Choupina, H.; Fernandes, J.M.; Rosas, M.J.; Vaz, R.; Cunha, J.P.S. Parkinson's Disease Assessment Based on Gait Analysis Using an Innovative RGB-D Camera System. In Proceedings of the 2014 36th Annual International Conference of the IEEE Engineering in Medicine and Biology Society, Chicago, IL, USA, 26–30 August 2014; Volume 2014, pp. 3126–3129.

15. Ťupa, O.; Procházka, A.; Vyšata, O.; Schätz, M.; Mareš, J.; Vališ, M.; Mařík, V. Motion Tracking and Gait Feature Estimation for Recognising Parkinson's Disease Using MS Kinect. *Biomed. Eng. Online* **2015**, *14*, 97. [CrossRef]

16. Borzì, L.; Varrecchia, M.; Sibille, S.; Olmo, G.; Artusi, C.A.; Fabbri, M.; Rizzone, M.G.; Romagnolo, A.; Zibetti, M.; Lopiano, L. Smartphone-Based Estimation of Item 3.8 of the MDS-UPDRS-III for Assessing Leg Agility in People with Parkinson's Disease. *IEEE Open J. Eng. Med. Biol.* **2020**, *1*, 140–147. [CrossRef]

17. Breit, S.; Spieker, S.; Schulz, J.B.; Gasser, T. Long-Term EMG Recordings Differentiate between Parkinsonian and Essential Tremor. *J. Neurol.* **2008**, *255*, 103–111. [CrossRef]

18. Mazzetta, I.; Gentile, P.; Pessione, M.; Suppa, A.; Zampogna, A.; Bianchini, E.; Irrera, F. Stand-Alone Wearable System for Ubiquitous Real-Time Monitoring of Muscle Activation Potentials. *Sensors* **2018**, *18*, 1748. [CrossRef]

19. Saggio, G.; Bocchetti, S.; Pinto, C.A.; Orengo, G.; Giannini, F. A Novel Application Method for Wearable Bend Sensors. In Proceedings of the 2nd International Symposium on Applied Sciences in Biomedical and Communication Technologies, Bratislava, Slovakia, 24–27 November 2009.

20. Saggio, G.; Quitadamo, L.R.; Albero, L. Development and Evaluation of a Novel Low-Cost Sensor-Based Knee Flexion Angle Measurement System. *Knee* **2014**, *21*, 896–901. [CrossRef]

21. Patrick, S.K.; Denington, A.A.; Gauthier, M.J.A.; Gillard, D.M.; Prochazka, A. Quantification of the UPDRS Rigidity Scale. *IEEE Trans. Neural Syst. Rehabil. Eng.* **2001**, *9*, 31–41. [CrossRef]

22. Di Lazzaro, G.; Ricci, M.; Saggio, G.; Costantini, G.; Schirinzi, T.; Alwardat, M.; Pietrosanti, L.; Patera, M.; Scalise, S.; Giannini, F.; et al. Technology-Based Therapy-Response and Prognostic Biomarkers in a Prospective Study of a de Novo Parkinson's Disease Cohort. *NPJ Park. Dis.* **2021**, *7*, 82. [CrossRef]

23. Suppa, A.; Kita, A.; Leodori, G.; Zampogna, A.; Nicolini, E.; Lorenzi, P.; Rao, R.; Irrera, F. L-DOPA and Freezing of Gait in Parkinson's Disease: Objective Assessment through a Wearable Wireless System. *Front. Neurol.* **2017**, *8*, 406. [CrossRef] [PubMed]

24. Zampogna, A.; Mileti, I.; Martelli, F.; Paoloni, M.; del Prete, Z.; Palermo, E.; Suppa, A. Early Balance Impairment in Parkinson's Disease: Evidence from Robot-Assisted Axial Rotations. *Clin. Neurophysiol.* **2021**, *132*, 2422–2430. [CrossRef]

25. Bobić, V.; Djurić-Jovičić, M.; Dragašević, N.; Popović, M.B.; Kostić, V.S.; Kvaščev, G. An Expert System for Quantification of Bradykinesia Based on Wearable Inertial Sensors. *Sensors* **2019**, *19*, 2644. [CrossRef] [PubMed]

26. Di Biase, L.; Summa, S.; Tosi, J.; Taffoni, F.; Marano, M.; Rizzo, A.C.; Vecchio, F.; Formica, D.; Di Lazzaro, V.; Di Pino, G.; et al. Quantitative Analysis of Bradykinesia and Rigidity in Parkinson's Disease. *Front. Neurol.* **2018**, *9*, 121. [CrossRef] [PubMed]

27. Dai, H.; Zhang, P.; Lueth, T.C. Quantitative Assessment of Parkinsonian Tremor Based on an Inertial Measurement Unit. *Sensors* **2015**, *15*, 25055–25071. [CrossRef]

28. Ricci, M.; di Lazzaro, G.; Pisani, A.; Mercuri, N.B.; Giannini, F.; Saggio, G. Assessment of Motor Impairments in Early Untreated Parkinson's Disease Patients: The Wearable Electronics Impact. *IEEE J. Biomed. Health Inform.* **2020**, *24*, 120–130. [CrossRef]

29. Zampieri, C.; Salarian, A.; Carlson-Kuhta, P.; Aminian, K.; Nutt, J.G.; Horak, F.B. The Instrumented Timed up and Go Test: Potential Outcome Measure for Disease Modifying Therapies in Parkinson's Disease. *J. Neurol. Neurosurg. Psychiatry* **2010**, *81*, 171–176. [CrossRef]

30. Lewek, M.D.; Poole, R.; Johnson, J.; Halawa, O.; Huang, X. Arm Swing Magnitude and Asymmetry during Gait in the Early Stages of Parkinson's Disease. *Gait Posture* **2010**, *31*, 256–260. [CrossRef]

31. El Arayshi, M.; Verrelli, C.M.; Saggio, G.; Iosa, M.; Gentile, A.E.; Chessa, L.; Ruggieri, M.; Polizzi, A. Performance Index for in Home Assessment of Motion Abilities in Ataxia Telangiectasia: A Pilot Study. *Appl. Sci.* **2022**, *12*, 4093. [CrossRef]

32. Saggio, G.; Tombolini, F.; Ruggiero, A. Technology-Based Complex Motor Tasks Assessment: A 6-DOF Inertial-Based System Versus a Gold-Standard Optoelectronic-Based One. *IEEE Sens. J.* **2021**, *21*, 1616–1624. [CrossRef]

33. Anderson, E.; Nutt, J. The Long-Duration Response to Levodopa: Phenomenology, Potential Mechanisms and Clinical Implications. *Park. Relat. Disord.* **2011**, *17*, 587–592. [CrossRef]

34. Zampogna, A.; Manoni, A.; Asci, F.; Liguori, C.; Irrera, F.; Suppa, A. Shedding Light on Nocturnal Movements in Parkinson's Disease: Evidence from Wearable Technologies. *Sensors* **2020**, *20*, 5171.

35. Colantonio, P.; Giannini, F.; Leuzzi, G.; Limiti, E. High Efficiency Low-Voltage Power Amplifier Design by Second-Harmonic Manipulation. *Int. J. RF Microw. Comput.-Aided Eng.* **2000**, *10*, 19–32. [CrossRef]

36. Colantonio, P.; Giannini, F.; Leuzzi, G.; Limiti, E. Harmonic Tuned PAs Design Criteria. *IEEE MTT-S Int. Microw. Symp. Dig.* **2002**, *3*, 1639–1642. [CrossRef]
37. Berardelli, A.; Sabra, A.F.; Hallett, M. Physiological Mechanisms of Rigidity in Parkinson's Disease. *J. Neurol. Neurosurg. Psychiatry* **1983**, *46*, 45–53. [CrossRef]
38. Ruonala, V.; Pekkonen, E.; Airaksinen, O.; Kankaanpää, M.; Karjalainen, P.A.; Rissanen, S.M. Levodopa-Induced Changes in Electromyographic Patterns in Patients with Advanced Parkinson's Disease. *Front. Neurol.* **2018**, *9*, 35. [CrossRef]

electronics

Article

Systolic Blood Pressure Estimation from PPG Signal Using ANN

Benedetta C. Casadei [1,*], Alessandro Gumiero [2], Giorgio Tantillo [2], Luigi Della Torre [2] and Gabriella Olmo [3]

[1] STMicroelectronics, 20010 Cornaredo, Italy
[2] STMicroelectronics, 20864 Agrate, Italy
[3] Department of Control and Computer Engineering, Politecnico di Torino, Corso Duca degli Abruzzi 24, 10129 Torino, Italy
* Correspondence: benedettacaterina.casadei@st.com

Abstract: High blood pressure is one of the most important precursors for Cardiovascular Diseases (CVDs), the most common cause of death in 2020, as reported by the World Health Organization (WHO). Moreover, many patients affected by neurodegenerative diseases (e.g., Parkinson's Disease) exhibit impaired autonomic control, with inversion of the normal circadian arterial pressure cycle, and consequent augmented cardiovascular and fall risk. For all these reasons, a continuous pressure monitoring of these patients could represent a significant prognostic factor, and help adjusting their therapy. However, the existing cuff-based methods cannot provide continuous blood pressure readings. Our work is inspired by the newest approaches based on the photoplethysmographic (PPG) signal only, which has been used to continuously estimate systolic blood pressure (SP), using artificial neural networks (ANN), in order to create more compact and wearable devices. Our first database was derived from the PhysioNet resource; we extracted PPG and arterial blood pressure (ABP) signals, collected at a sampling frequency of 125 Hz, in a hospital environment. It consists of 249,672 PPG periods and the relative SP values. The second database was collected at STMicroelectronics s.r.l., in Agrate Brianza, using the MORFEA3 wearable device and a digital cuff-based sphygmomanometer, as reference. The pre-processing phase, in order to remove noise and motion artifacts and to segment the signal into periods, was carried out on Matlab R2019b. The noise removal was one of the challenging parts of the study because of the inaccuracy of the PPG signal during everyday-life activity, and this is the reason why the MORFEA3 dataset was acquired in a controlled environment in a static position. Different solutions were implemented to choose the input features that best represent the period morphology. The first database was used to train the multilayer feed-forward neural network with a back-propagation model, whereas the second one was used to test it. The results obtained in this project are promising and match the Association for the Advancement of Medical Instruments (AAMI) and the British Hypertension Society (BHS) standards. They show a Mean Absolute Error of 3.85 mmHg with a Standard Deviation of 4.29 mmHg, under the AAMI standard, and reach the grade A under the BHS standard.

Keywords: cardiovascular diseases; blood pressure; hypertension; photoplethysmography; artificial neural networks; neurodegenerative disease; remote monitoring; telemonitoring; wearable sensor

Citation: Casadei, B.C.; Gumiero, A.; Tantillo, G.; Della Torre, L.; Olmo, G. Systolic Blood Pressure Estimation from PPG Signal Using ANN. *Electronics* **2022**, *11*, 2909. https://doi.org/10.3390/electronics11182909

Academic Editor: Maysam Abbod

Received: 1 July 2022
Accepted: 11 September 2022
Published: 14 September 2022

Publisher's Note: MDPI stays neutral with regard to jurisdictional claims in published maps and institutional affiliations.

1. Introduction

Thanks to the evolution of telecommunication technologies, the use of telemedicine (TLM) has spread rapidly in different areas. This paper focuses on *Telemonitoring*, which provides the acquisition and management of the patients' vital parameters from a remote location, so as to be continuously monitored at home. Blood pressure (BP) is one of the principal vital signs. The continuous and non-invasive remote monitoring of BP has been widely investigated in the past decades with the aim of preventing and treating hypertension. The idea of using PPG signal, collected in a optical non-invasive method, which is easy to obtain and can be used to detect blood volume fluctuations, has been explored. The amount of light scattered, reflected, or transmitted, calculated using the Beer–Lambert law, can provide an indication of changes in overall blood volume [1]. PPG, with respect to other biological signals, has a higher mobility, thus allowing continuous monitoring during everyday-life activity. Nevertheless, PPG has been widely used to measure blood oxygen; it has also been related to arterial blood pressure waveform [2], where Pulse Transit Time (PTT) and the Pulse Wave Velocity (PWV) have been calculated from the Electrocardiogram and the PPG signals. Other works also try to estimate BP deriving either the PTT or the PWV [3]. However, these methods require the use of two devices to acquire both electrocardiogram and photoplethysmogram, thus limiting the subject's freedom of movement. In order to overcome the above-mentioned synchronization problem, some studies have tried to determine the nonlinear relationship between PPG and ABP. In [4], 21 features and a multilayer feed-forward back propagation ANN with 21 input neurons and two output neurons, to estimate Systolic and Diastolic blood pressure, have been used, and satisfactory results have been achieved. In this work we implemented an ANN model able to estimate the SP starting from the PPG signal only. The paper is organized as follows. In Section 2, the state of the art is presented. In Section 3, the Materials and Methods are discussed. In Section 4, results are presented, and in Section 5, the conclusions are reported.

2. State of the Art

Several methods exist to measure BP, divided into two main categories: *invasive* blood pressure (IBP) and non-invasive blood pressure (NIBP) measurement systems. The first ones are used in the intensive care unit (ICU) and allow the beat-to-beat accurate monitoring of the patient's BP. However, they may cause hematoma, infections, and possible occurrence of thrombi. Being invasive, they can only be used in hospital.

NIBP measurement systems are divided into cuff-based and cuff-less. Cuff-based methods are routinely used in the clinical practice and at home by the patients themselves. The main disadvantages are the intermittent monitoring and the discomfort caused by the inflation of the cuff. Cuff-less methods are more recent and represent an efficient approach for continuous home monitoring. The most common cuff-less method is based on the Pulse Wave Velocity (PWV) analysis. That is because arterial stiffness increases the propagation velocity of the pressure wave, which implies an increase in blood pressure [5]. For example, in [2], PTT, defined as the time lag between ECG and PPG, was calculated to derive PWV, which was then correlated with blood pressure readings using a standard linear curve fitting algorithm.

Although in the literature there are many works demonstrating the huge potential of PTT to enable cuff-less BP measurement, they require a calibration stage and the use of two sensors placed at a known distance. In order to overcome these drawbacks, the idea of using pure PPG signal based method has taken off. Photoplethysmography is a simple and low-cost optical technology able to detect change in light absorption due to blood volume fluctuations. The system is composed of:

- *Light source*: light emitting diode (LED) is used to illuminate biological tissues;
- *Photo-detector*: it collects and converts light intensity variations into electric current.

PPG sensors are placed on the skin surface, and the amount of light detected by the PPG photo-detector provides important information about the performance of the vascular system. PPG waveform encompasses two different components. The physiological blood volume fluctuations, at each heart beat, are responsible of the pulsatile component (AC). On the other hand, the slow varying baseline (DC) is imputable to breathing, thermoregulation and sympathetic nervous system activity.

However, the correlation between PPG morphology and BP is nonlinear, and a morphological analysis on PPG signal is necessary in order to extract features useful to estimate the blood pressure. The today solutions are to first extract the morphological features from PPG and then estimates Systolic and Diastolic blood pressure. Nevertheless, the PPG signal is highly susceptible to motion artifacts, resulting in distorted signal and meaningless data zones [6], hence the interest in using ANN to correctly analyze it, removing uninformative signal portion. Deep Neural Networks have been widely used in classification, analysis, and biological signal segmentation, such as ECG and PPG [7]. In [8], Soltane et al. using an ANN has been able to categorize PPG signal in pathological and healthy, with a rate of success equals to 94.7%. Thanks to the advances of Deep Neural Networks, the PPG removing artifacts turns out to be not as challenging as it used to be. In [6], an unsupervised Convolutional Neural Network has been able to detect and remove artifacts in a PPG signal. In [4], an ANN is used to estimate systolic and diastolic blood pressure values, from twenty-one time domain features, extracted from 15,000 PPG periods (Figure 1). In [9], in addition to the time-domain features, frequency and complexity-analysis domain features were computed, so as to describe PPG periods morphology, complexity, amplitude and phase in the frequency domain. The performance of [9] has been evaluated on the MIMIC III database and on their everyday-life dataset collected at Jozef Stefan Institute (JSI). In [10], a PPG signal, a velocity plethysmogram (first derivative of PPG) and an accelerated plethysmogram signal (second derivative of PPG and an important indicator of arteriosclerosis), are fed into an autoencoder, to extract different features from the conventional ones, effective for the blood pressure estimation.

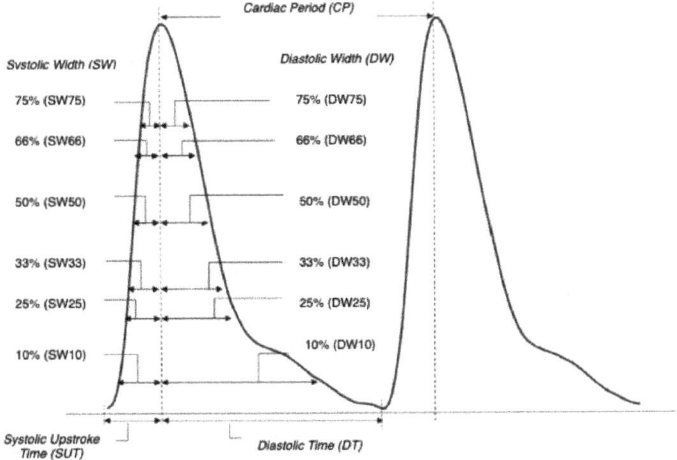

Figure 1. The 21 most common time domain features extracted in PPG from literature.

3. Materials and Methods

This project, developed at the *Remote Monitoring* group, STMicroelectronics in Agrate Brianza, Italy, is divided into two parts:

- In the first part, publicly available data have been used;
- In the second part, signals acquired by the *MORFEA3* (Figure 2) have been addressed (MORFEA3 is a prototype developed in the remote monitoring group in STMicroelectronics; it embeds different ST components, and allows a multispectral PPG reading).

Our first database, containing PPG and arterial blood pressure signals from 47 patients in ICU, was derived from *MIMIC III* (MIMIC III is a large, freely available clinical database). Signals were pre-processed in *Matlab R2019a*; a 4th order Butterworth low-pass filter with a cutoff frequency of 6.6 Hz was used to remove high-frequency noise from the ABP signal, and the *findpeaks* function was used to detect SP peaks. Each SP value was associated with the corresponding PPG period. The PPG signal was pre-processed using a band-pass filter with a bandwidth of 0.5 Hz and 7 Hz (The pulse wave frequency values of the PPG signal is in the range of 0.5–4.0 Hz [11]), and then segmented into one beat periods (Figure 3). The final dataset is composed as follows: each row contains all the period samples (since each period is made up of a different number of samples, it has been necessary to zero-pad all the periods, so as to have one hundred samples each and to preserve all the PPG period features, for both long and short periods), and the last column elements represent the SP values.

Figure 2. MORFEA3 board compared to a 1 euro cent coin.

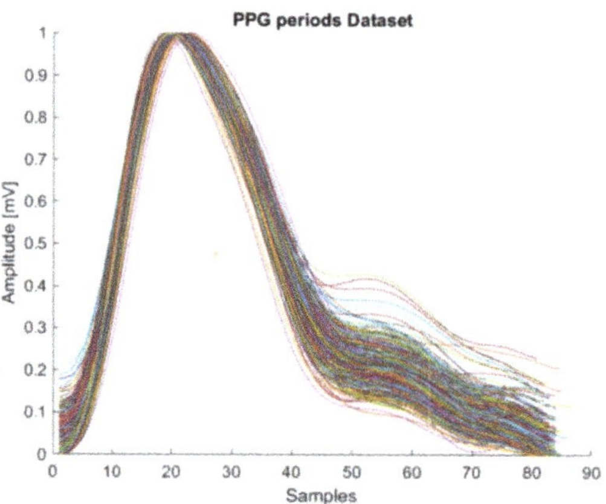

Figure 3. A portion of PPG periods dataset.

In order to estimate the SP, we implemented an ANN regression model, layer-by-layer, with *Keras*. An artificial neural network, a class of machine learning algorithms, one of the application of Artificial Intelligence (AI), is a mathematical model for predicting system performance inspired by the networks of neurons in the human brain. The ANN regression model implemented in this work was intended to predict the SP numerical value. The layers are as follows:

- The first layer is the input layer, composed of 100 neurons, and is equal to the number of samples of each PPG period.
- Dense layers were added as hidden layers; as their name suggests, all their neurons are fully connected to those in the next layer.
- The output layer consists of 1 neuron with a linear Activation Function, which directly provides the value of the weighted sum of the input.

To evaluate our NN model predictive performance, we adopted the *Mean Absolute Error* (MAE), which represents the absolute value of the average error (the difference between actual and predicted value):

$$MAE = \frac{\sum |y - y'|}{N} \tag{1}$$

where y represents the actual value, y' the predicted value, and N the number of sample points (it is more intuitive, since its values are expressed in the same unit of measurement of the input, and it is less sensitive to outliers than the *Root mean squared error* (RMSE)). Since the ultimate goal of this project is to embed the final regression model in a Micro-Controller Unit (MCU), in particular belonging to the *STM32L4+* family, it is necessary to evaluate its computational cost in terms of memory capacity required (Figure 4). To do this, the *STM32CubeMX* graphical tool was used.

Figure 4. Flash and RAM memory occupation percentage by the ANN regression model.

The first dataset used in our study, the *MIMIC III* one, contains periods with 80 mmHg < SP < 180 mmHg, corresponding to all the classes in the American Heart Association classification (reported in Table 1), except hypotension.

Table 1. American Heart Association hypertension guidelines classification.

Class Label	SP Range
Hypotension	<90
Desired	90–119
Pre-Hypertension	120–139
Stage 1 Hypertension	140–159
Stage 2 Hypertension	160–180

After the ANN regression model training phase using the MIMIC III dataset, PPG signal and NIBP were acquired at STMicroelectronics s.r.l. on 8 healthy subjects (5 males and 3 females), whose age ranges from 25 to 50 years old, during resting position, and the acquired data used for testing purposes (Figure 5).

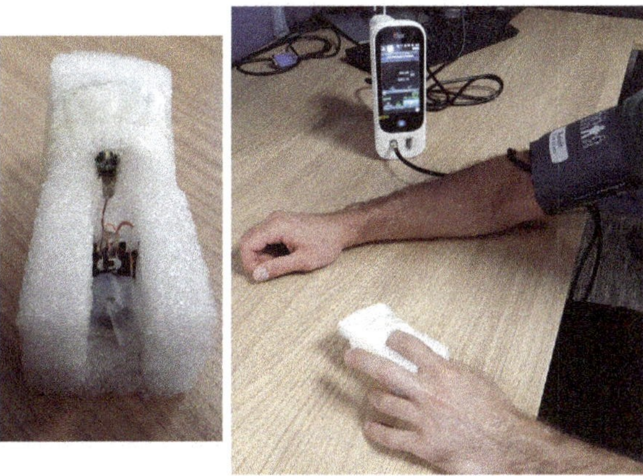

Figure 5. An example of BP and PPG setup for data acquisition on volunteers at STMicroelectronics.

BP was measured simultaneously using a sphygmomanometer wrapped around the left arm, and noted down on an Excel file, along with the acquisition time. The PPG signal was acquired from right hand finger using *MORFEA3* at a sampling frequency of 62.5 Hz. MORFEA3 uses a white LED as a white source and a spectrometer to separate the beam of white light into its components, in order to measure the green wavelength contribution. This wearable device, built with a printed circuit board (PCB), embeds a certain number of STMicroelectronics's sensors, such as a 3-axis accelerometer, two spectrometers VD6283, and a NTC thermistor managed by a MCU.

The MORFEA3 prototype interfaces with the outside world using a Bluetooth connection. To acquire the signals, it is placed within a suitable case, where it is possible to insert the right hand index fingertip, so as to keep the finger-device contact stable. After wearing the sphygmomanometer cuff, turning on the device and positioning the right hand finger inside the case, PPG signal and SP values are acquired. An example of filtered PPG signal acquired with the *MORFEA3* is represented in Figure 6.

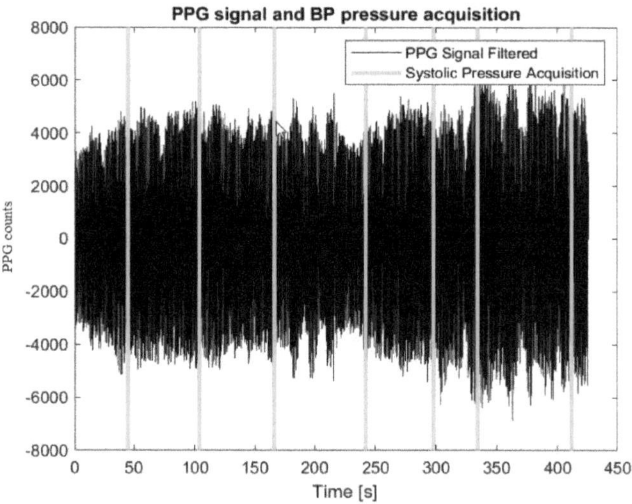

Figure 6. Filtered PPG MORFEA3 Signal with the SP acquisition instants superimposed.

The signal has been filtered with the same 4th order Butterworth filter used in the *MIMIC III* PPG signal pre-processing. Then, in order to make the *MORFEA3* PPG signal suitable for the chosen ANN model (this is because the number of input neurons of a model cannot be varied, and if we had segmented our signal without resampling, we would have obtained a shorter average length of the periods), it was resampled at 125 Hz. Once resampled, the PPG signal was segmented into periods, each associated to an SP value. This second dataset was composed of 6460 periods, with one hundred samples each, and, in the last column, the corresponding SP target. Starting from the discretization of the targets, the SP values, we realized that pathological values, such as Hypotension and Stage 1 and Stage 2 Hypertension, are not represented in our dataset. Moreover, our reference is not a continuous signal, as it was in the case of arterial pressure signal. Therefore, in order to obtain an SP value for each PPG periods, two consecutive SP records have been averaged between two acquisition instants.

In order to find the most suitable network architecture, we have made assumptions on the appropriate number of hidden layers and nodes. Our aim was to reach a good trade off between complexity and performance. First, we tried to use a basic NN model, so as to increase the number of hidden nodes in every step, until the best performance was reached. The first model used was inspired by [12] and is represented in Figure 7. The other hyperparameters are: learning rate equal to 0.007, sigmoid as Activation Function,

Adam as optimization algorithm, glorot uniform as initialization mode, Dropout rate equal to 0.2, batch size equal to 512, and number of epochs equal to 500.

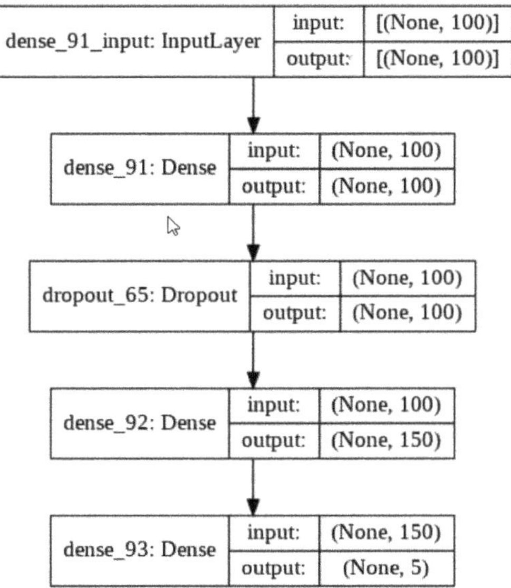

Figure 7. First ANN model.

To improve the model performance we started increasing the number of hidden layers. Starting from the simplest model, up to the analysis of more complex models, the performance over the *MIMIC III* test set improves, and then decreases, probably due to the overfitting phenomenon. Once the network architecture with the best generalization properties was identified, the model hyperparameters were tuned using the *Grid-Search* method (Grid-Search is a method used to find the optimal hyperparameters of a model which results in the most 'accurate' predictions) and the model performance over the test set was evaluated, with the purpose of finding the optimal combination of hyperparameters.

4. Results

For the purpose of this study, several NN architectures have been tested and various hyperparameters have been manipulated to compare the model outcome. The best hyperparameters combination, to obtain the best performance in terms of MAE, on the MIMIC III test set, is reported below:

- *Architecture*: [100 80 100 60];
- *Activation Function*: Softsign;
- *Optimization Algorithm*: Stochastic gradient descent (SGD);
- *Initialization Mode*: glorot uniform;
- *Learning Rate*: 0.007;
- *Dropout*: 0.2;
- *Batch Size*: 1024.

Once the most suitable network architecture and hyperparameters are selected, the regression model performance is evaluated over the MIMIC III test set and over our everyday-life MORFEA3 data set. The regression model performance obtained on the *MIMIC III* test set is presented in Table 2, and compared with the results obtained in [4], using an ANN with 21 input parameters, and with the results in [9], achieved using the hyperparameter tuned Ensemble of regression trees algorithm with RReliefF selected subset of features.

Table 2. The regression model results, in terms of MAE over the MIMIC III test set, compared with the results coming from literature. Data from [4,9].

	MIMIC III Test Set	**[4] Test Set**	**[9] Test Set**
MAE	2.99 mmHg	3.80 mmHg	4.90 mmHg
STD	3.37 mmHg	3.46 mmHg	6.59 mmHg

As mentioned before, in our everyday-life dataset, SP values in the range of Hypotension, Stage 1 Hypertension, and Stage 2 Hypertension have been not represented so it has been difficult to evaluate the regression model performance on the pathological classes. However, the performance obtained over the *MORFEA3* dataset was satisfactory and is reported in Table 3, and compared to the results obtained on the everyday life dataset collected at JSI in [9].

Table 3. The regression model results, in terms of MAE over the *MORFEA3* dataset, compared with the everyday-life dataset collected at JSI.

	MORFEA3 Test Set	**JSI Collected Test Set**
MAE	3.85 mmHg	7.87 mmHg
STD	4.29 mmHg	7.47 mmHg

Figures 8 and 9 show, respectively, the real SP values and the regression model predicted SP values on the *MIMIC III* test set and on the *MORFEA3* dataset. The predicted values referred to the values coming from the sphygmomanometer, which are, obviously, not continuous; however, we represented the values in a continuous manner (Figures 8 and 9), connecting the different estimation, because of simplicity.

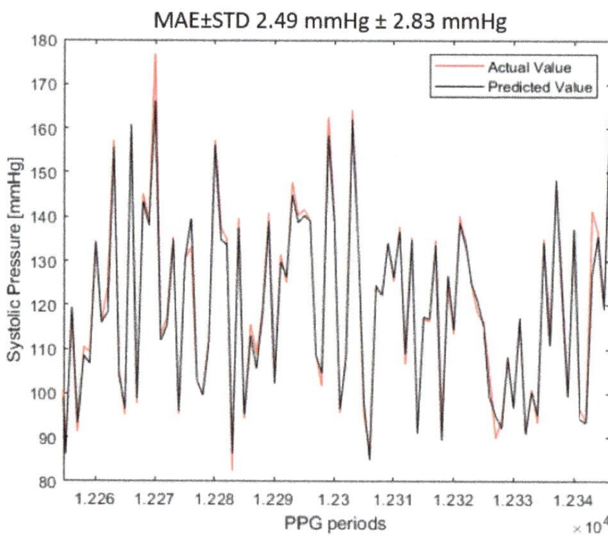

Figure 8. An example of actual and predicted SP values obtained on the *MIMIC III* test set.

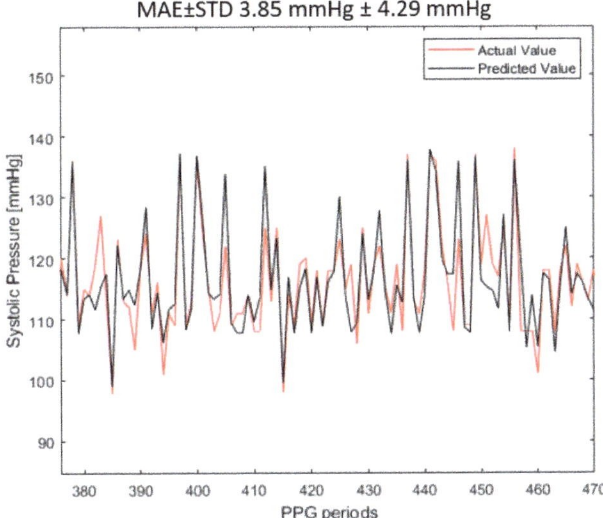

Figure 9. An example of actual and predicted SP values obtained on the *MORFEA3* test set.

The results obtained with the regression model presented in Table 2, are shown more in detail below (Table 4):

- MIMIC III Test set consisting of 74,902 PPG periods.
  - 83.48% of SP values were predicted with an absolute error lower than 5 mmHg;
  - 96.34% were predicted with an absolute error lower than 10 mmHg;
  - 98.88% were predicted with an absolute error lower than 15 mmHg.
- MORFEA3 dataset consisting of 6460 instances.
  - 74.58% of SP values were predicted with an absolute error lower than 5 mmHg;
  - 90.55% were predicted with an absolute error lower than 10 mmHg;
  - 97.27% were predicted with an absolute error lower than 15 mmHg.

Table 4. Percentage of absolute difference between actual and predicted SP values (mmHg).

	Error < 5 mmHg	Error < 10 mmHg	Error < 15 mmHg
MIMIC III Test set	83.48%	96.34%	98.88%
MORFEA3 Dataset	74.58%	90.55%	97.27%

These are very promising results because the obtained prediction errors are compatible with most clinical applications.

5. Conclusions

In recent decades, there has been a strong development in Telemedicine and in wearable systems. These allow continuous patients remote monitoring, without impairing the patient's freedom of movement, and communication at distance between patient and medical staff. In particular, our study aims the continuous monitoring of blood pressure, which is highly desirable in hypertension patients, since high blood pressure is considered as the main cause of CVDs, as well as in the prevention of the risk of dementia and in autonomic nervous system monitoring in most neurodegenerative diseases [13]. In the literature, photoplethysmography is considered a promising candidate for non-invasive and continuous BP monitoring [14].

In this project, the *MORFEA3* device was used to acquire, among others (accelerometric and temperature signals), the PPG signal in order to provide a real-time SP estimation, using an Artificial Neural Network. The regression model, with one hundred input neurons and one output neuron, has yielded stable results; it allows to evaluate the difference between actual SP values and predicted SP values. The results obtained on the *MORFEA3* dataset, in terms of MAE (3.85 mmHg $\leq$ 4.29 mmHg), satisfy the Association for the Advancement of Medical Instrumentation (MAE $\leq$5 mmHg with SD $\leq$ 8 mmHg for SP and DP) (AAMI) and the British Hypertension Society (BHS) standards (absolute difference between standard and test device $\leq$5 mmHg in the 60% of the readings, $\leq$10 mmHg in the 85% of the readings and $\leq$15 mmHg in the 95% of the readings) [15], as evidenced in Section IV, in Table 2 and in Table 4.

Although very promising, this work must be considered preliminary as it is affected by some limitations; in the AAMI and BHS protocols for General Population studies (they should include 85 subjects over 12 years; $\geq$30% males and $\geq$30% females and shall have $\geq$5% of SP values $\leq$100 mmHg, $\geq$5% with $\geq$160 mmHg, and $\geq$20% with $\geq$140 mmHg), 85 subjects are required, whereas in our work only 8 subjects have been tested. Additionally, we can state that our model turns out to be valid only for finger acquisition, since trying to estimate SP from wrist PPG signal, the model performance impairs. For this reason, it may be necessary to build another training set composed only of wrist PPG periods.

Author Contributions: Conceptualization, B.C.C., A.G. and L.D.T.; Data curation, B.C.C. and G.T.; Formal analysis, G.O.; Methodology, G.O.; Software, G.T.; Supervision, A.G. and L.D.T.; Writing—original draft, B.C.C.; Writing—review & editing, B.C.C., A.G., G.T., L.D.T. and G.O. All authors have read and agreed to the published version of the manuscript.

Funding: This research received no external funding.

Data Availability Statement: The MIMIC III freely available database, accessed on 17 of July 2022, used in the present study, can be found at: https://physionet.org/content/mimiciii/1.4/.

Conflicts of Interest: The authors declare no conflict of interest.

Abbreviations

The following abbreviations are used in this manuscript:

CVDs	Cardiovascular Diseases
WHO	World Health Organization
PPG	Photoplethysmogram
ANN	Artificial Neural Network
AAMI	Association for the Advancement of Medical Instruments
BHS	British Hypertension Society
STD	Standard Deviation
TLM	Telemedicine
BP	Blood Pressure
PTT	Pulse Transit Time
PWV	Pulse Wave Velocity
SP	Systolic Blood Pressure
IBP	Invasive Blood Pressure
NIBP	Non Invasive Blood Pressure
ICU	Intensive Care Unit
JSI	Jozef Stefan Institute

References

1. Ysehak Abay, T. Photoplethysmography for blood volumes and oxygenation changes during intermittent vascular occlusions. *J. Clin. Monit. Comput.* **2018**, *32*, 447–455. [CrossRef] [PubMed]
2. Mishra, B.; Thakkar, N. Cuffless blood pressure monitoring using PTT and PWV methods. In Proceedings of the International Conference on Recent Innovations in Signal Processing and Embedded Systems, RISE 2017, Bhopal, India, 27–29 October 2017; pp. 395–401. [CrossRef]
3. Xu, Y.; Zhang, W.; Wang, D.; Ping, P. A calibration method for cuffless continue blood pressure measurement using Gaussian normalized pulse transit time. In Proceedings of the 2018 IEEE International Instrumentation and Measurement Technology Conference (I2MTC), Houston, TX, USA, 14–17 May 2018; pp. 1–5.

4. Kurylyak, Y.; Lamonaca, F.; Grimaldi, D. A Neural Network-based method for continuous blood pressure estimation from a PPG signal. In Proceedings of the Conference Record—IEEE Instrumentation and Measurement Technology Conference, Minneapolis, MN, USA, 6–9 May 2013; pp. 280–283. [CrossRef]
5. Haddad, S.; Boukhayma, A.; Caizzone, A. Continuous PPG-Based Blood Pressure Monitoring Using Multi-Linear Regression. *IEEE J. Biomed. Health Informatics* **2022**, *26*, 2096–2105. [CrossRef] [PubMed]
6. Azar, J.; Makhoul, A.; Couturier, R.; Demerjian, J. Deep recurrent neural network-based autoencoder for photoplethysmogram artifacts filtering. *Comput. Electr. Eng.* **2021**, *92*, 107065. [CrossRef]
7. Esgalhado, F.; Fernandes, B.; Vassilenko, V.; Batista, A.; Russo, S. The Application of Deep Learning Algorithms for PPG Signal Processing and Classification. *Computers* **2021**, *10*, 158. [CrossRef]
8. Soltane, M.; Ismail, M.; Abidin, Z.; Rashid, A.; Ehsan, S.D. Artificial Neural Networks (ANN) Approach to PPG Signal Classification. *Int. J. Comput. Inf. Sci.* **2004**, *2*, 58–65.
9. Slapničar, G.; Luštrek, M.; Marinko, M. Continuous blood pressure estimation from PPG signal. *Informatica* **2018**, *42*, 33–42.
10. Shimazaki, S.; Bhuiyan, S.; Kawanaka, H.; Oguri, K. Features Extraction for Cuffless Blood Pressure Estimation by Autoencoder from Photoplethysmography. In Proceedings of the Annual International Conference of the IEEE Engineering in Medicine and Biology Society, EMBS, Honolulu, HI, USA, 18–21 July 2018; pp. 2857–2860. [CrossRef]
11. Moraes, J.L.; Rocha, M.X.; Vasconcelos, G.G.; Vasconcelos Filho, J.E.; de Albuquerque, V.H.C.; Alexandria, A.R. Advances in photopletysmography signal analysis for biomedical applications. *Sensors* **2018**, *18*, 1894. [CrossRef] [PubMed]
12. Augusti, F. Systolic Blood Pressure Evaluation from PPG Signal Using Artificial Neural Networks. Available online: https://webthesis.biblio.polito.it/17609/ (accessed on 30 June 2022).
13. Thorin, E. Hypertension and Alzheimer Disease Another Brick in the Wall of Awareness. *Hypertension* **2014**, *65*, 36–38. [CrossRef] [PubMed]
14. Kim, H.J.; Kang, W.H.; Lee, H.; Kim, N. Continuous Monitoring of Blood Pressure with Evidential Regression. *arXiv* **2021**, arXiv:2102.03542.
15. Stergiou, G.S.; Alpert, B.; Mieke, S.; Asmar, R.; Atkins, N.; Eckert, S.; Frick, G.; Friedman, B.; Graßl, T.; Ichikawa, T.; et al. A universal standard for the validation of blood pressure measuring devices: Association for the Advancement of Medical Instrumentation/European Society of Hypertension/International Organization for Standardization (AAMI/ESH/ISO) Collaboration Statement. *Hypertension* **2018**, *71*, 368–374. [CrossRef] [PubMed]

Article

Towards Posture and Gait Evaluation through Wearable-Based Biofeedback Technologies

Paola Cesari [1], Matteo Cristani [2,*], Florenc Demrozi [3,*], Francesco Pascucci [2], Pietro Maria Picotti [4], Graziano Pravadelli [2], Claudio Tomazzoli [2], Cristian Turetta [2], Tewabe Chekole Workneh [2] and Luca Zenti [1]

[1] Department of Neuroscience Biomedicine and Movement, University of Verona, 37129 Verona, Italy
[2] Department of Computer Science, University of Verona, 37129 Verona, Italy
[3] Department of Electrical Engineering and Computer Science, University of Stavanger, 4021 Stavanger, Norway
[4] LabofMove Research, 37122 Verona, Italy
* Correspondence: matteo.cristani@univr.it (M.C.); florenc.demrozi@uis.no (F.D.)

Abstract: In medicine and sport science, postural evaluation is an essential part of gait and posture correction. There are various instruments for quantifying the postural system's efficiency and determining postural stability which are considered state-of-the-art. However, such systems present many limitations related to accessibility, economic cost, size, intrusiveness, usability, and time-consuming set-up. To mitigate these limitations, this project aims to verify how wearable devices can be assembled and employed to provide feedback to human subjects for gait and posture improvement, which could be applied for sports performance or motor impairment rehabilitation (from neurodegenerative diseases, aging, or injuries). The project is divided into three parts: the first part provides experimental protocols for studying action anticipation and related processes involved in controlling posture and gait based on state-of-the-art instrumentation. The second part provides a biofeedback strategy for these measures concerning the design of a low-cost wearable system. Finally, the third provides algorithmic processing of the biofeedback to customize the feedback based on performance conditions, including individual variability. Here, we provide a detailed experimental design that distinguishes significant postural indicators through a conjunct architecture that integrates state-of-the-art postural and gait control instrumentation and a data collection and analysis framework based on low-cost devices and freely accessible machine learning techniques. Preliminary results on 12 subjects showed that the proposed methodology accurately recognized the phases of the defined motor tasks (i.e., rotate, in position, APAs, drop, and recover) with overall F1-scores of 89.6% and 92.4%, respectively, concerning subject-independent and subject-dependent testing setups.

Keywords: biofeedback; wearable sensors; neurodegenerative diseases; movement anticipation; machine learning

Citation: Cesari, P.; Cristani, M.; Demrozi, F.; Pascucci, F.; Picotti, P.M.; Pravadelli, G.; Tomazzoli, C.; Turetta, C.; Workneh, T.C.; Zenti, L. Towards Posture and Gait Evaluation through Wearable-Based Biofeedback Technologies. *Electronics* **2023**, *12*, 644. https://doi.org/10.3390/electronics12030644

Academic Editors: Jikui Luo and Riccardo Pernice

Received: 13 December 2022
Revised: 24 January 2023
Accepted: 26 January 2023
Published: 28 January 2023

1. Introduction

The control of the postural system is one of the fundamental neurophysiological mechanisms of the human body. It is fundamental to ensuring balance against gravity and fixing body orientation, and functions as a reference frame for perception–action coupling while efficiently dealing with the external world. Postural control is a dynamic process that requires sensory detection of body motions and integration of sensorimotor information within the central nervous system. In more detail, the central nervous system triggers the execution of appropriate musculoskeletal responses in order to obtain an equilibrium between destabilizing and stabilizing forces [1].

It has been shown, in the reference literature of physiatric medicine, that measurements of postural stability are critical for determining predictors of performance [2], for evaluating musculoskeletal injuries [3], for determining the effectiveness of physical training and rehabilitation treatments [4], and to provide injury prevention through the study of injury risk-factor analysis [3,5]. The body's motion is mainly based on the integration of the proprioceptive, visual, and vestibular inputs [6]. Afferent proprioceptive inputs are conveyed to different levels of the central nervous system [7–9]; however, most of them remain unconscious. The joint positions and movement sensing (kinaesthesia) are the expressions of the conscious component, but postural control is primarily based on the unconscious component [9]. Specifically for the antigravity movements, proprioceptive control represents the expression of the effectiveness of the stabilizing reflexes in controlling vertical stability [8]. In fact, antigravity movements are the activities that counteract gravity and postural instability with at least a phase of single-limb stance [6]. In this way, proprioceptive input represents the most relevant sensory system in the maintenance of static postural stability at all ages and fitness levels [10]. This topic is also relevant in neurodegenerative diseases such as Parkinson's disease (PD). For example, PD patients with postural instability have worse reactions to brief perturbations, more stance sway, and trouble switching between tasks. Moreover, quantifying balance changes in early and moderate-stage PD and the comparison to healthy subjects using clinical assessments of balance and musculoskeletal activation is paramount, primarily if performed through less invasive and costly systems [11,12].

Many tools to detect musculoskeletal activation have been used in sport and rehabilitative medicine. Mainly, electromyography (EMG) is employed for this purpose. However, EMGs are not yet widely used in combination with accelerometers for forecasting and customizing measures for the analysis of the human body's motion to achieve different goals. This is mainly due to the limited number of investigations that have been focusing on the nature of musculoskeletal response to a broad spectrum of stimuli able to identify the thresholds that establish the standard/ideal status of the postural system. Hence, there is a lack of low-cost technology that habilitates these measurements.

Furthermore, in the last decade, with the advent of the Internet of Things (IoT), embedded sensors have been integrated into personal devices such as smartphones and smartwatches. In several applications, sensors are integrated into clothes or other equipment/objects of daily life, becoming a central research topic due to their importance in many areas, including healthcare, interactive gaming, sports, and monitoring systems for general purposes in controlled and uncontrolled settings [13–15].

The primary purpose of the investigation we carried out in this project was to present a preliminary study on the design of a portable and reliable postural system prototype composed of HW and SW, adapted to diverse individual profiles concerning the performance viewpoint, from patients needing rehabilitation to top-level athletes. It is widely accepted in the community of psychiatric medicine that proper quantification of the postural system efficiency represents an essential assessment for improving the quality of life. However, most of the actual measurements are developed in a laboratory environment where natural movements are usually constrained by the instruments applied to subjects' bodies and the environment. This process is performed to distinguish, as precisely as possible, in the limits of the experimental setting, the roles of proprioception, visual and vestibular input using a low-cost and portable instrument. This habilitates the individuals to move freely and perform activities at home or in other uncontrolled environments (e.g., gyms or sports facilities) [5,16].

The technology integrated into such project includes state-of-the-art apparatus (i.e., force platform, EMG, and motion capture cameras) and wireless three-dimensional inertial units performing computation and data analysis over low-cost devices (i.e., sensors themselves, smartphones, and tablets) or cloud platforms.

In the project, we considered variants of these configurations' implementations to evaluate possible solutions to different settings. Subsequently, another essential goal will be the usability of the envisioned technology. Thus, the possibility to perform a comparative study with existing state-of-the-art validated systems is needed to obtain a well-ground assessment. Moreover, fundamental analysis of the specific characteristics of the exercises existing in the literature and a combinational calculation will define the subsequent exercise prescriptions; a multitude of exercise combinations will be available to satisfy the actual needs of the different clinical conditions. To identify these exercises, in future works, we are going to investigate the potential of reinforcement learning as a tool for customization (in order to adapt the response of the portable tool to the needs of the setting, such as medical diagnosis, performance test, training evaluation, and training progress measurements) and personalizing (in order to adapt to the individual variability of the applications mentioned above).

The project has a threefold scope:

(a) providing a theoretical foundation of a set of motor tasks employed to evaluate the performance of posture and gait in at least three contexts: rehabilitation, sports performance, and good health training at different ages;

(b) devising and testing a set of sensors that allow obtaining the above measures;

(c) defining a preliminary set of experiments and algorithms to identify individual profiles and compute, via machine learning methods, the correct set of functional exercises.

Finally, this paper presents the experimental apparatuses and describes the experimental design strictly related to the goals mentioned above. The experiments conducted were structured for a final purpose: obtaining an experimental system for identifying the correct patterns to be devised in the project's future developments.

The rest of the paper is organized as follows. Section 2 provides an overview of the relevant references to similar studies. Section 3 discusses the project's expected research outcome, and Section 4 introduces the overall experimental design. Section 5 presents a preliminary evaluation of the data collected by the designed setup. Finally, Section 6 presents some conclusions and sketches further work.

2. Related Work

Posture is studied in two aspects: (a) postural control and (b) postural orientation. The former involves studying the positional control of the body in space and orientation. Instead, the latter involves studying the relationships between body segments. A neutral state (also known as neutral posture) is observed when the upper trunk and head are at zero degrees concerning the vertebral column. Subjects deviating from this neutral posture are said to have low stability that can provoke accelerated intervertebral disc (IVD) degeneration, damage, and misalignment of vertebrae, producing nerve compression that can cause radicular manifestations, such as sensorimotor deficiencies and pain in the involved regions [17]. When considering the maintenance of the vertical posture in everyday life situations, postural control might become a complex task that requires the ability to anticipate and compensate postural strategies when fast actions are performed and when environmental perturbations are applied [18]. How individuals control their preparatory and compensatory postural adjustments is still under debate [19]. Several mechanisms help individuals to keep their posture when task conditions change due to self-inflicted perturbation (e.g., I am suddenly moving my upper arm forward [20] or when somebody is pushing me [21]). These mechanisms are represented by changes in the activation levels of postural muscles called early postural adjustments (EPAs), starting up to 1000ms prior to the impact [22], and anticipatory postural adjustments (APAs), starting 0–150ms prior to the impact [23]. The primary role played by the EPAs is to adjust the posture and facilitate action planning. Typical examples are seen in preparation for making a step [24] or to avoid contact with an approaching object [25]. On the other hand, the function of APAs is to generate forces that act against an effect (mechanical) of a predictable perturbation [26]. Here we concentrate on these ecological motor tasks where individuals are challenged to control posture when facing a highly dynamic situation. The tasks selected in this experimental design involve sequences of actions that require the maintenance of stable posture while standing on an unstable proprioceptive platform and receiving in an unexpected or expected way a perturbation requiring sudden balance recovery. These motor tasks will help unveil the individual strategies adopted given the individual's level of skill. Based on the literature background, a single limb stance is regularly used to examine the postural system [16,27–29],

EMG, electrocardiography (ECG), and inertial sensors integrated into wearables are emerging as promising low-cost and easily usable solutions in everyday life [13] and health care contexts [30]. Inertial measurement unit (IMU)-based movement identification can be achieved by statistical classification or be threshold-based [31]. Such statistical methods utilize supervised machine learning, which links features of a movement to possible movement states in terms of the observation's possibility [32]. Many of these studies are devoted explicitly to disabled people with diminished gait/posture abilities. This holds for multiple sclerosis patients [33] and Parkinson's disease sufferers [30,34,35]. The ability to monitor the gait of multiple sclerosis patients and provide correct biofeedback can help prevent falls and detect freezing (an aspect that can be fruitful also for Parkinson's patients) [33,36]. Prototype systems often include integrated sensors located on the ankles to track gait movements. Body sensors are positioned near the cervical vertebra or on the shoulders to monitor body posture [13]. Many systems can also measure parameters that might be difficult to provide manually, such as the maximum acceleration of the patients during standing up, or the time it takes from sitting to standing [37,38].

Moreover, the current diffusion of machine learning methods employed in gait, posture analysis, and feedback is not comprehensive, but a few significant results have already been achieved. A relevant group of investigations has been designed for decoding algorithms for brain–machine interactions (BMIs) that use the spiking activity as their control signal [39]. These approaches are powerful in devising usable technologies. Specifically, feedback for reinforcement-learning-based brain–machine interfaces using confidence metrics has been addressed [40]. Some studies show how to derive the required evaluative feedback from a biological source, using both the feedback's quantity and quality, and incorporate it into reinforcement learning controller architecture to maximize performance. Analogously, the Berlin BCI has developed an accurate system that works from the first session in BCI-naive Subjects [41].

An overview of the various steps in the brain–computer interface (BCI) cycle, i.e., the loop from the measurement of brain activity, classification of data, feedback to the subject, and the effect of feedback on brain activity, is the focus of [30,42]. On the other hand, the role of technology for accelerated motor learning in sports is investigated in [43]. Finally, parallel man–machine training in ECG-based cursor control development is the subject of [30,44]. Some references should be given to smart environment previous investigations as a foundation of the method developed here, emphasizing some development related to energy management [45] and concerning the design of energy-efficient transmission protocols for wireless body area networks [46,47]. However, the systems mentioned above present many limitations related to accessibility, economic cost, size, battery life, intrusiveness, and usability (i.e., controlled and uncontrolled home or working context) environments.

3. Expected Research Outcomes of the Project

This section illustrates the project goals, reporting the most suitable application scenarios of the technology we envision and an overview of the presented architecture design. For better comprehension, the goals are presented from a top-down perspective.

(a) basic application scenarios;
(b) envisioned technology;
(c) system architecture workflow.

3.1. Application Scenarios

Three different potential application scenarios have been devised under the supervision and collaboration of psychiatric medicine personal and sports training experts. Scenario 1 was a controlled environment enriched with a set of sensors to the extent that it makes this environment smart. Scenario 2 was set without specifying whether the performed activities were to be carried out indoors or outdoors. It is legitimate to suppose that the environmental setting shall be relatively poor regarding available interactions, including the potential unavailability of an Internet connection. Scenario 3 identified variations determined by post-traumatic rehabilitation, personalized performance control over the evolution of illnesses with harmful consequences on the patient's stability, and training of athletes with special needs.

Scenario 1: Diagnostic Evaluation

In the context of a psychiatric medical practice, a patient who suffers from postural instability due to a traumatic event (e.g., car accident) or neurodegenerative disorder is visited by medical personnel. The diagnostic process is assisted with the envisioned technology. The patient is asked to execute a sequence of three exercises: single-stance stability test, forward movement of arms with weights, and step-on gait on a free-range. During the exercises, the patient wears a jacket equipped with a set of sensors and interacts with a visual focus tool that helps her to identify a fixed point at a given distance. The jacket interacts with an application that works on the cloud, measures the reaction time or other variables, including anticipation's effectiveness in the movements, and provides the operator the possibility of marking progress in performance quality based on a fixed threshold that the operator can define concerning age, sex, and the clinical condition of the patient. The whole process is recorded on video, and the instrumental measures are saved on the patient's profile.

Scenario 2: Sport performance benchmarking

An athlete training for a sporting event is monitored by her coach. He provides her with a performance benchmark in line with the event requirements and expectations. The athlete has a given training period for preparing for the event and a performance level she has to accomplish in order to be competitive. The performance benchmarks have been defined by the coach based on the athletic preparation path of the athlete. While following the coach's requests, the athlete executes some training exercises while wearing the jacket described in Scenario 1. Every athlete's exercise is compared against the benchmark performance and consequently identified in terms of a negative gradient concerning the benchmark itself.

Scenario 3: Rehabilitation Follow-Up

During the rehabilitation period, a patient wearing the jacket described in Scenario 1 attends a program consisting of a series of exercises. Each step in the series requires comparing the performance with the provided reference benchmark defined by the psychiatrist during the diagnostic process. The patient measures are the same as in Scenario 2 and represent how a patient uses the jacket in a medical context.

3.2. Envisioned Technology

Our project's aim is to design an accessible, low-cost, small, dedicated low-cost technological solution (i.e., Gait and Posture Smart Jacket (GPSJack)) suitable for the previously introduced application scenarios. Figure 1 shows a graphic representation of the envisioned technology where the subject is wearing the GPSJack and is immersed in the Gait and Posture Smart Environment (GPSEnv). Numbers 1–5 mark the sensors attached to the jacket, and number 6 marks the tablet application used by the top-level user. Medical and potentially sports training staff can interact with the GPSJack/GPSEnv through a tablet to guarantee total portability and versatility of the envisioned technological solutions.

Figure 1. Visual description of the GPSJack technology and application scenario.

Therefore, the GPSJack system represents the low-cost inertial sensor-based system, which is the project's final goal. GPSEnv represents the adaption of the existing state-of-the-art technologies to the presented scenarios for validation purposes. The final GSP-Jack/GPSEnv architecture will provide to medical/sports staff the capability to perform the following actions:

- register a GPSJack/GPSEnv administrator;
- start setup of a GPSJack/GPSEnv;
- registration of a new user;
- definition of the profile of a registered user;
- eliminate a profile;
- execution of a base test for a registered user;
- registration of a new exercise;
- registration of a new benchmark for an existing exercise;
- assignment of a benchmark for a given exercise to a specific user;
- execution of an exercise;
- registration of a sequence of exercises as a training path;
- association of benchmark values for a training path;
- assignment of benchmark values for a training path to a given user;
- visualization of a single test progresses along a temporal interval;
- visualization of progresses with respect to a given benchmark;

Based on the above-defined functions, several background software instruments are required. The technologies for managing and analyzing the data from the sensors, which we may name the GPSJack Framework, have been envisioned. Nevertheless, machine-learning-based algorithms will guide the personalization of the benchmark process by employing intelligent reinforcement learning methods. Finally, since the proposed technology has its main applications in healthcare, it will provide, in addition to the classical data protection techniques, a physical protection layer that, based on the radio signals' propagation patterns, will habilitate the possibility of utilizing the tablet *if and only if* the tablet is under a certain distance (e.g., 5 m) from the GPSJack nodes.

3.3. Architecture Workflow

The system's architecture is composed of several modules, each one with a single responsibility, the logic model of which is reported in Figure 2.

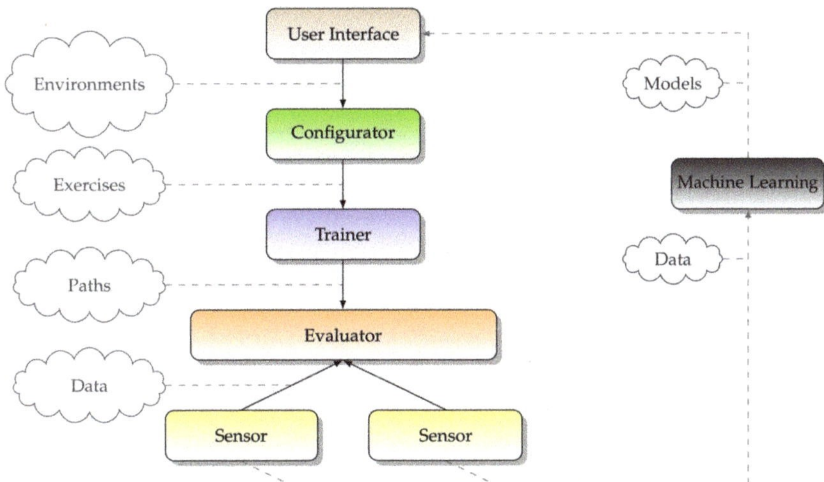

Figure 2. Logic model of system architecture.

Every module is related to at least another one:

User Interface: allows the user to input "environments" and provides visualization of all data/results returned by the Machine Learning module;

Configurator: takes as input a set of environments and gives as output a set of feasible "exercises" to be used by the Trainer, according to user input and machine learning algorithms' indications;

Trainer: uses the exercises and determines the "should be" paths, meaning which exercises shall be executed and with which environment;

Evaluator: responsible for the evaluation of data and automatic comparison with benchmarks;

Sensor: deals with wearable sensors, collecting and normalizing data;

Machine Learning: gathers data and induces models.

The User Interface's output is an ordered set of "environments", which are one of the inputs of the Configurator. The Configurator continuously computes, and at given times uses models from the Machine Learning module to produce a set of feasible "exercises". These exercises are to be given to the Trainer, whose outputs are chains of exercises, also named "paths", to be executed by patients (or athletes). At different times, different paths are possible due to the work of the Trainer. The data are then gathered from the Sensor module and sent to the Evaluator to be stored, visualized, and compared with benchmarks. They are also used to devise possible "paths" and exercises to be delivered as hints to the user. This is the responsibility of Machine Learning, which acts as a feedback generator for the whole system, enabling the system to enhance performances continuously.

4. Methodology Design Workflow

This section presents the overall project information concerning hardware composing the GPSEnv and GPSJack, and software regarding edge computation and the preliminary data analysis pipeline. To achieve the system's architectural requirements, we have designed a four-step method, illustrated in Figure 3.

- the first step consists of the identification of the most suitable motor tasks;
- the second step consists of the description and design of the used data collection systems composing the GPSEnv/GPSJack;
- the third step exploits the collected data analysis workflow;

-    the fourth step exploits the modeling of individuals' gait and posture invariants by machine learning.

Figure 3. GPSJack/GPSEnv system design workflow.

4.1. Identification of Motor Tasks

The motor tasks should contain specific characteristics. We aimed to select tasks that challenge postural stability and require the ability to foresee and anticipate the consequences of actions given the presence of a sudden perturbation that might change posture from a stable to an unstable state. In this way, we will be able to train fundamental motor skills such as action adaptation, compensation, and anticipation, while, on the other hand, measuring the performance of such skills. We will describe a typical trial involving a sequence of movements that satisfies the task requirements stated above:

1.   The subject stands on an unstable proprioceptive board (Figure 4) while holding a heavy ball. The task is to rotate at an angle of 30° or 60° or 90°, both right and left, while keeping his feet still and rotating only the torso, without losing equilibrium. Body balance is evaluated by analyzing the center of pressure (COP) migration. This sequence of activities identifies two distinct motor task phases: (a) rotate and (b) in position.

2.   Once in position, the subject is asked to drop the ball quickly (in the 150 ms preceding this action, the anticipatory postural adjustments (APAs) can be identified and analyzed) captured by the EMG and the force platform. This sequence of activities identifies two distinct motor task phases: (c) APAs and (d) drop.

3.   After the loss of balance (due to the fast drop of the ball), the subject is asked to regain the balance as fast as possible. The analysis of COP migration for defining the balance recovery after the perturbation is considered. This activity identifies one motor task phase: (d) recover.

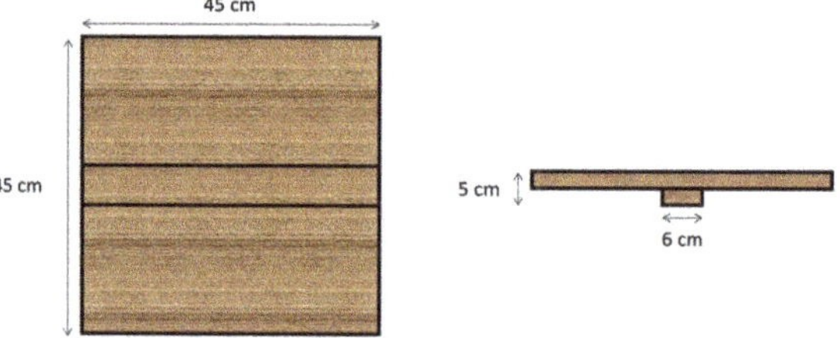

Figure 4. The wooden board: bottom and lateral views.

We analyzed different experimental conditions to define the departure from the standard measurements by considering the same task while changing the biomechanical and perceptual conditions and testing different populations ranging from elite athletes to individuals affected by neuromuscular diseases. Figure 5 presents an overview of the force platform x-axis data of a motor task. As shown, it is composed by five different movements: (a) rotate, (b) in position, (c) APAs, (d) drop, and (e) recover. The task starts at most 10 s after the emission of an audio signal. Such a signal is later used during the manual synchronization of the GPSJack and GPSEnv data streams.

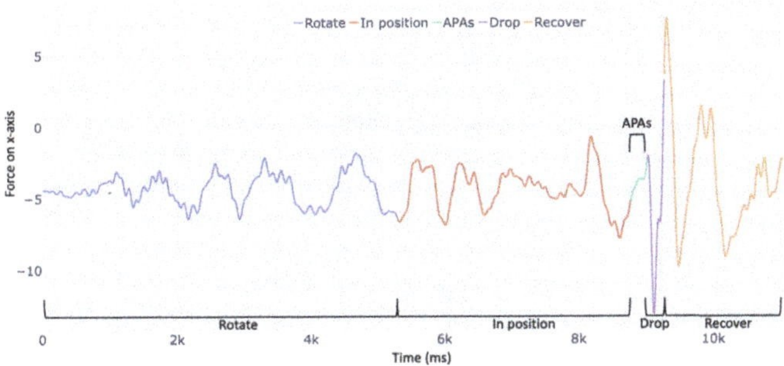

Figure 5. Overview of motor task phases: (**blue**) rotate, (**red**) in position, (**green**) APAs, (**purple**) drop, and (**yellow**) recover.

Data Collection Procedure

To validate the GPSJack system, subjects of different ages, genders, and motor backgrounds performed the introduced protocol of exercises. This protocol aimed to analyze the tested subjects' gait and posture using state-of-the-art instrumentation to identify the anticipatory movements and minimum requirements that the GPSJack nodes (as shown in Figure 1) will have to implement.

Figure 4 presents the proprioceptive board. Area dimensions were 0.45 m × 0.45 m, and height was 0.025 m. On the bottom of the surface, the board was touching the ground utilizing a beam glued along the board mid-line, having the same length as the wooden board, and being 0.025 m in height and 0.06 m in width. Figure 6, on the left, presents how the board was used during the exercise.

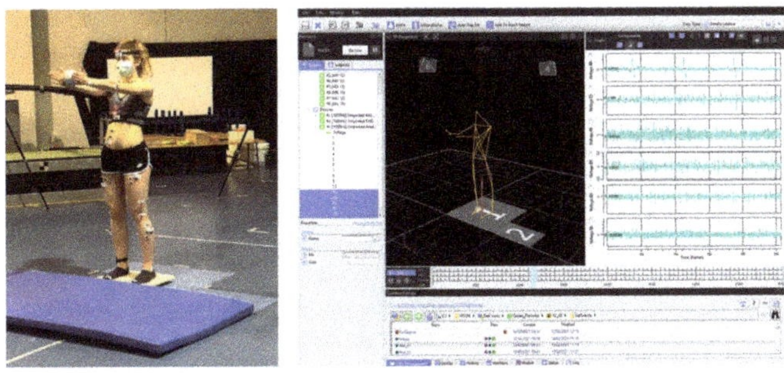

Figure 6. Experimental setup: the task.

4.2. Data Collection Systems

This section presents the instruments involved in the design of the GPSEnv and GPSJack systems. GPSEnv is defined as a state-of-the-art apparatus. GPSJack was designed to be a low-cost and long-battery-life system on chip (SoC).

4.2.1. GPSEnv Apparatus

It performs gait and posture analyses based on the combination of three different instruments:

- Force platform;
- Surface EMGs;
- Motion capture cameras.

Force Platform

The forces in three orthogonal directions, along with the COP migration, are measured by a force platform (https://tinyurl.com/rhkktv4 accessed on 10 December 2022), coupled with a 6-channel strain gauge amplifier (https://tinyurl.com/zy2efpn8 accessed on 10 December 2022), with a sampling frequency of 1000 Hz, presenting a size of 0.9 m × 0.9 m. The AMTI Biomechanics Force Platform, model BP900900, features composite construction, resulting in a low-mass instrument with excellent frequency response. Specifically designed for the precise measurement of ground reaction forces, the BP900900 measures the three orthogonal force components, the moments about the three axes, and the center of pressure in the horizontal plane producing a total of eight outputs. The high sensitivity, low crosstalk, excellent repeatability, and long term stability of this platform make it ideal for research and clinical studies. The BP900900 is easy to use and is available with 1000, 2000, or 4000 pound (4450, 8900, or 17,800 Newtons) vertical capacity. This force platform is shown in Figure 6, on the left part, under the subject's feet, and in right image, represented virtually as platform 1.

Surface electromyography (EMG)

The surface EMG activity of sixteen postural muscles, on both sides of the body, is recorded using electrocardiographic electrodes located on the subject's body, as shown by the red markers in Figure 7. The muscles used are: the rectus abdominis (RA), erector spinae (ES), rector femoris (RF), biceps femoris (BF), vastus lateralis (VL), tensor fasciae latae (TL), tibialis anterior (TA), and soleus (SO). Guidelines from the http://www.seniam. org/ accessed on 10 December 2022 (Surface ElectroMyoGraphy for the Non-Invasive Assessment of Muscles) are used to guarantee consistency in the muscles' anatomical localization. The https://fccid.io/VH6ZWTX07/User-Manual/User-Manual-903877 accessed on 10 December 2022 EMG system, produced by Aurion S.r.l., is used to collect and amplify EMG signals at a sampling rate of 1000 Hz.

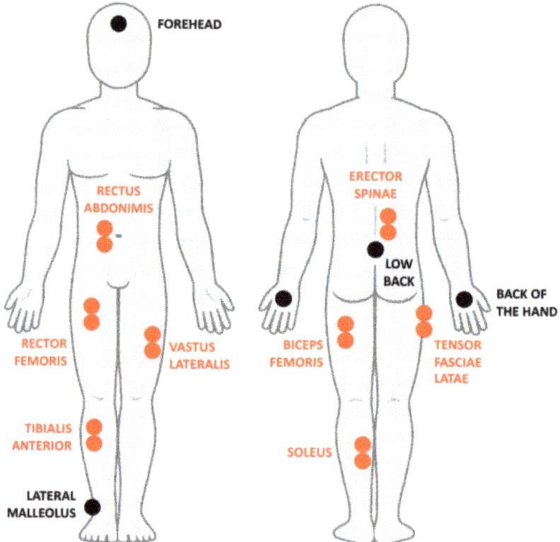

Figure 7. Locations of retro-reflective markers (black) and EMG electrodes (red). EMGs were placed on both sides of the body.

Motion Capture Cameras

Concerning the kinematic analysis, five retro-reflective markers are attached to each subject. Markers' positions are located on the subject's body as shown by the black markers in Figure 7.

The markers are placed on the backs of both hands, on the forehead, on the lowerback, and on the lateral malleoulus on the dominant leg. The position, velocity, and acceleration of every marker are recorded at a sampling rate of 200 Hz, using eight motion capture cameras (https://www.evl.uic.edu/sjames/mocap/resources/Doc/MXhardware_Reference.pdf accessed on 10 December 2022), featuring multiple high-speed processors that perform real-time proprietary image processing and Vicon Nexus 2.6 software.

4.2.2. GPSJack Apparatus

The GPSJack prototype uses the nRF52840 system on chip (SoC), built over the 32-bit ARM CortexTM-M4 CPU with a floating-point unit running at 64 MHz. The nRF52840 is the most advanced member of the Nordic Semiconductor nRF52 Series SoC family (https://www.nordicsemi.com/Products/Low-power-short-range-wireless/nRF52840 accessed on 10 December 2022). It is fully multi-protocol and capable of supporting Bluetooth 5, Bluetooth mesh, Thread, Zigbee, 802.15.4, ANT, and 2.4 GHz proprietary stacks [48]. Furthermore, the nRF52840 uses a sophisticated on-chip adaptive power management system achieving exceptionally low energy consumption.

This SoC interfaces with various electronic devices capable of perceiving different types of measurement of the movement context. In the setup we discuss here, the nRF52840 defines the core of each GPSJack node. On the same HW board, each SoC core communicates with various sensors, such as an accelerometer, gyroscope, and magnetometer. However, since the design of every single node of the GPSJack from scratch would require further targeted effort, and since the goal of the project is to show the suitability of the system in recognizing the different phases of the defined task, we made use of the Nordic Thingy 52 IoT Sensors kit shown in Figure 8. In future developments, the Nordic Thingy 52 can be replaced by dedicated data collection nodes based on the nRF52840 SoC, presenting reduced dimensions and integrating only sensors relevant to the scenario. It transmits data to/from its sensors and actuators to a receiver implemented through a PC, single board computing (SBC) (e.g., Raspberry pi 4 or Odroid H2+), or a mobile application running on a tablet or smartphone [48]. Extended device characteristics are listed in the following:

- Dimensions: 5 cm × 5 cm × 1.5 cm, weight: 47 g;
- Motion-tracking sensors: nine axis motion sensor including 3-axis gyroscope, 3-axis accelerometer, and 3-axis magnetometer;
  - Sampling frequency: up to 200 Hz;
  - Full scale: up to 16 g for accelerometer and up to 2000 dps for gyroscope;
- Battery: rechargeable Li-Po battery with 1440 mAh capacity;
- Microprocessor: 64 Mhz Cortex M4 MCU;
- Communication: Bluetooth Low Energy (BLE);
- Cost: 38 Dollars.

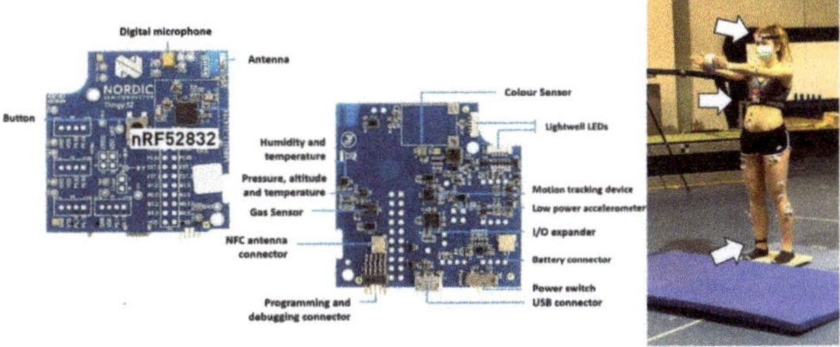

Figure 8. Nordic Thingy 52 board (**on the left**) and its usage in the data collection setup (**on the right**).

Nodes positions and number are not definitive, since their positions directly depend on the preliminary study carried out with the previously introduced instrumentation. The GPSJack sampling frequency, the number of nodes, and the positions will be adjusted (reduced) based on the previous phase's outcome, thereby reducing the battery consumption of the overall system.

Moreover, the designed GPSJack system can be composed of up to 11 different nodes that collect data and communicate with a single tablet or SBC device. Based on the performed tests, the designed GPSJack (configured as shown in Figure 1, 5 nodes and 1 data aggregator) can communicate, without data loss, with the tablet device at a maximal distance of 45 m. Concerning the battery life, the GPSJack data collection nodes can compute for more than 48 h at a sampling frequency of 200 Hz.

Figure 9 presents an overview of the main characteristics of the GPSJack android mobile application running over a tablet device, presenting data collection and visualization.

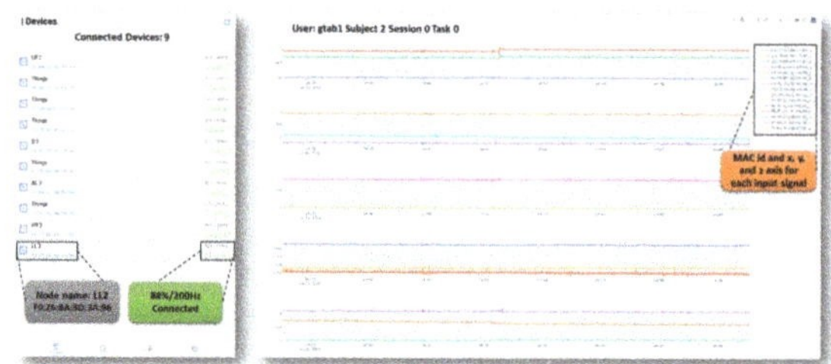

Figure 9. GPSJack Android mobile application running on the tablet: node connection and data visualization.

The raw data (aka, time series), perceived by the GPSJack prototype (at a maximal sampling frequency of 200 Hz) and the existing in-laboratory architecture (perceived at a frequency of 200–1000 Hz), will be pre-treated by applying different data processing steps.

Finally, the GPSJack android mobile application executing on a tablet device has the ability to video-record at 60 FPS the performed tasks. Such recording is synchronized with the data stream perceived by the data collection nodes (i.e., Nordic Thingy 52).

4.2.3. GPSJack/GPSEnv Synchronization

The GPSJack and the GPSEnv data streams present different timestamps and are not synchronized. The synchronization is manually performed offline, using the audio start signal emitted by the Vicon system and the video recording of the task performed by the GPSJack system. The annotator identifies the precise timestamp of the GPSJack system where the start audio signal is emitted by the GPSEnv system. In particular, the annotator identifies the precise video frame during which the signal is emitted (i.e., a granularity of 16 ms). In future developments, the aim will be to automatically synchronize the GPSJack/GPSEnv data streams using existing solutions [49], thereby excluding the time-consuming offline synchronization process.

4.3. Data Analysis

This section presents, starting with the data collected by the mentioned instruments (i.e., EMGs, force platform, kinematic, and GPSJack), the main data processing steps, performed with different processing methods in a defined order [50]. This workflow is presented in Figure 10.

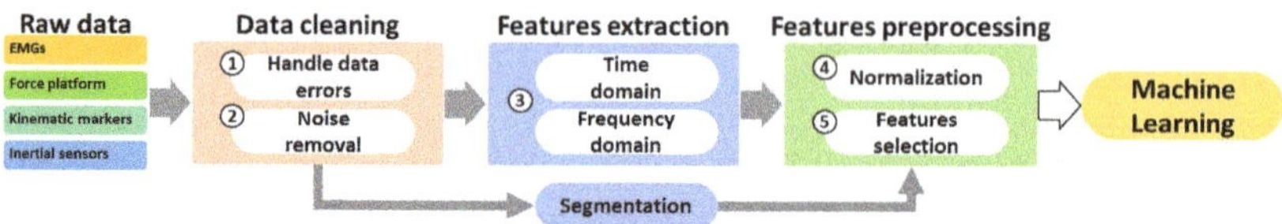

Figure 10. Data analysis workflow.

4.3.1. Data Cleaning

This phase emphasizes data patterns by reducing their dependence on environmental/HW noise and the data collection architecture, which often leads to data loss or corruption during transmission. In particular, dedicated data cleaning techniques must handle missing or corrupted data to maintain the time series structure and information. In this step, the corrupted and missing data issue is handled by applying an interpolation data-filling method that replaces such data with a value that follows the time series's previous and consequent pattern [51]. The noise's impact is reduced by applying a 4-order low-pass filter with a cut-off frequency of 20 Hz. We do not apply any data cleaning method to the existing in-lab architecture, since the architecture's proprietary software already performs such a step.

4.3.2. Feature Extraction

Furthermore, since standard pattern-recognition models are not always suitable for raw data, the machine learning training phase is anticipated by a feature extraction step during which time-series are represented as a set of features in the time and/or frequency domain [52,53]. We make use of the https://github.com/fraunhoferportugal/tsfel accessed on 10 December 2022 [54] to represent each time window of 150 ms (equal to the APAs movement duration) perceived by the mentioned instrumentation in a set of 160 features in the frequency and time domains.

Table 1 shows the most commonly extracted time (e.g., min, max, mean, std, etc.) and frequency (e.g., Fast Fourier Transform (FFT), Discrete Fourier Transform (DFT), Discrete Wavelet Transform (DWT), etc.) domain features that are usually combined, further increasing the recognition accuracy. Other features, extracted by using other frameworks or handcrafted (extracted manually), can be used, since the proposed workflow is modular. Thus, we can easily substitute each module.

Table 1. Most used time and frequency-domain features.

Time Domain Features	Frequency Domain Features
(1) maximum, (2) minimum, (3) mean, (4) standard deviation, (5) root mean square, (6) range, (7) median, (8) skewness, (9) kurtosis, (10) time-weighted variance, (11) interquartile range, (12) empirical cumulative density function, (13) percentiles (10, 25, 75, and 90), (14) sum of values above or below percentile (10, 25, 75, and 90), (15) square sum of values above or below percentile (10, 25, 75, and 90), (16) number of crossings above or below percentile (10, 25, 75, and 90), (17) mean amplitude deviation, (18) mean power deviation, (19) signal magnitude area, (20) signal vector magnitude, (21) covariance, (22) simple moving average of sum of range of a signal, (23) sum of range of a signal, (24) sum of standard deviation of a signal, (25) maximum slope of simple moving average of sum of variances of a signal, (26) autoregression.	(1) Fast Fourier Transform (FFT) coefficients, (2) Discrete Fourier Transform (DFT), (3) Discrete Wavelet Transform (DWT), (4) first dominant frequency, (5) ratio between the power at the dominant frequency and the total power, (6) ratio between the power at frequencies higher than 3.5 Hz and the total power, (7) two signal fragmentation features, (8) DC component in FFT spectrum, (10) energy spectrum, (11) entropy spectrum, (12) sum of the wavelet coefficients, (13) squared sum of the wavelet coefficients and energy of the wavelet coefficients, (14) auto-correlation, (15) mean-crossing rate, (16) spectral entropy, (17) spectral energy, (18) wavelet entropy values, (19) mean frequency, (20) energy band.

BUsers could also decide not to apply the feature extraction step and use the date in the form provided by the previous block applying a standard data segmentation phase. In such a case, the feature selection algorithm is not applied [13,50].

4.3.3. Preprocessing of Features

The extracted features could present a wide range of values that will govern the training process, but such features are not those that primarily represent the dataset's characteristics or the final pattern-recognition model's accuracy. Data normalization transforms multi-scaled data to the same scale, and all variables equally influence the model, improving the learning algorithm's stability and performance [55]. Our methodology makes use of the robust scaling normalization technique that scales each feature of the dataset by subtracting the median ($Q_2(x)$) of this feature and then dividing by the interquartile range (IQR) ($Q_3(x) - Q_1(x)$). This scaler is robust to outliers, in contrast with the other scalers that arehighly affected by outliers. When working with datasets in which different features are used to represent every single sample, the datasets perform independent normalization for every feature.

Moreover, a large number of features does not imply high recognition quality, since they can positively or negatively impact the recognition process. Therefore, feature selection techniques identify features that positively and negatively influence the recognition process, reducing the model's dependence on irrelevant features. The exclusion of a certain number of features decreases the training process's complexity, since a smaller dataset generally requires less training time. The main benefits of such techniques are (i) reducing overfitting by eliminating redundant data, which consequently reduces noise-related issues; (ii) improving accuracy, since misleading data are eliminated; (iii) reducing training time due to fewer data points; and iv) raising interest in certain features demonstrating higher importance [55,56]. We use the tree-based feature selection technique to compute impurity-based feature importances, discarding irrelevant features in cooperation with other feature selection techniques.

In conclusion, the execution of this series of data treatment steps transforms the raw data, subject to noise and errors, to an optimal number of features in the time and frequency domains. This features set will be used by the machine learning models in the fourth phase of the methodology, as shown in Figure 3. The features-preprocessing step is anticipated by the hold/leave-out validation techniques, generating the training and testing datasets. Then, the training dataset will be preprocessed as mentioned above, and then the testing dataset is preprocessed based on the training dataset's requirements.

5. Preliminary Experimental Evaluation

Following the experimental design of Section 4, we collected data from 12 different male subjects, whose physical characteristic are shown in Table 2.

Table 2. Subjects' characteristics.

Subject	Age (Years)	Height (cm)	Weight (kg)
22	24	182	72.6
23	24	183	61.1
24	22	186	97.7
25	28	176	65.1
26	28	172	69.7
28	23	170	64.7
29	24	174	69.0
30	28	176	78.0
31	25	187	86.9
32	23	189	78.1
33	23	176	66.9
34	19	175	72.8
Min.	19	170	61.1
Max.	28	189	97.7
Avg.	24.76	178.65	73.71
Std.	2.78	6.15	9.97

At the end of the data collection phase, each subject had performed a total of 84 7 tasks (i.e., do not rotate, rotate of an angle of $30°$ or $60°$ or $90°$, both right and left) $\times$ 2 statuses (i.e., stable/unstable) $\times$ 6 repetitions) data collection sessions. Each session involved the five movements phases (i.e., rotate, in position, APAs, drop, and recover) described in Section 4.1 and shown in Figure 5.

In this preliminary evaluation, the collected data are divided based on the data collection technologies we utilized (force platform, EMGs, or acceleration data) into three different datasets, as shown in Table 3.

Table 3. Overview of the three types of datasets analyzed.

Dataset ID	Time [ms]	Force Platform (#1)	EMG's (#10)	Acceleration (#9)	Sampling Frequency [Hz]
A	✓	✓	✓		1000
B	✓			✓	200
C	✓	✓	✓	✓	200

(# val) number of sensors.

Subsequently, for each dataset type (i.e., A, B, and C), we applied the data-processing pipeline defined in Section 4.3 and shown in Figure 10. In particular, the data, in segmentation and feature representation forms were segmented in time windows of 150 ms as the hypothetical duration of the APA movement phase), were used to train three different machine learning models, whose performances in recognizing the movement phases of Figure 5 were measured in terms of accuracy A), precision (P), recall (R), and F1-score ($F1$), defined as follows [57]:

$$A = \frac{tp+tn}{tp+tn+fp+fn} \qquad P = \frac{tp}{tp+fp}$$

$$R = \frac{tp}{tp+fn} \qquad F1 = 2 \times \frac{P \times R}{P+R}$$

Here, tp represents the number of true positives, tn represents the number of true negatives, fn represents the number of false negatives, and fp represents the number of false positives.

5.1. Preliminary Results

Tables 4 and 5 present the results of our preliminary analysis. Table 4 presents the results for datasets A, B, and C from all subjects simultaneously by performing a k-fold test (k = 5) on all subjects' data. Table 5 presents the results for datasets A, B, and C for every single subject by performing a k-fold test (k = 5) on each subject's data.

Table 4. Average results for segmentation and feature-extraction data representation of all 12 subjects' data.

| | Segmentation | | | | | | | | | | | | Feature Extraction | | | | | | | | | | | |
| | A | | | | B | | | | C | | | | A | | | | B | | | | C | | | |
Model	A	P	R	F1	A	P	R	F1	A	P	R	F1	A	P	R	F1	A	P	R	F1	A	P	R	F1
k-NN	58.1	56.2	58.1	55.2	47.4	45.0	47.4	44.0	57.3	55.5	57.3	54.4	42.7	39.4	42.7	40.2	38.9	36.8	38.9	35.8	71.4	67.7	71.4	69.2
RF	77.7	80.2	77.7	76.0	54.5	55.3	54.5	48.3	88.8	89.3	88.8	88.2	84.0	84.8	84.0	83.5	63.0	68.6	63.0	59.5	89.9	90.2	89.9	89.6
LDA	49.2	48.9	49.2	47.3	42.5	64.7	42.5	25.7	49.5	49.2	49.5	48.4	36.8	34.6	34.5	31.45	61.3	67.6	61.3	56.9	65.1	67.4	65.0	63.6

The results of Table 4 show that the random forest performed the best on all three datasets (i.e., A, B, and C), and for both data-treatment types (i.e., segmentation and feature extraction). Moreover, when differentiating by dataset type, the results show that the conjunct information of dataset A (i.e., EMGs and Force Platform) and dataset B (i.e., acceleration values) was put to use significantly better than when used separately. In particular, in terms of F1-score, there was an increment of 12.2% from dataset A to dataset C and 39.9% from dataset B to dataset C. Such results indicate that the acceleration provides precious information concerning the recognition of the studied activities. Overall, the achieved F1-scores for all subjects using the random forest model and dataset C were 88.2% and 89.6%, respectively, in segmentation and feature representation modes.

Table 5. Results on segmentation and feature extraction data representation for each subject.

| Subject | Model | Segmentation | | | | | | | | | | | | Feature Extraction | | | | | | | | | | | |
| | | A | | | | B | | | | C | | | | A | | | | B | | | | C | | | |
		A	P	R	F1	A	P	R	F1	A	P	R	F1	A	P	R	F1	A	P	R	F1	A	P	R	F1
P22	k-NN	56.8	55.6	56.8	54.4	40.0	42.1	40.0	39.0	56.5	55.3	56.5	54.2	39.4	37.4	39.4	37.7	34.5	31.9	34.5	32.6	44.5	60.5	44.5	48.9
	RF	83.6	84.8	83.6	82.7	54.1	55.9	54.1	49.2	84.9	85.7	84.9	84.1	83.6	84.8	83.6	83.0	61.5	63.3	61.5	59.4	83.9	85.1	83.9	83.2
	LDA	59.5	59.5	59.5	59.4	32.1	30.8	32.1	31.2	54.8	56.4	54.8	55.5	39.3	76.0	39.3	40.0	3.0	8.3	3.0	0.4	44.5	60.5	44.5	48.9
P23	k-NN	66.7	67.3	66.7	64.3	42.1	42.8	42.1	37.4	65.8	63.5	65.8	63.0	49.1	46.8	49.1	47.8	37.8	34.7	37.8	35.0	82.9	81.4	82.8	80.3
	RF	86.1	87.7	86.1	85.6	48.9	47.9	48.9	43.5	99.8	99.8	99.8	99.8	87.6	88.0	87.6	87.2	62.4	64.9	62.4	60.3	100	100	100	100
	LDA	61.1	61.9	61.1	61.4	37.8	33.7	37.8	33.4	98.6	98.7	98.6	98.6	30.5	0.1	3.1	0.01	3.6	12.9	3.6	1.2	97.4	97.5	97.4	97.4
P24	k-NN	63.7	64.0	63.7	62.6	46.1	43.9	46.1	41.0	64.4	63.7	64.4	63.1	46.1	44.2	46.1	44.9	40.9	38.0	40.9	37.2	65.8	61.6	65.8	63.5
	RF	89.4	90.8	89.4	88.8	56.2	60.6	56.2	49.8	88.1	89.6	88.1	87.3	91.4	91.7	91.4	91.1	65.3	70.5	65.3	62.3	92.1	92.3	92.1	91.7
	LDA	59.7	60.8	59.7	60.0	35.0	32.3	35.0	33.8	52.8	57.2	52.8	54.5	58.2	66.6	58.2	57.1	30.0	34.4	30.0	15.0	61.8	64.3	61.8	62.9
P25	k-NN	63.3	62.8	63.3	60.9	49.5	50.9	49.5	46.5	63.4	63.0	63.4	61.1	48.0	45.4	48.0	46.4	37.6	37.6	37.6	34.9	68.3	65.0	68.3	66.2
	RF	85.0	85.6	85.2	84.7	55.9	63.4	55.9	52.1	90.1	90.3	90.1	89.7	86.1	85.9	86.1	85.8	65.2	72.3	65.2	62.4	90.1	90.0	90.1	89.7
	LDA	55.9	56.1	55.9	55.7	37.5	35.2	37.5	35.6	54.1	54.4	54.1	54.1	20.0	7.8	20.0	10.5	3.7	11.6	3.7	1.9	40.0	46.4	40.0	31.8
P26	k-NN	65.0	64.1	65.0	62.3	48.1	46.4	48.1	43.1	65.0	64.1	65.0	62.3	49.7	46.7	49.7	47.5	37.3	32.9	37.3	34.2	80.4	76.9	80.4	77.7
	RF	86.3	86.5	86.3	85.9	58.7	61.5	58.7	52.7	99.8	99.8	99.8	99.8	88.4	88.3	88.4	88.1	65.3	65.2	65.3	60.8	100	100	100	100
	LDA	57.9	58.6	57.9	58.2	41.7	38.7	41.7	38.6	98.3	98.4	98.3	98.3	41.3	25.7	41.3	31.0	31.4	11.2	31.4	15.8	98.3	98.4	98.3	98.3
P28	k-NN	65.5	64.3	65.5	62.8	46.7	46.2	46.7	41.7	65.5	64.3	65.6	62.8	52.3	49.5	52.3	50.2	38.7	33.6	38.7	35.0	64.0	60.7	64.1	62.2
	RF	84.2	85.4	84.2	83.1	56.7	60.7	56.7	49.7	85.8	86.5	85.8	84.8	86.1	86.8	86.1	85.4	63.4	67.1	63.4	59.0	85.5	86.1	85.5	84.8
	LDA	58.8	59.3	58.8	59.0	46.7	41.4	46.7	39.1	53.6	56.2	53.5	54.5	56.6	62.7	56.6	60.6	3.4	17.9	3.4	0.6	50.8	57.0	50.1	52.5
P29	k-NN	61.0	60.1	61.0	58.1	46.8	47.5	46.8	42.7	61.1	60.6	61.1	58.4	49.0	46.0	49.0	46.7	39.6	37.9	39.6	36.5	69.4	65.4	69.4	67.0
	RF	85.0	87.0	85.0	84.3	59.3	59.5	59.3	51.4	84.8	86.5	84.8	84.0	86.1	87.4	86.1	85.6	65.1	68.4	65.1	58.9	87.8	88.8	87.8	87.3
	LDA	61.7	61.3	61.7	61.3	39.0	35.1	39.0	35.9	54.6	57.4	54.6	55.4	59.9	65.6	59.9	61.5	46.9	22.9	46.9	30.5	49.2	54.9	49.2	50.3
P30	k-NN	59.2	62.0	59.2	55.9	51.4	54.5	51.4	43.5	59.7	59.3	59.7	56.5	42.6	39.8	42.6	40.4	37.9	34.9	37.9	34.6	77.8	75.3	77.8	74.2
	RF	84.6	85.8	84.6	83.6	61.8	65.0	61.8	57.2	99.8	99.8	99.8	99.8	86.4	87.1	86.4	85.9	69.9	72.8	69.9	67.3	99.8	99.8	99.8	99.8
	LDA	53.4	55.3	53.4	54.1	41.1	38.3	41.1	38.9	99.0	99.0	99.0	99.0	11.9	45.4	11.9	13.4	30.1	17.5	30.1	14.9	98.7	98.8	98.7	98.8
P31	k-NN	58.2	55.4	58.2	54.8	45.8	44.3	45.8	38.3	55.6	52.6	55.6	52.5	42.9	39.8	42.9	39.4	38.9	37.1	38.9	35.7	77.8	72.7	78.8	74.4
	RF	81.2	83.1	81.2	80.2	53.7	59.6	53.6	47.5	99.8	99.8	99.8	99.8	86.4	87.3	86.4	85.8	62.0	67.2	62.0	58.9	99.9	99.9	99.9	99.9
	LDA	59.8	60.0	59.8	59.7	44.1	41.1	44.1	40.4	98.9	98.9	98.9	98.9	65.3	66.1	65.3	65.3	16.8	66.1	16.8	20.5	98.4	98.4	98.4	98.4
P32	k-NN	68.3	67.8	68.3	65.1	30.7	36.8	30.7	30.2	68.3	67.8	68.3	65.0	56.3	53.7	56.3	54.3	41.0	37.5	41.1	38.2	71.2	67.5	71.2	68.7
	RF	87.2	87.3	87.2	86.1	58.5	63.6	58.5	53.2	89.3	87.0	89.3	87.9	88.3	87.5	88.3	87.5	67.4	66.1	67.4	64.4	88.6	87.9	88.6	87.9
	LDA	56.8	56.7	56.8	56.7	42.3	37.5	42.3	37.9	53.6	55.5	53.6	54.4	15.3	3.9	15.3	06.2	18.6	20.3	18.6	7.0	42.2	25.4	42.2	28.8
P33	k-NN	69.2	65.5	69.2	66.5	45.0	46.1	45.0	43.4	69.4	65.3	69.4	66.7	53.2	49.0	53.2	50.3	38.1	35.0	38.1	34.8	70.4	66.5	70.4	67.7
	RF	89.8	90.3	89.8	89.2	53.9	54.8	53.9	48.2	92.0	92.5	92.0	91.3	91.4	91.2	91.4	90.9	63.9	64.7	63.9	59.6	93.8	94.0	93.8	93.2
	LDA	60.0	60.3	60.0	60.1	38.9	37.7	38.9	37.3	54.2	56.4	54.2	55.0	14.3	45.8	14.3	18.3	3.3	35.1	3.3	0.8	50.0	57.6	50.0	46.9
P34	k-NN	73.4	73.3	73.4	71.6	52.8	51.4	52.8	45.6	73.9	73.9	73.9	72.0	57.6	54.1	57.7	55.6	40.5	36.4	40.5	37.2	76.4	73.0	76.4	74.5
	RF	91.6	91.8	91.6	91.3	58.8	67.9	58.8	52.1	93.8	94.1	93.8	93.4	91.7	91.6	91.6	91.5	66.2	70.1	66.2	60.5	92.4	92.3	92.4	92.2
	LDA	63.3	63.5	63.3	63.4	43.1	37.6	43.1	39.0	57.7	58.9	57.7	58.2	28.8	9.1	28.8	13.4	28.9	26.8	28.9	13.7	49.5	49.2	49.5	48.4
k-NN	Min	56.8	55.4	56.8	54.4	30.7	36.8	30.7	30.2	55.6	52.6	55.6	52.5	39.4	37.4	39.4	37.7	34.5	31.9	34.5	32.6	44.5	60.5	44.5	48.9
	Max	69.2	67.8	69.2	66.5	51.4	54.5	51.4	46.5	69.4	67.8	69.4	66.7	56.3	53.7	56.3	54.3	41	38	41.1	38.2	82.9	81.4	82.8	80.3
	Avg	63.4	62.6	63.4	60.7	44.7	45.6	44.7	40.6	63.2	61.8	63.2	60.5	48.1	45.3	48.1	46.0	38.4	35.6	38.4	35.3	70.2	68.5	70.3	68.3
RF	Min	81.2	83.1	81.2	80.2	48.9	47.9	48.9	43.5	84.8	85.7	84.8	84.0	83.6	84.8	83.6	83	61.5	63.3	61.5	58.9	83.9	85.1	83.9	83.2
	Max	89.4	90.8	89.4	88.8	61.8	65	61.8	57.2	99.8	99.8	99.8	99.8	91.4	91.7	91.4	91.1	69.9	72.8	69.9	67.3	100	100	100	100
	Avg	85.3	86.4	85.3	84.5	56.4	59.8	56.4	50.6	92.2	92.5	92.2	91.7	87.0	87.5	87.0	86.5	64.8	67.8	64.8	61.4	92.8	93.0	92.8	92.4
LDA	Min	53.4	55.3	53.4	54.1	32.1	30.8	32.1	31.2	52.8	54.4	52.8	54.1	11.9	0.1	3.1	0.01	3.0	8.3	3.0	0.4	40.0	25.4	40.0	28.8
	Max	61.7	61.9	61.7	61.4	46.7	41.4	46.7	40.4	99.0	99.0	99.0	99.0	65.3	76.0	65.3	65.3	46.9	66.1	46.9	30.5	98.7	98.8	98.7	98.8
	Avg	58.5	59.0	58.5	58.6	39.7	36.4	39.7	36.5	71.8	73.2	71.8	72.3	39.8	42.0	37.1	34.6	18.8	22.3	18.8	10.8	68.1	70.2	68.1	66.8

Table 5 shows that when training and testing the models with one specific subject, the models' performances are subject-dependent. As shown from the statistics of each model for all subjects (bottom of the table), the RF model performed much better (F1 score > 91.6%) in both segmentation and feature representation modes, showing that the recognition accuracy, in terms of F1-score, is on average 5% higher than when training and testing with all 12 subjects' data (see Table 4). Again, differentiating by dataset type, the results show that the conjunct information of dataset A (i.e., EMGs and Force Platform) and dataset B (i.e., acceleration values) performed significantly better than when used separately. In particular, on average, there was an increment of 7.3% in the F1-score from dataset A to dataset C and 31.1% from dataset B to dataset C.

5.2. GPSJack Evaluation

Concerning the suitability and principal characteristics of the GPSJack system, this section presents its evaluation in terms of RAM, storage, CPU, battery consumption, and data loss. In particular, a Samsung Galaxy Tab A7 with the following HW/SW characteristics has been tested in the setup described below.

- **Processor:** Qualcomm Snapdragon Octo-Core (4 × 2 GHz + 4 × 1.8 GHz);
- **Operating system:** Android 10;
- **RAM:** 3 GB;
- **Storage:** 32 GB;
- **Connectivity:** Bluetooth 5.0, Wi-Fi a/b/g/n/ac Dual Band;
- **Battery:** 7040 mAh;
- **Dimension:** 247.6 × 157.4 × 7 mm per 476 g;
- **Price:** $140.

Three Nordic Thingy 52 were connected to the Samsung Galaxy for three consecutive hours. The setup was tested once for each sampling frequency: (i) 50 Hz, (ii) 100 Hz, and (iii) 200 Hz. Table 6 presents the evaluation results obtained using the Android Studio Profiler suit.

Table 6. Data aggregators' profiling using Android Studio 4.1.2.

	Galaxy Tab A7		
Frequency (Hz)	**50**	**100**	**200**
RAM (MB/h)	147	180	224
Storage (Mb/h)	50	102	185
CPU (%)	31	40	46
Battery (mAh)	312	325	342
Data loss (%)	0	0	0

As observed, the designed system can work on various setups with no data loss and low storage, RAM, and CPU usage. Moreover, its battery consumption allows a data collection phase of almost 8 h. Based on the tests performed during the project, the Nordic Thingy 52 can efficiently compute for more than 48 consecutive hours at 200 Hz.

5.3. Discussion

The results of these preliminary experiments clearly show that the conjugate of EMG, Force Platform, and acceleration data performs considerably better than their separate utilization. This improvement enables a better understanding and in-depth study of human motion. In fact, the GPSJack prototype has good potential to capture relevant information, enabling the possibility to recognize the studied motion classes with an average F1-score of 89.6% when using all the subjects' data at once. Furthermore, when tested on single subjects, the F1-score ranged from a minimum of 83.2% to a maximum of 100%, outperforming the usage of only one of the aforementioned data sources. Even though the collected acceleration data present precious information, further work must be conducted to increase the overall performance and reduce the dependence on state-of-the-art technology. This can be done by: (i) implementing more complex recognition models than the used k-NN, RF, and LDA; and (ii) exploiting the utilization of a larger number and different positions of data collection nodes on the human body. Nevertheless, the information generated by the EMG and Force Platforms is paramount; thus, a possible next step in addition to those mentioned above would be the integration of EGM sensors into the same GPSJack nodes.

6. Conclusions and Further Development

This paper has dealt with defining the experimental design of the "Biofeedback Wearable and Environmental Technologies for Postural Correction" project. We illustrated the target technology, described the project's evaluation workflow (i.e., state-of-the-art instruments and low-cost wearable sensors, data processing flow, and machine learning-based analysis), and provided a high-level description of the context in which the envisioned technologies are forecasted to operate. In particular, we devised a methodology investigating how to build the parameters that allow the physiatric medical staff to evaluate the patient. Three challenging motor tasks were identified to, on the one hand, train fundamental motor skills such as action adaptation, compensation, and anticipation, and on the other hand, to measure the performance levels fir such skills. To measure the quality of the performed motor tasks, we evaluated a low-cost body area network (aka GPSJack). GPSJack uses at most eleven data collection nodes integrating an accelerometer, gyroscope, and magnetometer sensors and can compute for 48 h at a sampling frequency of 200 Hz. Moreover, we used an android mobile application that works as a data aggregator and controller for the GPSJack system, through which the user can observe the collected data and the posture and gait quality indicators. Nevertheless, in conjunction with the GPSJack, we integrated the GPSEnv based on state-of-the-art gait and posture evaluation systems.

Tests on data collected from 12 subjects for a total of 84 data collection sessions each showed that the designed system could highly accurately recognize the phases of the defined motor tasks. In particular, in a subject-independent setup, we achieved an F1-score of 89.6% in recognizing the five studied movement states (i.e., rotate, in position, APAs, drop, and recover). With a subject-dependent setup, the F1-score ranged from 100% for subjects 23 and 26 to 83.2% for subject 22. These results show that the acceleration information that we will add to the state-of-the-art systems significantly increases recognition capabilities.

It is widely accepted in the community of psychiatric medicine that proper quantification of the postural system's efficiency represents an essential assessment for improving quality of life in the elderly, patients with neurological pathologies, and athletes. Moreover, since most of the actual measurements are made in a laboratory environment where natural movements are constrained by the instruments applied to subjects' bodies and the environment, a system usable in uncontrolled and unconstrained environments (e.g., home, gym, or sports facilities) habilitates the individuals to move freely in their natural environment and perform the required motor tasks. Thus, the designed system will evaluate profiles from the performance viewpoint of individuals ranging from patients undergoing rehabilitation to top-level elite athletes in controlled and uncontrolled environments.

Since the final goal of the project is performing the defined task in uncontrolled environments and using only the acceleration information provided by a system such as GPSJack, the next step will concern the exploitation of the data collection phase while making use of a large number of nodes positioned on different body parts and of more complex pattern-recognition models than the used k-NN, RF, and LDA. In particular, deep learning models such as recurrent neural networks (RNN) and long short-term memory (LSTM) have shown optimal results in such fields. Moreover, since the information captured by the EMG sensors is paramount, integrating an EMG sensor into the GPSJack nodes should be considered.

Author Contributions: Conceptualization, P.C., G.P., P.M.P. and M.C.; methodology, P.C., G.P., F.D., P.M.P. and M.C.; software, T.C.W., F.D., C.T. (Claudio Tomazzoli), C.T. (Cristian Turetta) and F.P.; validation, T.C.W., F.D., L.Z., C.T. (Claudio Tomazzoli), C.T. (Cristian Turetta) and F.P.; formal analysis, P.C., G.P., F.D., P.M.P. and M.C.; investigation, All Authors; writing—original draft preparation, All authors; writing—review and editing, F.D., T.C.W., M.C., G.P. and P.C.; supervision, P.C., G.P. and M.C.; project administration, P.C., G.P., P.M.P. and M.C.; funding acquisition, P.C., G.P., P.M.P. and M.C. All authors have read and agreed to the published version of the manuscript.

Funding: This research was funded by Regione del Veneto grant number 1695-0007-1463-2019.

Data Availability Statement: Data are available at request

References

1. Sell, T.C. An examination, correlation, and comparison of static and dynamic measures of postural stability in healthy, physically active adults. *Phys. Ther. Sport* **2012**, *13*, 80–86. [CrossRef] [PubMed]
2. Sell, T.; Tsai, Y.S.; Smoliga, J.; Myers, J.; Lephart, S. Strength, flexibility, and balance characteristics of highly proficient golfers. *J. Strength Cond. Res.* **2007**, *21*, 1166–1171.

3. Riva, D.; Bianchi, R.; Rocca, F.; Mamo, C. Proprioceptive Training and Injury Prevention in a Professional Men's Basketball Team: A Six-Year Prospective Study. *J. Strength Cond. Res.* **2016**, *30*, 461–475. [CrossRef] [PubMed]
4. Riva, D.; Rossitto, F.; Battocchio, L. Postural muscle atrophy prevention and recovery and bone remodelling through high frequency proprioception for astronauts. *Acta Astronaut.* **2009**, *65*, 813–819. [CrossRef]
5. Ageberg, E.; Roberts, D.; Holmstrom, E.; Friden, T. Balance in Single-Limb Stance in Patients with Anterior Cruciate Ligament Injury: Relation to Knee Laxity, Proprioception, Muscle Strength, and Subjective Function. *Am. J. Sport. Med.* **2005**, *33*, 1527–1537. [CrossRef]
6. Rosengren, K.; Riva, D.; Mamo, C.; Fanì, M.; Saccavino, P.; Rocca, F.; Momenté, M.; Fratta, M. Single Stance Stability and Proprioceptive Control in Older Adults Living at Home: Gender and Age Differences. *J. Aging Res.* **2013**, *2013*, 561695. [CrossRef]
7. Kobayashi, H.; Hayashi, Y.; Higashino, K.; Saito, A.; Kunihiro, T.; Kanzaki, J.; Goto, F. Dynamic and static subjective visual vertical with aging. *Auris Nasus Larynx* **2002**, *29*, 325–328. [CrossRef]
8. Sherrington, C.S. *The Integrative Action of the Nervous System*; Cambridge University Press: Cambridge, UK, 2016. [CrossRef]
9. Riemann, B.L.; Lephart, S.M. The sensorimotor system, part I: The physiologic basis of functional joint stability. *J. Athl. Train.* **2002**, *37*, 71–79.
10. Cohen, H.; Heaton, L.G.; Congdon, S.L.; Jenkins, H.A. Changes in Sensory Organization Test Scores with Age. *Age Ageing* **1996**, *25*, 39–44. [CrossRef]
11. Feller, K.J.; Peterka, R.J.; Horak, F.B. Sensory re-weighting for postural control in Parkinson's disease. *Front. Hum. Neurosci.* **2019**, *13*, 126. [CrossRef]
12. Kamieniarz, A.; Michalska, J.; Marszałek, W.; Stania, M.; Słomka, K.J.; Gorzkowska, A.; Juras, G.; Okun, M.S.; Christou, E.A. Detection of postural control in early Parkinson's disease: Clinical testing vs. modulation of center of pressure. *PLoS ONE* **2021**, *16*, e0245353. [CrossRef] [PubMed]
13. Demrozi, F.; Pravadelli, G.; Bihorac, A.; Rashidi, P. Human Activity Recognition Using Inertial, Physiological and Environmental Sensors: A Comprehensive Survey. *IEEE Access* **2020**, *8*, 210816–210836. [CrossRef] [PubMed]
14. Demrozi, F.; Pravadelli, G.; Tighe, P.J.; Bihorac, A.; Rashidi, P. Joint Distribution and Transitions of Pain and Activity in Critically Ill Patients. In Proceedings of the 2020 42nd Annual International Conference of the IEEE Engineering in Medicine & Biology Society (EMBC), Montreal, QC, Canada, 20–24 July 2020; pp. 4534–4538.
15. Mikos, V.; Heng, C.H.; Tay, A.; Yen, S.C.; Chia, N.S.Y.; Koh, K.M.L.; Tan, D.M.L.; Au, W.L. A wearable, patient-adaptive freezing of gait detection system for biofeedback cueing in Parkinson's disease. *IEEE Trans. Biomed. Circuits Syst.* **2019**, *13*, 503–515. [CrossRef]
16. Verhagen, E.; Bobbert, M.; Inklaar, M.; van Kalken, M.; van der Beek, A.; Bouter, L.; van Mechelen, W. The effect of a balance training programme on centre of pressure excursion in one-leg stance. *Clin. Biomech.* **2005**, *20*, 1094–1100. [CrossRef]
17. Yoong, N.K.M.; Perring, J.; Mobbs, R.J. Commercial postural devices: A review. *Sensors* **2019**, *19*, 5128. [CrossRef]
18. Piscitelli, D.; Falaki, A.; Solnik, S.; Latash, M.L. Anticipatory postural adjustments and anticipatory synergy adjustments: Preparing to a postural perturbation with predictable and unpredictable direction. *Exp. Brain Res.* **2017**, *235*, 713–730. [CrossRef]
19. Latash, M.L. Muscle coactivation: Definitions, mechanisms, and functions. *J. Neurophysiol.* **2018**, *120*, 88–104. [CrossRef]
20. Aruin, A.S.; Forrest, W.R.; Latash, M.L. Anticipatory postural adjustments in conditions of postural instability. *Electroencephalogr. Clin. Neurophysiol. Mot. Control.* **1998**, *109*, 350–359. [CrossRef]
21. Bertucco, M.; Nardello, F.; Magris, R.; Cesari, P.; Latash, M.L. Postural Adjustments during Interactions with an Active Partner. *Neuroscience* **2021**, *463*, 14–29. [CrossRef]
22. Krishnan, V.; Aruin, A.S.; Latash, M.L. Two stages and three components of the postural preparation to action. *Exp. Brain Res.* **2011**, *212*, 47–63. [CrossRef]
23. Massion, J. Movement, posture and equilibrium: Interaction and coordination. *Prog. Neurobiol.* **1992**, *38*, 35–56. [CrossRef] [PubMed]
24. Crenna, P.; Frigo, C. A motor programme for the initiation of forward-oriented movements in humans. *J. Physiol.* **1991**, *437*, 635–653. [CrossRef]
25. Krishnan, V.; Latash, M.L.; Aruin, A.S. Early and late components of feed-forward postural adjustments to predictable perturbations. *Clin. Neurophysiol.* **2012**, *123*, 1016–1026. [CrossRef] [PubMed]
26. Cordo, P.; Nashner, L.M. Properties of postural adjustments associated with rapid arm movements. *J. Neurophysiol.* **1982**, *47*, 287–302. [CrossRef] [PubMed]
27. Clark, R.A.; Pua, Y.H.; Oliveira, C.C.; Bower, K.J.; Thilarajah, S.; McGaw, R.; Hasanki, K.; Mentiplay, B.F. Reliability and concurrent validity of the Microsoft Xbox One Kinect for assessment of standing balance and postural control. *Gait Posture* **2015**, *42*, 210–213. [CrossRef] [PubMed]
28. Hertel, J.; Olmsted-Kramer, L.C.; Challis, J.H. Time-to-Boundary Measures of Postural Control during Single Leg Quiet Standing. *J. Appl. Biomech.* **2006**, *22*, 67–73. [CrossRef] [PubMed]
29. Messier, S.P.; Royer, T.D.; Craven, T.E.; O'Toole, M.L.; Burns, R.; Ettinger, W.H. Long-Term Exercise and its Effect on Balance in Older, Osteoarthritic Adults: Results from the Fitness, Arthritis, and Seniors Trial (FAST). *J. Am. Geriatr. Soc.* 48, 131–138. [CrossRef]
30. De Venuto, D.; Mezzina, G. Multi-Sensing System for Parkinson's Disease Stage Assessment based on FPGA-embedded Serial SVM Classifier. *IEEE Des. Test* **2019**, *38*, 44–51. [CrossRef]

31. McKinney, Z.; Heberer, K.; Nowroozi, B.N.; Greenberg, M.; Fowler, E.; Grundfest, W. Pilot evaluation of wearable tactile biofeedback system for gait rehabilitation in peripheral neuropathy. In Proceedings of the 2014 IEEE Haptics Symposium (HAPTICS), Houston, TX, USA, 23–26 February 2014; pp. 135–140.
32. Baldini, G.; Steri, G.; Dimc, F.; Giuliani, R.; Kamnik, R. Experimental identification of smartphones using fingerprints of built-in micro-electro mechanical systems (MEMS). *Sensors* **2016**, *16*, 818. [CrossRef]
33. McGinnis, R.S.; Mahadevan, N.; Moon, Y.; Seagers, K.; Sheth, N.; Wright, J.A., Jr.; DiCristofaro, S.; Silva, I.; Jortberg, E.; Ceruolo, M.; others. A machine learning approach for gait speed estimation using skin-mounted wearable sensors: From healthy controls to individuals with multiple sclerosis. *PLoS ONE* **2017**, *12*, e0178366. [CrossRef]
34. Demrozi, F.; Bacchin, R.; Tamburin, S.; Cristani, M.; Pravadelli, G. Towards a wearable system for predicting the freezing of gait in people affected by Parkinson's disease. *IEEE J. Biomed. Health Inform.* **2019**, *24*, 2444–2451. [CrossRef] [PubMed]
35. Borzì, L.; Mazzetta, I.; Zampogna, A.; Suppa, A.; Olmo, G.; Irrera, F. Prediction of freezing of gait in Parkinson's disease using wearables and machine learning. *Sensors* **2021**, *21*, 614. [CrossRef] [PubMed]
36. Angelini, L.; Hodgkinson, W.; Smith, C.; Dodd, J.M.; Sharrack, B.; Mazzà, C.; Paling, D. Wearable sensors can reliably quantify gait alterations associated with disability in people with progressive multiple sclerosis in a clinical setting. *J. Neurol.* **2020**, *267*, 2897–2909. [CrossRef] [PubMed]
37. Bolam, S.M.; Batinica, B.; Yeung, T.C.; Weaver, S.; Cantamessa, A.; Vanderboor, T.C.; Yeung, S.; Munro, J.T.; Fernandez, J.W.; Besier, T.F.; others. Remote patient monitoring with wearable sensors following knee arthroplasty. *Sensors* **2021**, *21*, 5143. [CrossRef]
38. Pandhare, V.; Miller, M.; Vogl, G.W.; Lee, J. Ball Screw Health Monitoring with Inertial Sensors. *IEEE Trans. Ind. Inform.* **2022**. [CrossRef]
39. Li, Z.; Zhao, S.; Duan, J.; Su, C.Y.; Yang, C.; Zhao, X. Human cooperative wheelchair with brain–machine interaction based on shared control strategy. *IEEE/ASME Trans. Mechatron.* **2016**, *22*, 185–195. [CrossRef]
40. Prins, N.W.; Sanchez, J.C.; Prasad, A. Feedback for reinforcement learning based brain–machine interfaces using confidence metrics. *J. Neural Eng.* **2017**, *14*, 036016. [CrossRef]
41. Blankertz, B.; Losch, F.; Krauledat, M.; Dornhege, G.; Curio, G.; Müller, K. The Berlin Brain-Computer Interface: Accurate performance from first-session in BCI-naive subjects. *IEEE Trans. Biomed. Eng.* **2008**, *55*, 2452–2462. [CrossRef]
42. van Gerven, M.; Farquhar, J.; Schaefer, R.; Vlek, R.; Geuze, J.; Nijholt, A.; Ramsey, N.; Haselager, P.; Vuurpijl, L.; Gielen, S.; Desain, P. The brain–computer interface cycle. *J. Neural Eng.* **2009**, *6*, 041001. [CrossRef]
43. Pustišek, M.; Wei, Y.; Sun, Y.; Umek, A.; Kos, A. The role of technology for accelerated motor learning in sport. *Pers. Ubiquitous Comput.* **2021**, *25*, 969–978. [CrossRef]
44. Kostov, A.; Polak, M. Parallel man-machine training in development of EEG-based cursor control. *IEEE Trans. Rehabil. Eng.* **2000**, *8*, 203–205. [CrossRef] [PubMed]
45. Tomazzoli, C.; Cristani, M.; Karafili, E.; Olivieri, F. Non-monotonic reasoning rules for energy efficiency. *J. Ambient. Intell. Smart Environ.* **2017**, *9*, 345–360. [CrossRef]
46. Sodhro, A.H.; Li, Y.; Shah, M.A. Energy-efficient adaptive transmission power control for wireless body area networks. *IET Commun.* **2016**, *10*, 81–90. [CrossRef]
47. Sodhro, A.H.; Sangaiah, A.K.; Sodhro, G.H.; Lohano, S.; Pirbhulal, S. An energy-efficient algorithm for wearable electrocardiogram signal processing in ubiquitous healthcare applications. *Sensors* **2018**, *18*, 923. [CrossRef] [PubMed]
48. Turetta, C.; Demrozi, F.; Pravadelli, G. A freely available system for human activity recognition based on a low-cost body area network. In Proceedings of the 2022 IEEE 46th Annual Computers, Software, and Applications Conference (COMPSAC), Online, 27 June–1 July 2022; pp. 395–400.
49. Coviello, G.; Florio, A.; Avitabile, G.; Talarico, C.; Wang-Roveda, J.M. Distributed full synchronized system for global health monitoring based on flsa. *IEEE Trans. Biomed. Circuits Syst.* **2022**, *16*, 600–608. [CrossRef]
50. Demrozi, F.; Turetta, C.; Pravadelli, G. B-HAR: An open-source baseline framework for in depth study of human activity recognition datasets and workflows. *arXiv* **2021**, arXiv:2101.10870.
51. Pratama, I.; Permanasari, A.E.; Ardiyanto, I.; Indrayani, R. A review of missing values handling methods on time-series data. In Proceedings of the 2016 International Conference on Information Technology Systems and Innovation (ICITSI), Bali, Indonesia, 24–27 October 2016; pp. 1–6.
52. Guyon, I.; Gunn, S.; Nikravesh, M.; Zadeh, L.A. *Feature Extraction: Foundations and Applications*; Springer: Cham, Switzerland, 2008; Volume 207.
53. Storcheus, D.; Rostamizadeh, A.; Kumar, S. A survey of modern questions and challenges in feature extraction. *Feature Extraction: Modern Questions and Challenges*; Microtome Publishing: Brookline, MA, USA, 2015; pp. 1–18.
54. Barandas, M.; Folgado, D.; Fernandes, L.; Santos, S.; Abreu, M.; Bota, P.; Liu, H.; Schultz, T.; Gamboa, H. TSFEL: Time Series Feature Extraction Library. *SoftwareX* **2020**, *11*, 100456. [CrossRef]
55. Bishop, C.M. *Pattern Recognition and Machine Learning*; Springer: Cham, Switzerland, 2006.

56. Scikit Learn. Available online: https://scikit-learn.org/stable/modules/feature_selection.html (accessed on 2 November 2020).
57. Powers, D.M. Evaluation: From precision, recall and F-measure to ROC, informedness, markedness and correlation. *arXiv* **2020**, arXiv:2010.16061.

Article

An Efficient Machine Learning Approach for Diagnosing Parkinson's Disease by Utilizing Voice Features

Arti Rana [1], Ankur Dumka [2,3], Rajesh Singh [4,5], Mamoon Rashid [6,7,*], Nazir Ahmad [8] and Manoj Kumar Panda [9]

1   Computer Science & Engineering, Veer Madho Singh Bhandari Uttarakhand Technical University, Dehradun 248007, India
2   Department of Computer Science and Engineering, Women Institute of Technology, Dehradun 248007, India
3   Department of Computer Science & Engineering, Graphic Era Deemed to be University, Dehradun 248001, India
4   Division of Research and Innovation, Uttaranchal Institute of Technology, Uttaranchal University, Dehradun 248007, India
5   Department of Project Management, Universidad Internacional Iberoamericana, Campeche 24560, Mexico
6   Department of Computer Engineering, Faculty of Science and Technology, Vishwakarma University, Pune 411048, India
7   Research Center of Excellence for Health Informatics, Vishwakarma University, Pune 411048, India
8   Department of Information System, College of Applied Sciences, King Khalid University, Muhayel 61913, Saudi Arabia
9   Department of Electrical Engineering, G.B. Pant Institute of Engineering and Technology, Pauri 246194, India
*   Correspondence: mamoon.rashid@vupune.ac.in

Citation: Rana, A.; Dumka, A.; Singh, R.; Rashid, M.; Ahmad, N.; Panda, M.K. An Efficient Machine Learning Approach for Diagnosing Parkinson's Disease by Utilizing Voice Features. *Electronics* **2022**, *11*, 3782. https://doi.org/10.3390/electronics11223782

Academic Editors: Gabriella Olmo, Florenc Demrozi and Yu Zhang

Received: 13 October 2022
Accepted: 16 November 2022
Published: 17 November 2022

Publisher's Note: MDPI stays neutral with regard to jurisdictional claims in published maps and institutional affiliations.

Abstract: Parkinson's disease (PD) is a neurodegenerative disease that impacts the neural, physiological, and behavioral systems of the brain, in which mild variations in the initial phases of the disease make precise diagnosis difficult. The general symptoms of this disease are slow movements known as 'bradykinesia'. The symptoms of this disease appear in middle age and the severity increases as one gets older. One of the earliest signs of PD is a speech disorder. This research proposed the effectiveness of using supervised classification algorithms, such as support vector machine (SVM), naïve Bayes, k-nearest neighbor (K-NN), and artificial neural network (ANN) with the subjective disease where the proposed diagnosis method consists of feature selection based on the filter method, the wrapper method, and classification processes. Since just a few clinical test features would be required for the diagnosis, a method such as this might reduce the time and expense associated with PD screening. The suggested strategy was compared to PD diagnostic techniques previously put forward and well-known classifiers. The experimental outcomes show that the accuracy of SVM is 87.17%, naïve Bayes is 74.11%, ANN is 96.7%, and KNN is 87.17%, and it is concluded that the ANN is the most accurate one with the highest accuracy. The obtained results were compared with those of previous studies, and it has been observed that the proposed work offers comparable and better results.

Keywords: ANN; KNN; machine learning (ML); naïve Bayes classification; Parkinson's disease; SVM

1. Introduction

Parkinson's disease, commonly known as Tremor, is affected by a reduction in dopamine levels in the brain which damages a person's motion functions, or physical functioning. It is one of the world's most common diseases. Intermittent neurological signs and symptoms result from these lesions, which get worse as the disease progresses [1]. Because aging causes changes in our brains, such as loss of synaptic connections and changes in neurotransmitters and neurohormones, this condition is more frequent among the elderly. With the passage of time, the neurons in a person's body begin to die and become inimitable. The consequences of neurological problems and the falling dopamine levels in the patient's body show gradually, making them difficult to detect until the patient's condition requires medical treatment [2].

However, the symptoms and severity levels are different for individuals. Major symptoms of this disease are deficiency in speech, short-term memory loss, loss of balance, and unbalanced posture [1].

Every year, 8.5 million individual cases of this disease are registered worldwide, as per the World Health Organization (WHO) report in 2019 [3]. The chance of developing this disease rises with age; currently, there are 4% of sufferers worldwide under 50 years of age. This disease is the most widespread neurodegenerative disease in the world after Alzheimer's disease, impacting millions of people [4,5]. Therapy for this disease is still in its initial stages, and doctors can only assist patients in alleviating the symptoms of the disease [6]. However, there are no definite diagnostics for this disease, and the diagnosis is largely dependent on the medical history of the patient [1]. As invasive procedures are typically used for diagnosis and therapy, which are both expensive and demanding [7], a reasonably straightforward and accurate way to diagnose this disease looks very relevant.

1.1. Machine Learning-Based Detection of Parkinson's Disease

Over the past few decades, researchers have looked at a new way of detecting this disease through ML techniques, a subset of artificial intelligence (AI). Clinical personnel might better recognize these disease patients by combining traditional diagnostic indications with ML.

As walking is the most common activity in every person's day-to-day life, it has been linked to physical as well as neurological disorders. This disease, for example, has been identifiable using gait (mobility) data. Gait analysis approaches offer advantages such as being non-intrusive and having the potential to be extensively used in residential settings [8]. Few researchers have attempted to combine ML methods to make the procedure autonomous and possible to do offline [9].

Furthermore, persons with the subject disease in its early stages might experience speech problems [10]. These include dysphonia (weak vocal fluency), repetitious echoes (a tiny assortment of audio variations), and hypophonia (vocal musculature disharmony) [7,11]. Information from human aural emissions might be detected and evaluated using a computing unit [12,13].

1.2. Research Problem and Motivation

Early PD detection in PD patients is a crucial challenge. Even if their health deteriorates, people can enhance their quality of life if they receive an early diagnosis. Another issue is that the diagnosis of PD requires a number of steps, including gathering a thorough neurological history from the patient and examining their motor abilities in various environments.

The majority of recent studies deal with the homo dataset (text, speech, video, or image). Problems with dataset modification and multi-data handling procedures have been highlighted in the suggested study. The effectiveness of disease prediction is regulated as a result of the examination of a particular dataset. More real-time solutions are made possible by the use of machine learning-based techniques for multivariate data processing. The multi-variate vocal data analysis (MVDA) is driven to provide multiple dataset attribute-based Parkinson's disease identification utilizing machine learning approaches. This study examines the potential for improving multi-variate and multimodal data processing, which aids in raising the disease detection rate. The existing research simultaneously concentrates on various ML-based techniques such as support vector machines, naïve Bayes, K-NN, and artificial neural network evaluations of Parkinson's data based on voice features. The MVDA employs extensive datasets and machine learning approaches to improve disease identification based on these works. The incorporation of numerous patients' multivariate acoustic characteristics in the proposed MVDA is encouraged. The subjective disease has been diagnosed with the help of proposed machine learning techniques under the MVDA system.

1.3. Contribution

This research article covers the techniques of machine learning which are implemented in the auditory analysis of speech to diagnose this disease. The benefits and shortcomings of these algorithms in detecting the disease are thoroughly contrasted, and existing comparative studies' potential drawbacks are explored. The accuracy of ANN in speech analysis for diagnosis is the finest among different classifiers; however, the assumption is to enhance and adapt to the difficulties that may come from the data. Using the naïve Bayes classifier with suitable pre-processing might result in greater average accuracy. The main contribution of this paper is as follows:

a. To identify which machine learning algorithms, such as SVM, KNN, naive Bayes, and ANN, offer the most accurate classifications and diagnosis of Parkinson's disease.
b. To develop statistical evaluations for the diagnosis of Parkinson's disease in order to identify the frequency at which the best training and test results will be acquired, and consequently to assist in upcoming literature-based research.
c. The proposed system has used an ANN classifier to attain the maximum classification accuracy when compared to the approaches used in earlier research.
d. In order to improve the prediction of PD, a comprehensive methodology was employed to explore the effectiveness and efficiency of various feature selection approaches.
e. The proposed model is examined with four machine learning methods, including SVM, naive Bayes, k-NN, and ANN, as well as with earlier and more current studies on PD detection.

1.4. Structure of Proposed Work

The structure of the study is as follows: Section 2 describes the related research survey. Section 3 discusses the methodology used to achieve the proposed objective. Section 4 defines the materials and methods. Section 5 examines the experiment and results. Section 6 discusses the comparative study and discussion. Finally, Section 7 concludes the proposed work.

2. Related Works

In order to distinguish PD cases from healthy controls, a variety of modern machine learning algorithms, including support vector machines, artificial neural networks, logistic regression, naïve Bayes, etc., have been successfully used. In this study, numerous databases, including Web of Science, Elsevier, MDPI, Scopus, Science Direct, IEEE Xplore, Springer, and Google Scholar, were utilized to survey relevant papers on Parkinson's disease.

In a survey by [14], the authors used KNN, SVM, and discrimination-function-based (DBF) classifiers for the diagnosis of PD. In their study, they used several parameters such as jitter, fundamental frequency, pitch, shimmer, and other statistical measures. The best accuracy among these classifiers was obtained from KNN with a 93.83% accuracy rate and it also provided good performance in other parameters, such as sensitivity, specificity, and error rate.

The authors in [15] used a convolution neural network classifier applied to speech classification datasets. The accuracy reached throughout the training phase, which was over 77%, makes the results optimistic. In accordance with the works mentioned above, [16] examined a variety of classifiers to identify individuals who were likely to have Parkinson's disease. They used 40 participants for their investigation, including 20 PD patients and 20 healthy controls. According to the experimental findings, the naive Bayes classifier has a detection accuracy of 65%, with a sensitivity rate of 63.6% and a specificity rate of 66.6%, respectively. In [17], the authors used three types of classifiers based on KNN, SVM, and multilayer perceptron (MLP) to diagnose Parkinson's disease. Among all these ML classifiers, SVM using an RBF kernel outperformed with an overall classification accuracy rate of 85.294%.

A summary of the most recent deep learning methods for audio signal processing is given in another work by [18]. The works that have been examined include convolution neural networks as well as other long short-term memory architecture models and audio-specific neural network models. Similar to the previous studies, [19] detected PD using naive Bayes and other machine learning approaches. In their method, relevant features were extracted from the voice signal of PD patients and healthy control subjects using signal processing techniques. The naive Bayes algorithm shows a 69.24% detection accuracy and 96.02% precision rate for the 22 voice characteristics. In [20], the authors suggested a technique for detecting Parkinson's disease using SVM on shifted delta cepstral (SDC) and single frequency filtering cepstral coefficients (SFFCC) features extracted from speech signals of PD patients and healthy controls. Comparing the standard MFCC + SDC features to the SDC + SFFCC features, performance increases of 9% were observed. The 73.33% detection accuracy with a 73.32% F1-score was displayed by the conventional SVM on SDC + SFCC features. In addition to the naive Bayes classifier, several additional supervised methods, including but not restricted to well-known deep learning methods, have been suggested to identify PD patients among healthy controls.

In a survey conducted by [21], the authors examined two recognizing decision forests i.e., SysFor and ForestPA, along with the most widely used random forest classifier, which has been utilized as a Parkinson's detector. In their study, as compared to SysFor and ForestPA, random forest's average detection accuracy on incremental trees showed 93.58%. For the purpose of classifying Parkinson's disease through sets of acoustic vocal (voice) characteristics, the authors [22] suggested two frameworks based on CNN. Both frameworks are used for the mixing of different feature sets, although they combine feature sets in different ways. While the second framework provides feature sets to the parallel input levels that are directly connected to convolution layers, the first framework first combines several feature sets before passing them as inputs to the nine-layered CNN.

AI is assisting physicians in better diagnosing and treating diseases such as postoperative hypotension, and more advanced future models may have even more widespread medical uses. The evolutionary step in the creation of therapeutic pathways and adherence is machine learning. The real benefit of machine learning, however, is that it enables provider organizations to use information about the patient population from their own systems of record to create therapeutic pathways that are unique to their procedures, clientele, and physicians [23].

The vocal biomarkers and the description of the Aachen aphasia database, which contains recordings and transcriptions of therapy sessions, were covered in [24]. The authors also discussed how the biomarkers and the database could be used to build a recognition system that automatically maps pathological speech to aphasia type and severity.

In [25], the authors examined the suggested technique using a dataset of 288 audio files from 96 patients, including 48 healthy controls and 48 participants with cognitive impairment. The suggested method outperformed techniques based on manual transcription and speech annotation, with classification results that were comparable to those of the most advanced neuropsychological screening tests and an accuracy rate of 90.57%.

In [26], the authors intended to enlighten on the early indicators of major depressive relapse, which were discreetly measured using remote measurement technologies (RMT).

RMT has the potential to alter how depression and other long-term disorders are evaluated and handled if it is found to be acceptable to patients and other important stakeholders and capable of providing clinically meaningful information predicting future deterioration.

It can be seen from the reviews above that all the research that has been carried out is only restricted to a small number of datasets. The above previous works inspired us to try a new methodology. In this study, we experimented with several feature selection methods before comparing the results with various machine learning classifiers. Table 1 illustrates the review of ML techniques used to diagnose major symptoms of PD i.e., speech recording, handwriting pattern, and gait features, where data were collected from the UCI machine learning repository, the University of Oxford (UO), and other resources for 20 studies.

Table 1. Comparative Studies of Machine Learning Approaches to diagnose Parkinson's Disease.

Reference	Feature	Machine Learning Algorithms Used	Objective	Tools Used	Source of Data	No. of Subjects	Outcomes
Sakar et al., 2019 [27]	Speech	Naïve Bayes, Logistic Regression, SVM (RBF and Linear), KNN, Random Forest, MLP	Classification of PD from HC	JupyterLab with python programming language	Collected from participants	252, 188 PD + 64 HC	Highest accuracy obtained from SVM (RBF)—86%
Yasar A. et al., 2019 [28]	Speech	Artificial Neural Network	Classification of PD from HC	MATLAB	Collected from participants	80, 40 PD + 40 HC	Accuracy of ANN—94.93%
Avuçlu, E., Elen, A, 2020 [29]	Speech	KNN, Random Forest, Naïve Bayes, SVM	Classification of PD from HC	JupyterLab with python programming language	UCI machine learning repository	31, 23 PD + 8 HC	Accuracy from Naïve Bayes—70.26%
Marar et al., 2018 [30]	Speech	Naïve Bayes, ANN, KNN, Random Forest, SVM, Logistic Regression, Decision Tree (DT)	Classification of PD from HC	R programming	Collected from participants	31, 23 PD + 8 HC	Highest accuracy obtained from ANN—94.87%
Sheibani R et al., 2019 [31]	Speech	Ensemble Based Method	Classification of PD from HC	JupyterLab with python programming language	UCI machine learning repository	31, 23 PD + 8 HC	Accuracy obtained from ensemble learning—90.6%,
John M. Tracy et al., 2020 [32]	Speech	Logistic Regression (L2-Regularized), Random Forest, Gradient Boosted Trees	Classification of PD from HC	Python	mPower database	2289, 246 PD + 2023 HC	Highest accuracy obtained from gradient boosted trees Recall—79.7%, Precision—90.1%, F1-score—83.6%
Cibulka et al., 2019 [33]	Handwriting Patterns	Random Forest	Classification of PD from HC	Not mentioned	Collected from participants	270, 150 PD + 120 HC	Classification error for rs11240569, rs708727, rs823156 is 49.6%, 44.8%, 49.3% respectively.
Hsu S-Y et al., 2019 [34]	Handwriting Patterns	SVM with RBF Kernel, Logistic Regression	Classification of PD from HC	Weka	PACS	202, 94 Severe PD + 102 mild PD + 6 HC	Highest accuracy obtained from SVM-RBF 83.2% having sensitivity 82.8%, specificity 100%
Drotár, P et al., 2016 [35]	Handwriting Patterns	K-NN, Ensemble AdaBoostClassifier, Support Vector Machine	Classification of PD from HC	Python [scikit-learn library]	PaHaW database	37 PD and 38 HC	Accuracy—81.3%

Table 1. *Cont.*

Reference	Feature	Machine Learning Algorithms Used	Objective	Tools Used	Source of Data	No. of Subjects	Outcomes
Fabian Maass et al., 2020 [36]	Handwriting Patterns	SVM	Classification of PD from HC	Weka	UCI machine learning repository	157, 82 PD + 68 HC +7 Normal Pressure Hydrocephalus (NPH)	sensitivity-80%, and specificity—83%
J. Mucha et al., 2018 [37]	Handwriting Patterns	Random Forest Classifier	Classification of PD from HC	Python Programming	PaHaW database	69, 33 PD + 36 HC	Obtained classification accuracy-90% with sensitivity 89%, and specificity 91%
Wenzel et al., 2019 [38]	Handwriting Patterns	CNN	Classification of PD from HC	MATLAB	PPMI database	645, 438 PD + 207 HC	Accuracy-97.2%
Segovia, F. et al., 2019 [39]	Handwriting Patterns	SVM with 10 Cross Validation	Classification of PD from HC	Python programming	Virgen De La Victoria Hospital, Malaga, Spain	189, 95 PD + 94 HC	Accuracy-94.25%
Ye, Q. et al., 2018 [40]	Gait	Least Square (LS)—SVM, Particle Swarm Optimization (PSO)	Classification of PD, ALS, HD from HC	Not mentioned	Neurology Outpatient Clinic at Massachusetts General Hospital, Boston, MA, USA	64, 15 PD + 16 HC + 13 (Amyotrophic lateralsclerosis disease (ALS)) + 20 (Huntington's disease (HD))	Accuracy to diagnose PD from HC- 90.32%, Accuracy to diagnose HD from HC-94.44%, Accuracy to diagnose ALS from HC- 93.10%
Klomsae, A et al., 2018 [41]	Gait	Fuzzy KNN	Classification of PD, ALS, HD from HC	Not mentioned	Neurology Outpatient Clinic at Massachusetts General Hospital, Boston, MA, USA	64, 15 PD + 20 HD + 13 ALS + 16 HC	Accuracy to diagnose PD from HC- 96.43%, Accuracy to diagnose HD from HC-97.22%, Accuracy to diagnose ALS from HC-96.88%
J. P. Félix et al., 2019 [42]	Gait	SVM, KNN, Naïve Bayes, LDA, Decision Tree	Classification of PD from HC	MATLAB R2017a	Neurology Outpatient Clinic at Massachusetts General Hospital, Boston, MA, USA	31, 15 PD + 16 HC	Highest accuracy obtained from SVM, KNN, and decision tree- 96.8%
Andrei et al., 2019 [43]	Gait	SVM	Classification of PD from HC	Not mentioned	Laboratory for Gait and Neurodynamics	166, 93 PD + 73 HC	Accuracy-100%
Priya SJ et al., 2021 [44]	Gait	ANN	Classification of PD from HC	MATLAB R2018b	Laboratory for Gait and Neurodynamics	166, 93 PD + 73 HC	Accuracy-96.28%
Oğul, et al., 2020 [45]	Gait	ANN	Classification of PD from HC	MATLAB	Laboratory for Gait and Neurodynamics	166, 93 PD + 73 HC	Classification accuracy-98.3%
Li B et al., 2020 [46]	Gait	Deep CNN	Classification of PD from HC	Not mentioned	Collected from participants	20, 10 PD + 10 HC	Accuracy-91.9%

3. Proposed Work

The proposed ML model uses an SVM, naïve Bayes, KNN, and ANN algorithm in the core. These algorithms are widely used in the literature since they are easy to use and only need a small number of parameters to be tuned. There are several processes involved in developing a model to detect PD from voice recordings. In the first phase, relevant features are extracted from the dataset for better understanding. In the second phase, machine learning techniques are applied to classify healthy as well as PD patients, which are dependent on acoustic features to predict the outputs in the form of visual representation of graphs and percentage of accuracy score tables. Finally, in the third phase, there is a difference between the entire machine learning classifier models to predict the best accuracy score. The complete technical process of the proposed work is represented in Figure 1. The proposed methodology is shown to be better than the other methodologies with respect to computational cost since few voice features were used instead of heavy feature extraction processes such as MRI, motion sensors, or handwriting assessments. Additionally, the performances of different popular classifiers were evaluated, and the best classifier was found to be ANN for PD diagnosis problems.

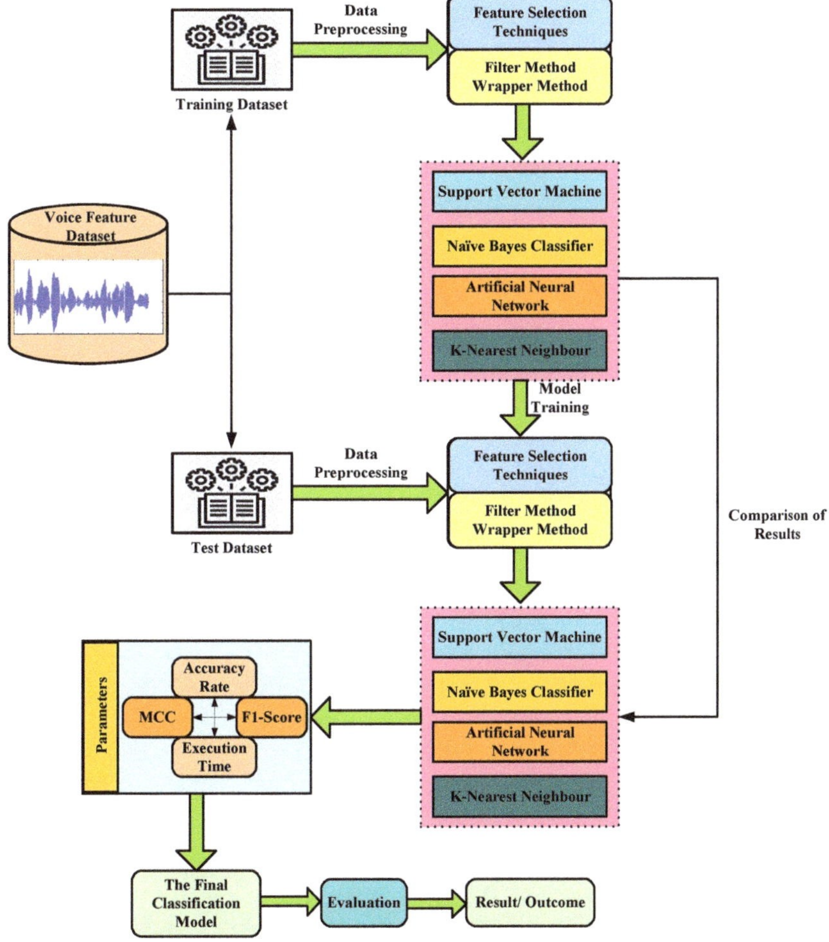

Figure 1. Diagram of the flowchart of the proposed work.

Feature Selection

Due to many available features, feature selection is a frequent approach used to minimize the dimension of data in machine learning based on voice analysis. As demonstrated in Figure 2, all feature selection algorithms have the same aim of reducing redundancy and increasing relevance, which improves the accuracy of the disease's diagnosis. Prior to supplying the data to the classifier, a variety of feature selection strategies were used. The filter-based strategies take into account the importance of the characteristics. As a result, they are stable and scalable and have a low level of complexity [47,48]. The major drawback of this method is that, especially when the data are flowing in a stream, it may overlook certain useful aspects [49]. Both univariate and multivariate techniques based on filters are possible [50]. According to statistically based criteria such as information gain (IG) [51–53], the univariate approaches analyze attributes. Multivariate approaches calculate feature dependence before ranking the feature. In addition, a widely utilized statistical technique for data analysis is principle component analysis (PCA). By choosing a collection of features that accurately reflects the entire data set, PCA can minimize the size of the data sets. The initial variables' principal components are the components with the largest variance value since PCA is a conversion technique. Following that, the other principal components are arranged in descending order of variance values [54]. Additionally, the wrapper-based algorithms assess the quality of the chosen features based on the learning classifier's performance.

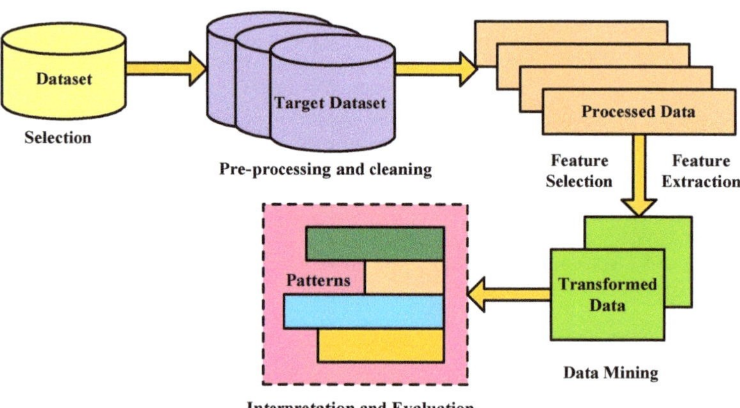

Figure 2. Feature Selection and Feature Extraction from Dataset.

In the pre-processing section, the whole procedure for filter techniques takes place independent of the model. The models are skipped by the filter. Filter methods primarily consider the data's distribution and correlation and internal relationships. As a result, filter techniques have the advantage of being simple and quick to compute. Because of their simplicity and quick computing speed, filter approaches are commonly used in the diagnosis of this disease. Some popular filtering methods are listed below. The minimum redundancy and maximum relevancy (mRMR) method selects characteristics that are far apart but have a strong "correlation" with the classification variable.

The wrapper method decides whether to have or reject a feature depending on a classifier's working change [55]. The wrapper method takes certain classifiers into account and provides a well-tailored subset. As a result, wrapper methods have a lower chance of finding the local maximum. Due to its huge gain in performance, the wrapper approach is popular among ML diagnostics. However, it has drawbacks such as being prone to overfitting and being computationally costly. Wrapper-based feature selection techniques use a classifier to build ML models with different predictor variables and select the variable subset that leads to the best model.

In contrast, filter-based methods are statistical techniques independent of a learning algorithm used to compute the correlation between the predictor and independent variables. The predictor variables are scored according to their relevance to the target variable. The variables with higher scores are then used to build the ML model. Therefore, this research aims to use a filter-based feature selection method, to identify the most relevant features for improved PD detection.

4. Materials and Methods

4.1. Dataset

The dataset of recorded speech signals was obtained from Max Little of the University of Oxford [56,57]. Table 2 contains the details of the dataset. This dataset has an assortment of acoustic speech measures from 195 persons, where 147 persons have Parkinson's disease. All the attributes in the dataset characterize an individual voice measure, and each tuple represents a total number of voice recordings made by these people. The objective of the dataset is to differentiate fit persons compared to the unhealthy using the "status" column, which is set to negative for fit persons and positive for those having the disease.

Table 2. Detail of Parkinson's Dataset.

Dataset Characteristic	Multivariate
No. of Instances	197
Attributes Characteristic	Real
No. of Attributes	23
Missing Values	N/A
Made by	Max Little of the University of Oxford
Associated Tasks	Classification
Types of Classification	Binary {0 for healthy and 1 for PD patient}

4.2. Parkinson's Disease Diagnosis Based on Voice Analysis and Machine Learning

Some studies have concentrated on the acoustic level or the fluctuations in fundamental frequency (F0) caused by vocal activities. The effects of power spectral analysis of F0 phonation in persons with sensorineural audibility loss and the disease have been examined in [58–60]. F0's rhythm was unique in the incidence and amplitude of the diseases. Further, the study demonstrated that the F0 analysis can be a useful tool for neurological diseases under investigation. The autocorrelation function approach was used to find the basic frequencies of speech transmissions. According to the concept, Parkinsonian dysprosody is frequently described as a simple neuro-motor disorder.

The understanding and generation of pitch characteristics in a group of patients were examined to confirm the idea. Conventional medications, such as LDOPA, define that in the early stages of PD, LDOPA is a very effective treatment of subjective disease [61]. In [62], the authors use deep learning to categorize the patient's speech data as "severe" and "not severe". The evaluation measures employed in this study were the unified Parkinson's disease ranking scale (UPDRS). The motor UPDRS examines the patient's motor ability on a 0–108 scale, while the entire UPDRS provides a range of scores from 0 to 1766.

4.3. Classification of Parkinson's Disease with ML Classifier

In this technique, we'll use an ML classifier to classify the disease. First, we select a target variable of patient health status and measure the number of patients in this report. We visualize the data graphically after assessing the health status of a patient. Two types of datasets were developed: 80% of the dataset was used for training and 20% for the testing dataset. In the following Figure 3, the score of 0 represents the healthy persons in the sample, whose count is 48, and 1 represents the patients with Parkinson's disease, whose count is 147. The count of Parkinson's disease patients in the dataset: 147 out of 195 (75.38%). The count of healthy persons in the dataset: 48 out of 195 (24.62%).

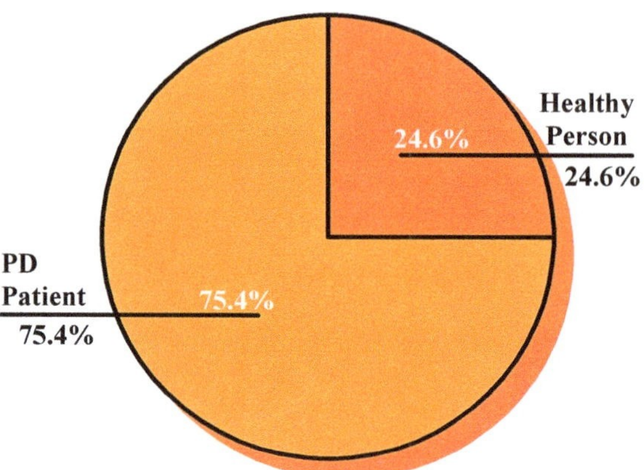

Figure 3. Health Status of PD Patient.

4.4. Building of Machine Learning Techniques with Classifier Evaluation Metrics

By using different types of classifiers, it becomes easy to detect the disease. Classification sensitivity, Matthews's correlation coefficient (MCC), accuracy, specificity, F-score (F-measure), and other measurement parameters are used to distinguish it. Each of these measurement criteria includes a formula for calculating it and determining which classifier is the most qualitatively appropriate for the analysis. It is requisite to focus on the confusion matrix before developing these criteria [63]. The confusion matrix of the multi-class classifier is shown in Figure 4.

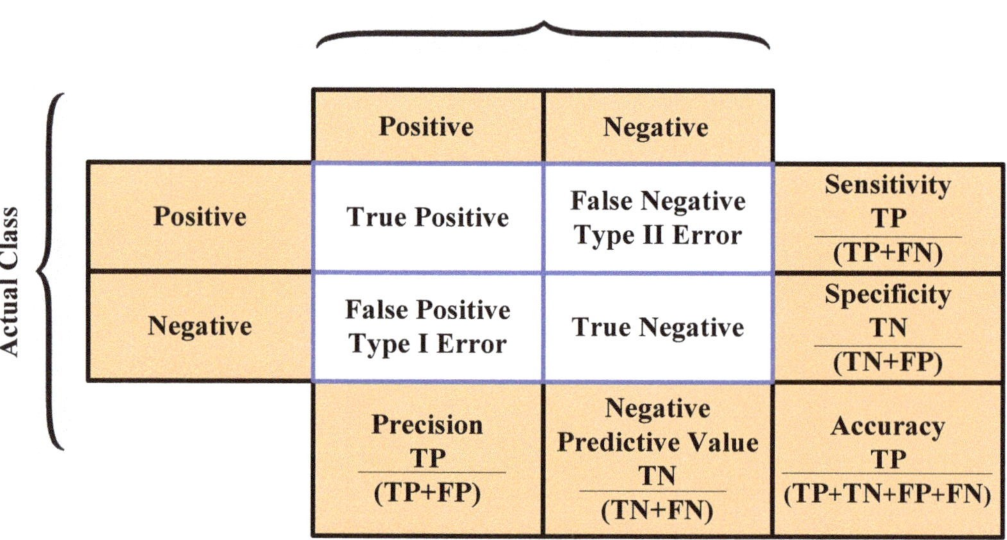

Figure 4. Confusion Matrix with Sensitivity, Specificity, Accuracy, and Precision value.

F1-Score: It represents the accuracy of a model on a given dataset which is also known as F-Score as shown in Equation (1):

$$F - Score = 2 * \frac{precision * sensitivity}{precision + sensitivity} \tag{1}$$

MCC: It is utilized for model evaluation to evaluate the quality of the binary and multi-class classifications as shown in Equation (2). It is based on true-negative, true-positive, and false-negative, false-positive. It lies between -1 to 1 which is defined as follows:

$$MCC = \frac{TP * TN - FP * FN}{\sqrt{(TP + FP)(TP + FN)(TN + FP)(TN + FN)}} \tag{2}$$

(-1): Contradiction between prediction and observation
(0): No better than random prediction
(1): Perfect classifier (accurate prediction).

5. Experiments and Results

The proposed work is implemented in Python 3.7: JupyterLab. Here we detail the experimental setup and the results of the four machine learning classification methods.

5.1. SVM-Classifier

SVM is one of the most prevalent classifier models because it provides accurate as well as highly robust results. The fundamental goal of SVM is to classify the training data by separating the classes while executing a multiple-class learning activity. It allows for the best classification performance on training data and accurately classifies patterns from the data [64]. The training procedure uses a sequential minimization strategy, and classification accuracy is shown to be higher in SVM due to its greater generalization ability [65]. The linear SVM is calculated by using the following Equation (3).

$$y = f(x) = w^T x - b \tag{3}$$

where x represents the data, y represents the class label, w represents the weight of vector orthogonal to the decision hyper-plane, b represents the offset of the hyper-plane and T shows the transpose operator [66].

In this study, we use the sklearn library in the SVM-classifier module for the classification of the given dataset. Table 3 represents the results that are generated by using the SVM classifier (Figure 5). Figure 6 represents the confusion matrix with the true positive, true negative, false positive, and false negative value of a PD person by using the SVM classifier.

Table 3. SVM Classifier.

Name	Results
Accuracy Score of test data	87.17%
Accuracy Score of training data	88.46%
Execution Time	0.03111 s
F1-score	66.19%
MCC	56.59%

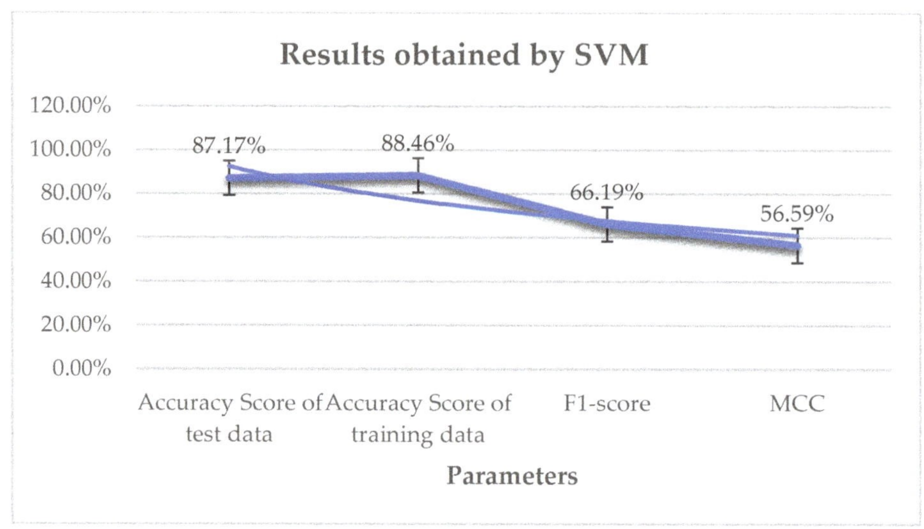

Figure 5. Results obtained by SVM.

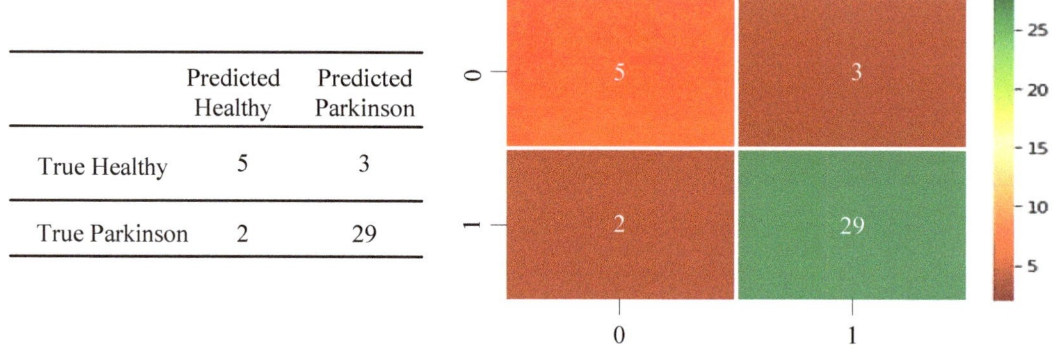

	Predicted Healthy	Predicted Parkinson
True Healthy	5	3
True Parkinson	2	29

Figure 6. Confusion Matrix and Heatmap of SVM Classifier.

5.2. Naïve Bayes Classifier

Another main essential category method of ML is the naive Bayes classifier technique. It provides effective classification and learning and the majority of results are acquired through the naïve Bayes method [67]. Naïve Bayes, based on Bayes' theorem, determines the likelihood of an event occurring depending on the event's circumstances. For instance, variations in the voice are common in people with the disease; hence, these symptoms are linked to the prediction for diagnosis of this disease. The naive variation of the theorem extends and simplifies the original Bayes theorem, which gives a mechanism for determining the probability of a target occurrence. To estimate the likelihood of the medical condition, the data comprise numerous speech signal variants. The sklearn Gaussian naive Bayes algorithm is used to provide the classifier module for the execution of the naïve Bayes categorization. The result of the classifier is shown in Table 4 and graphical representation is illustrated in Figure 7.

Table 4. Naïve Bayes Classifier Results.

Name	Results
Accuracy Rate of test data	74.11%
Accuracy Rate of training data	76.23%
Execution Time	0.0323 s
F1-score	86.74%
MCC	66.56%

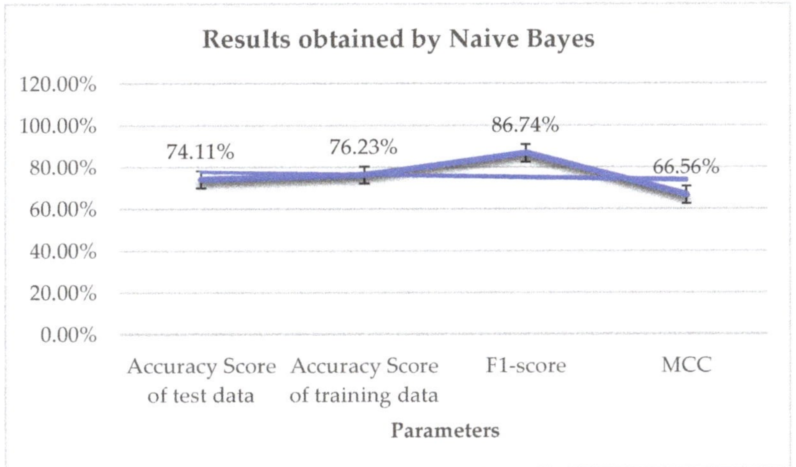

Figure 7. Results obtained by Naïve Bayes.

5.3. Artificial Neural Network

ANN is a subfield of deep neural networks that predict how the human brain works. In general, there is a significant distinction between the human brain and ANN. The brain has 'n' number of parallel neurons, whereas the machine only has a finite sum of processors. Additionally, neurons are meeker and more relaxed than computer processors. Another major disparity between computer systems and the brain is the ability to process information on a larger scale. Neurons are made up of synapses or networks that operate together [64,68]. In this article, the main aim is to classify the functionality of ANN techniques in the early detection of this disease which is built on the subsequent phases:

i.   Identifying the responsibility and function of ANN in the detection of this disease.
ii. Making observations on labels and features of datasets.
iii. Grouping the types of the studied disease centered on their symptoms.
iv. Examining the accurate outcomes.

These outcomes can be further used in the medical sector as direction for developers considering ANN deployment to enhance the civic health potential as a reaction to the studied disease [69].

In the experiment of an artificial neural network, the dataset was split into two parts i.e., the training dataset (80%) and the test dataset (20%). The classification results of the artificial neural network were found to be very high in the form of the average accuracy score which was the highest among all the classification methods, i.e., 96.7% shown in Table 5 and graphical representation is shown in Figure 8.

Table 5. Artificial Neural Network Classifier Outcome.

Title	Results
Accuracy Rate of test data	96.7%
Accuracy Rate of training data	97.4%
Execution Time	0.025 s
F1-Score	87.01%
MCC	70.11%

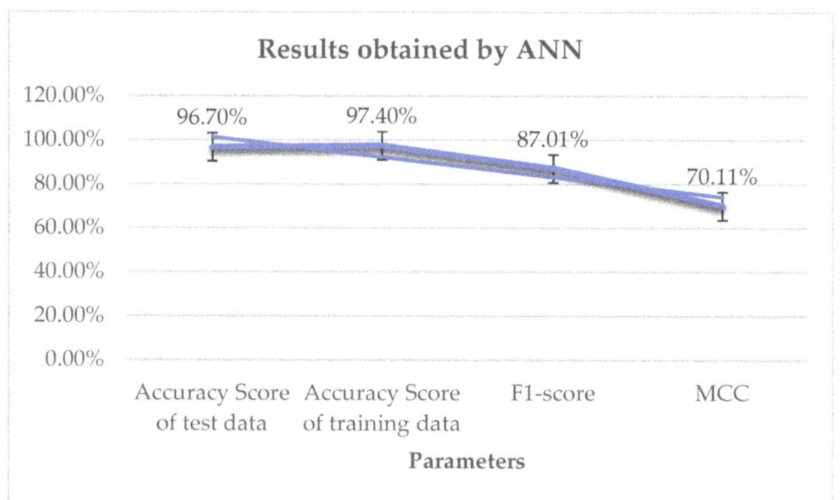

Figure 8. Results obtained by ANN.

5.4. K-Nearest Neighbor

The KNN technique is costly while presenting with a huge training dataset since it has been used most of the time in pattern recognition. KNN is the base concept of learning by analogy utilized to categorize the nearest neighbors. It is accomplished by comparing closely similar training tuples to the provided test tuple. As a result, "n" characteristics are utilized to recognize training tuples in which each tuple corresponds to a distinct point in the n-dimensional space. The KNN classifier's responsibility in the event of an unlabeled tuple is to explore the pattern space for all k training tuples that are close together [64]. This study aims to identify the accuracy rate of detecting the subject disease. To find out the difference between affected patients and healthy persons, the KNN algorithm is used. In terms of accuracy, experimental data reveal that the ANN classifier outperformed the KNN classifier on average. The results of the KNN classifier are shown in Table 6 with the accuracy rate of the training and test datasets, F1-score, and MCC illustrated in Figure 9.

Table 6. KNN Classifier Results.

Name	Results
Accuracy Rate of test data	87.17%
Accuracy Rate of training data	88.46%
Execution Time	0.03111 s
F1-score	71%
MCC	65.02%

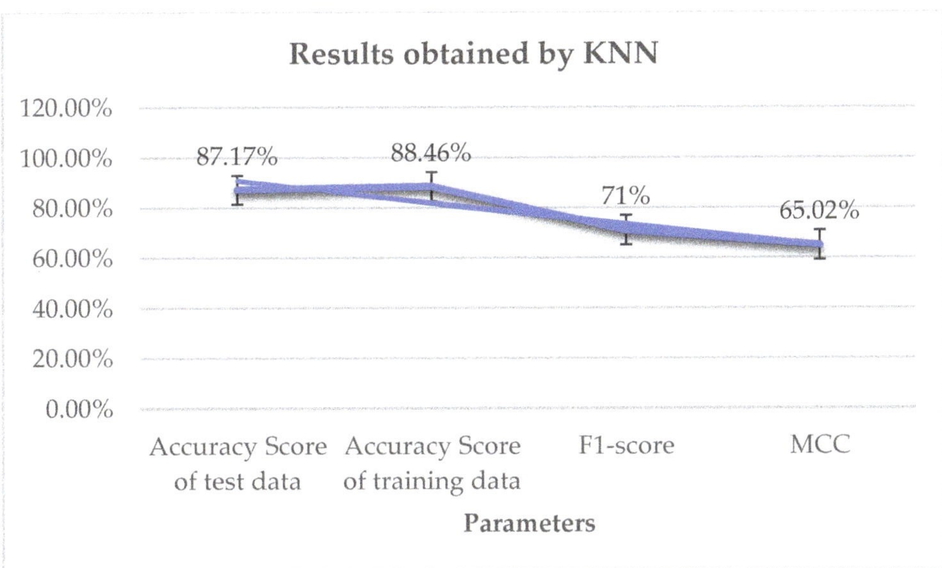

Figure 9. Results obtained by KNN.

5.5. Summary of Evaluation Results

The performance of all the classifier models used in the experiment for the disease's prediction is illustrated in Table 7. The artificial neural network classifier scores the highest accuracy rate followed by SVM, naïve Bayes, and KNN. Figure 10 shows the graphical representation of the results obtained by these four ML classifiers based on various parameters. Table 7 illustrates that SVM attained the average accuracy for the training and test datasets, which are 88.46% and 87.17% respectively, F1-score (66.19%), and MCC (56.59%), sensitivity and specificity 62.5% and 93.54%, respectively. In addition, the naïve Bayes achieved the average accuracy for the training and test datasets, F1-score, MCC, sensitivity, and specificity, which are 76.23%, 74.11%, 86.74%, 66.56%, 84%, and 79.76% respectively.

Table 7. An overview of evaluation results.

| | Performance Measure | | | | | |
| | Accuracy | | F1-Score | MCC | Sensitivity | Specificity |
	Training Dataset	Test Dataset				
SVM	88.46%	87.17%	66.19%	56.59%	62.5%	93.54%
Naïve Bayes	76.23%	74.11%	86.74%	66.56%	84%	79.76%
KNN	88.46%	87.17%	71%	65.02%	60.0%	93.54%
ANN	97.4%	96.7%	87.01%	70.11%	92.42%	91.25%

It has been observed that the results obtained by the SVM and KNN have the same values for all the parameters except MCC (65.02 %) and sensitivity (60%). Finally, the best accuracy was obtained by the ANN where the results of parameters such as accuracy of the training and test datasets, F1-score, MCC, sensitivity, and specificity are 97.4%, 96.7%, 64.55%, 87.01%, 70.11%, 92.42%, and 91.25%, respectively. Overall, the results of our experiments show that ANN outperforms SVM, naive Bayes, and KNN.

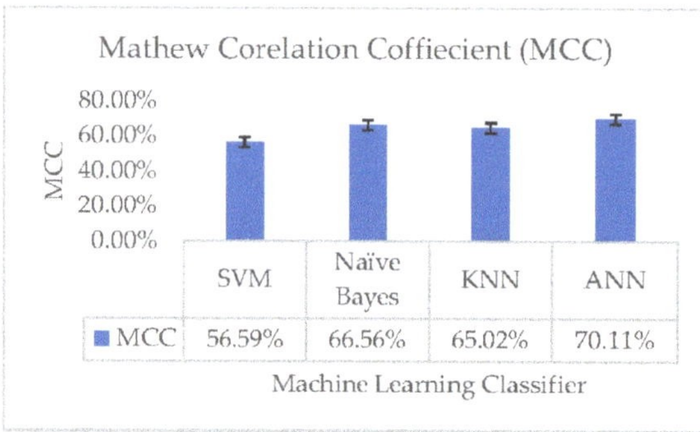

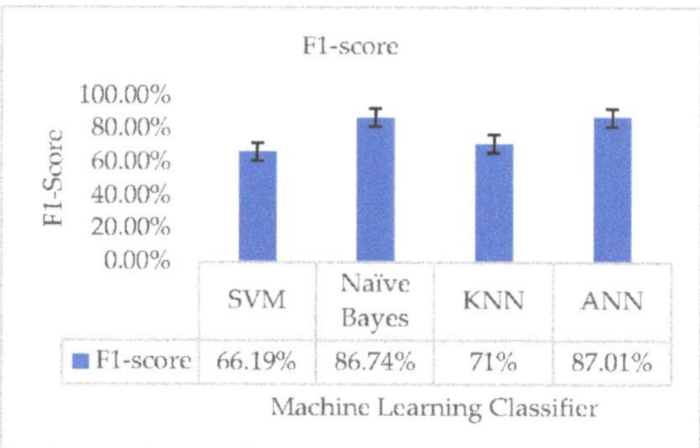

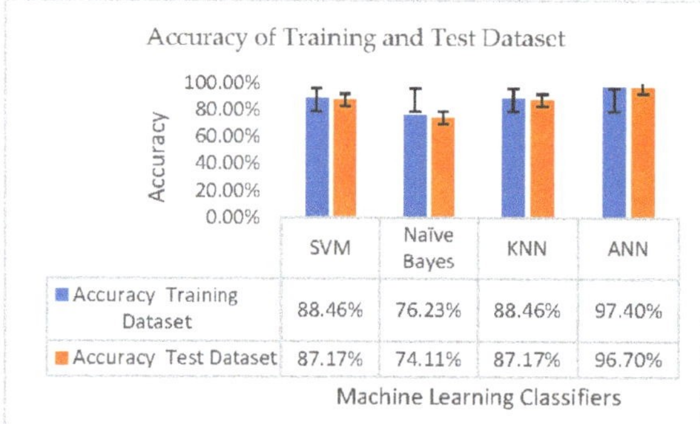

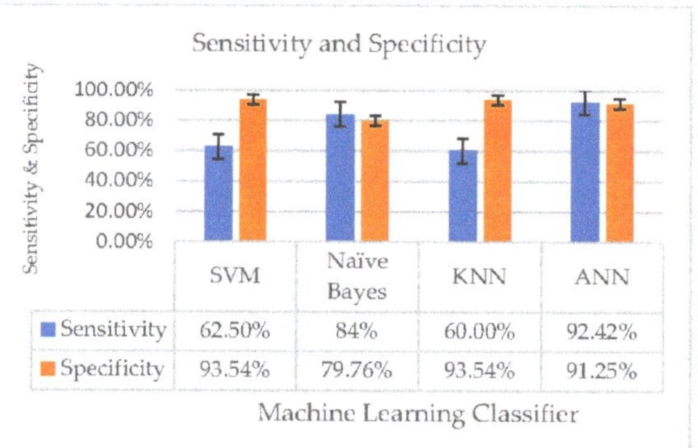

Figure 10. Graphical representation of distributions of performance measures for all classifiers.

6. Comparative Study and Discussion

This section examines the efficient comparative result analysis of the proposed technique with other conventional machine learning techniques. The comparison of the proposed study with previously published research is shown in Table 8.

Table 8. Performance Comparison with previous studies.

Reference	Basis	Machine Learning Classifier	Accuracy	Sensitivity	Specificity
Sakar et al. [70]	Speech	SVM and KNN	68.45%	60	50
Vadovsk'y and Parali [71]	Speech	C4.5 + C5.0 + randomforest + CART	66.5	NA	NA
Ouhmida, A. [72]	Speech	SVM, K-NN, Decision Tree	98.26% (AUC)	NA	NA
Mabrouk et al., [73]	Speech	Random forest, SVM, MLP, KNN	78.4% (SVM), 82.2% (KNN)	NA	NA
Benba et al. [74]	Speech	HFCC-SVM	87.5%	90%	85%
Proposed Work	Speech	SVM, naïve Bayes, KNN and ANN	87.17%, 74.11%, 87.17%, and 96.7%	62.5%, 84%, 60%, and 92.42%	93.54%, 79.76%, 93.54%, and 91.25%

As per the comparative analysis, the proposed model (using four machine learning algorithms) shows better results obtained as compared to all other experimental machine learning models and the existing state of the art. In the proposed study, the best result was achieved by ANN with 96.7% accuracy, which is higher than the other experimental algorithms. The authors of [49] collected 20 PD and 20 HC speech datasets using high-quality recording equipment and used KNN and SVM to analyze the datasets in order to detect PD. KNN and SVM classifiers performed with accuracy rates of 59.52% (LOSO) and 68.45% (LOSO), respectively. In addition to [50], the authors used various algorithms such as C4.5, C5.0, random forest, and CART based on decision trees. The authors experimented on 40 individuals' records, where 50% were affected with the subjective disease and 50% were HC. For this study, the highest average model accuracy of 66.5% was attained. ANN was used by [51] to identify PD. The dataset was obtained from the University of California, Irvine's machine learning library. A total of 45 attributes were chosen as input values and one outcome for the categorization using the MATLAB tool. With an accuracy of 94.93%, their suggested model was able to differentiate healthy individuals from PD subjects. In [52], the authors used random forest, SVM, MLP, and KNN classifiers for the detection of PD patients from HC. The result obtained from this study was 78.4% and 82.2% for the SVM and KNN classifiers, respectively. In a study by [53], the authors examined the comparison between the patients with PD (PWP) and healthy controls (HC) based on a variety of speech samples. In their study, human factor cepstral coefficients (HFCC) were applied. The extracted HFCC was used to generate the average voice print for each voice recording. For the classification, SVM was used with a variety of kernels, including RBF, polynomial, linear, and MLP. The SVM's linear kernel allowed for the highest accuracy of 87.5%.

In addition to the comparisons mentioned above, the performance of the proposed methodology is compared with related ML methods for PD analysis in various scenarios and with various types of evaluated PD datasets. The proposed technique outperformed other similar contributions of ML methods in terms of performance for diagnosing PD, as seen in the above table, and is thus superior to them.

7. Conclusions

Automated ML techniques may classify PD from HC and predict the outcome using non-invasive speech biomarkers as features. With noisy and high-dimensional data, our study compares the performance of multiple machine learning classifiers for disease detection. Accuracy at the clinical level is feasible with careful feature selection. In this paper, we compared ML classifiers: SVM with an accuracy of 87.17%, naïve Bayes' classifier with an accuracy of 74.11%, ANN with an accuracy of 96.7%, and KNN with an accuracy of 87.17%. We used these techniques to distinguish between affected patients and healthy people. The disease is diagnosed using human speech signals. The acquired results demonstrate how feature selection techniques work well with ML classifiers, especially when working with voice data where it is possible to extract a large number of phonetic characteristics. The proposed early diagnosis approach makes it possible to detect PD with high accuracy in its early stages and the subjective disease's severe symptoms can be prevented. Many categorization algorithms are being used in the medical imaging area to obtain the best level of accuracy. This research may be used in different machine learning methods and datasets to improve classifier performance and reach the maximum accuracy score. In order to improve the accuracy of the models created, future efforts will make use of the already-existing recordings and add to the number of existing attributes. In order to compare the collected data, various different records processing software that are available online may also be used.

Author Contributions: Conceptualization, A.R.; methodology, A.R. and A.D.; validation, A.R. and M.R.; formal analysis, A.R. and N.A.; writing—original draft preparation, A.R.; writing—review and editing, M.K.P. and M.R.; supervision, A.D. and R.S. All authors have read and agreed to the published version of the manuscript.

Funding: There was no external funding received for this article.

Institutional Review Board Statement: Not applicable.

Informed Consent Statement: Not applicable.

Data Availability Statement: Data in this research paper will be shared upon request made to the first author.

Conflicts of Interest: The authors declare no conflict of interest.

References

1. DeMaagd, G.; Philip, A. Parkinson's Disease and Its Management: Part 1: Disease Entity, Risk Factors, Pathophysiology, Clinical Presentation, and Diagnosis. *Pharm. Ther.* **2015**, *40*, 504–532.
2. Rizek, P.; Kumar, N.; Jog, M.S. An update on the diagnosis and treatment of Parkinson disease. *CMAJ* **2016**, *188*, 1157–1165. [CrossRef] [PubMed]
3. Available online: https://www.who.int/news-room/fact-sheets/detail/parkinson-disease (accessed on 30 October 2022).
4. De Rijk, M.C.; Launer, L.J.; Berger, K.; Breteler, M.M.; Dartigues, J.F.; Baldereschi, M.; Fratiglioni, L.; Lobo, A.; Martinez-Lage, J.; Trenkwalder, C.; et al. Prevalence of Parkinson's disease in Europe: A collaborative study of population-based cohorts. Neurologic Diseases in the Elderly Research Group. *Neurology* **2000**, *54* (Suppl. S5), S21–S23. [PubMed]
5. Cantürk, İ.; Karabiber, F. A machine learning system for the diagnosis of Parkinson's disease from speech signals and its application to multiple speech signal types. *Arab. J. Sci. Eng.* **2016**, *41*, 5049–5059. [CrossRef]
6. Singh, N.; Pillay, V.; Choonara, Y.E. Advances in the treatment of Parkinson's disease. *Prog. Neurobiol.* **2007**, *81*, 29–44. [CrossRef]
7. Rana, A.; Rawat, A.S.; Bijalwan, A.; Bahuguna, H. Application of multi-layer (perceptron) artificial neural network in the diagnosis system: A systematic review. In Proceedings of the 2018 International Conference on Research in Intelligent and Computing in Engineering (RICE), San Salvador, El Salvador, 22–24 August 2018; pp. 1–6.
8. Lakany, H. Extracting a diagnostic gait signature. *Pattern Recognit.* **2008**, *41*, 1627–1637. [CrossRef]
9. Figueiredo, J.; Santos, C.P.; Moreno, J.C. Automatic recognition of gait patterns in human motor disorders using machine learning: A review. *Med. Eng. Phys.* **2018**, *53*, 1–12. [CrossRef]
10. Hazan, H.; Hilu, D.; Manevitz, L.; Ramig, L.O.; Sapir, S. Early diagnosis of Parkinson's disease via machine learning on speech data. In Proceedings of the 2012 IEEE 27th Convention of Electrical and Electronics Engineers in Israel, Eilat, Israel, 14–17 November 2012; pp. 1–4. [CrossRef]
11. Karan, B.; Sahu, S.S.; Mahto, K. Parkinson disease prediction using intrinsic mode function based features from speech signal. *Biocybern. Biomed. Eng.* **2019**, *40*, 249–264. [CrossRef]
12. Frid, A.; Safra, E.J.; Hazan, H.; Lokey, L.L.; Hilu, D.; Manevitz, L.; Ramig, L.O.; Sapir, S. Computational diagnosis of Parkinson's Disease directly from natural speech using machine learning techniques. In Proceedings of the 2014 IEEE International Conference on Software Science, Technology and Engineering, Washington, DC, USA, 11–12 June 2014; pp. 50–53.
13. Rawat, A.S.; Rana, A.; Kumar, A.; Bagwari, A. Application of multi layer artificial neural network in the diagnosis system: A systematic review. *IAES Int. J. Artif. Intell.* **2018**, *7*, 138. [CrossRef]
14. KarimiRouzbahani, H.; Daliri, M.R. Diagnosis of Parkinson's Disease in Human Using Voice Signals. *BCN* **2011**, *2*, 12–20.
15. Khamparia, A.; Gupta, D.; Nguyen, N.G.; Khanna, A.; Pandey, B.; Tiwari, P. Sound Classification Using Convolutional Neural Network and Tensor Deep Stacking Network. *IEEE Access* **2019**, *7*, 7717–7727. [CrossRef]
16. Bourouhou, A.; Jilbab, A.; Nacir, C.; Hammouch, A. Comparison of classification methods to detect the parkinson disease. In Proceedings of the 2016 International Conference on Electrical and Information Technologies (ICEIT), Tangiers, Morocco, 4–7 May 2016; pp. 421–424.
17. Sharma, A.; Giri, R.N. Automatic Recognition of Parkinson's Disease via Artificial Neural Network and Support Vector Machine. *Int. J. Innov. Technol. Explor. Eng. (IJITEE)* **2014**, *4*, 7.
18. Purwins, H.; Li, B.; Virtanen, T.; Schluter, J.; Chang, S.-Y.; Sainath, T.N. Deep Learning for Audio Signal Processing. *IEEE J. Sel. Top. Signal Process.* **2019**, *13*, 206–219. [CrossRef]
19. Zhang, L.; Qu, Y.; Jin, B.; Jing, L.; Gao, Z.; Liang, Z. An Intelligent Mobile-Enabled System for Diagnosing Parkinson Disease: Development and Validation of a Speech Impairment Detection System. *JMIR Public Health Surveill.* **2020**, *8*, e18689. [CrossRef]
20. Kadiri, S.R.; Kethireddy, R.; Alku, P. Parkinson's Disease Detection from Speech Using Single Frequency Filtering Cepstral Coefficients. In Proceedings of the Interspeech 2020, Shanghai, China, 25–29 October 2020. [CrossRef]
21. Pramanik, M.; Pradhan, R.; Nandy, P.; Bhoi, A.K.; Barsocchi, P. Machine Learning Methods with Decision Forests for Parkinson's Detection. *Appl. Sci.* **2021**, *11*, 581. [CrossRef]
22. Gunduz, H. Deep Learning-Based Parkinson's Disease Classification Using Vocal Feature Sets. *IEEE Access* **2019**, *7*, 115540–115551. [CrossRef]
23. Available online: https://www.dataversity.net/improving-clinical-insights-machine-learning/# (accessed on 27 August 2022).
24. Kohlschein, C.; Schmitt, M.; Schuller, B.; Jeschke, S.; Werner, C.J. A machine learning based system for the automatic evaluation of aphasia speech. In Proceedings of the 2017 IEEE 19th International Conference on e-Health Networking, Applications and Services (Healthcom), Dalian, China, 12–15 October 2017; pp. 1–6. [CrossRef]
25. Bertini, F.; Allevi, D.; Lutero, G.; Montesi, D.; Calzà, L. Automatic Speech Classifier for Mild Cognitive Impairment and Early Dementia. *ACM Trans. Comput. Healthc.* **2022**, *3*, 1–11. [CrossRef]
26. Matcham, F.; On Behalf of the RADAR-CNS Consortium; Pietro, C.B.D.S.; Bulgari, V.; de Girolamo, G.; Dobson, R.; Eriksson, H.; Folarin, A.A.; Haro, J.M.; Kerz, M.; et al. Remote assessment of disease and relapse in major depressive disorder (RADAR-MDD): A multi-centre prospective cohort study protocol. *BMC Psychiatry* **2019**, *19*, 72. [CrossRef]
27. Sakar, C.O.; Serbes, G.; Gunduz, A.; Tunc, H.C.; Nizam, H.; Sakar, B.E.; Tutuncu, M.; Aydin, T.; Isenkul, M.E.; Apaydin, H. A comparative analysis of speech signal processing algorithms for Parkinson's disease classification and the use of the tunable-factor wavelet transform. *Appl. Soft Comput.* **2019**, *74*, 255–263. [CrossRef]

28. Yasar, A.; Saritas, I.; Sahman, M.A.; Cinar, A.C. Classification of Parkinson disease data with artificial neural networks. In Proceedings of the IOP Conference Series: Materials Science and Engineering, Wuhan, China, 10–12 October 2019; Volume 675, p. 012031.
29. Avuçlu, E.; Elen, A. Evaluation of train and test performance of machine learning algorithms and Parkinson diagnosis with statistical measurements. *Med. Biol. Eng. Comput.* **2020**, *58*, 2775–2788. [CrossRef]
30. Marar, S.; Swain, D.; Hiwarkar, V.; Motwani, N.; Awari, A. Predicting the occurrence of Parkinson's Disease using various Classification Models. In Proceedings of the 2018 International Conference on Advanced Computation and Telecommunication (ICACAT), Bhopal, India, 28–29 December 2018; pp. 1–5.
31. Nikookar, E.; Sheibani, R.; Alavi, S.E. An ensemble method for diagnosis of Parkinson's disease based on voice measurements. *J. Med. Signals Sens.* **2019**, *9*, 221–226. [CrossRef] [PubMed]
32. Tracy, J.M.; Özkanca, Y.; Atkins, D.C.; Ghomi, R.H. Investigating voice as a biomarker: Deep phenotyping methods for early detection of Parkinson's disease. *J. Biomed. Inform.* **2019**, *104*, 103362. [CrossRef] [PubMed]
33. Cibulka, M.; Brodnanova, M.; Grendar, M.; Grofik, M.; Kurca, E.; Pilchova, I.; Osina, O.; Tatarkova, Z.; Dobrota, D.; Kolisek, M. SNPs rs11240569, rs708727, and rs823156 in SLC41A1 Do Not Discriminate Between Slovak Patients with Idiopathic Parkinson's Disease and Healthy Controls: Statistics and Machine-Learning Evidence. *Int. J. Mol. Sci.* **2019**, *20*, 4688. [CrossRef] [PubMed]
34. Hsu, S.-Y.; Lin, H.-C.; Chen, T.-B.; Du, W.-C.; Hsu, Y.-H.; Wu, Y.-C.; Tu, P.-W.; Huang, Y.-H.; Chen, H.-Y. Feasible Classified Models for Parkinson Disease from 99mTc-TRODAT-1 SPECT Imaging. *Sensors* **2019**, *19*, 1740. [CrossRef] [PubMed]
35. Drotár, P.; Mekyska, J.; Rektorová, I.; Masarová, L.; Smékal, Z.; Faundez-Zanuy, M. Evaluation of handwriting kinematics and pressure for differential diagnosis of Parkinson's disease. *Artif. Intell. Med.* **2016**, *67*, 39–46. [CrossRef]
36. Maass, F.; Michalke, B.; Willkommen, D.; Leha, A.; Schulte, C.; Tönges, L.; Mollenhauer, B.; Trenkwalder, C.; Rückamp, D.; Börger, M.; et al. Elemental fingerprint: Reassessment of a cerebrospinal fluid biomarker for Parkinson's disease. *Neurobiol. Dis.* **2019**, *134*, 104677. [CrossRef]
37. Mucha, J.; Mekyska, J.; Faundez-Zanuy, M.; Lopez-De-Ipina, K.; Zvoncak, V.; Galaz, Z.; Kiska, T.; Smekal, Z.; Brabenec, L.; Rektorova, I. Advanced Parkinson's Disease Dysgraphia Analysis Based on Fractional Derivatives of Online Handwriting. In Proceedings of the 2018 10th International Congress on Ultra Modern Telecommunications and Control Systems and Workshops (ICUMT), Moscow, Russia, 5–9 November 2018; pp. 1–6. [CrossRef]
38. Wenzel, M.; Milletari, F.; Krüger, J.; Lange, C.; Schenk, M.; Apostolova, I.; Klutmann, S.; Ehrenburg, M.; Buchert, R. Automatic classification of dopamine transporter SPECT: Deep convolutional neural networks can be trained to be robust with respect to variable image characteristics. *Eur. J. Pediatr.* **2019**, *46*, 2800–2811. [CrossRef]
39. Segovia, F.; Gorriz, J.M.; Ramirez, J.; Martinez-Murcia, F.J.; Castillo-Barnes, D. Assisted Diagnosis of Parkinsonism Based on the Striatal Morphology. *Int. J. Neural Syst.* **2019**, *29*, 1950011. [CrossRef]
40. Ye, Q.; Xia, Y.; Yao, Z. Classification of Gait Patterns in Patients with Neurodegenerative Disease Using Adaptive Neuro-Fuzzy Inference System. *Comput. Math. Methods Med.* **2018**, *2018*, 9831252. [CrossRef]
41. Klomsae, A.; Auephanwiriyakul, S.; Theera-Umpon, N. (2018). String grammar unsupervised possibilistic fuzzy c-medians for gait pattern classification in patients with neurodegenerative diseases. *Comput. Intell. Neurosci.* **2018**, *2018*, 1869565. [CrossRef]
42. Felix, J.P.; Vieira, F.H.T.; Cardoso, A.A.; Ferreira, M.V.G.; Franco, R.A.P.; Ribeiro, M.A.; Araujo, S.G.; Correa, H.P.; Carneiro, M.L. A Parkinson's Disease Classification Method: An Approach Using Gait Dynamics and Detrended Fluctuation Analysis. In Proceedings of the 2019 IEEE Canadian Conference of Electrical and Computer Engineering (CCECE), Edmonton, AB, Canada, 5–8 May 2019. [CrossRef]
43. Andrei, A.-G.; Tautan, A.-M.; Ionescu, B. Parkinson's Disease Detection from Gait Patterns. In Proceedings of the 2019 E-Health and Bioengineering Conference (EHB), Iasi, Romania, 21–23 November 2019; pp. 1–4. [CrossRef]
44. Priya, S.J.; Rani, A.J.; Subathra, M.S.P.; Mohammed, M.A.; Damaševičius, R.; Ubendran, N. Local Pattern Transformation Based Feature Extraction for Recognition of Parkinson's Disease Based on Gait Signals. *Diagnostics* **2021**, *11*, 1395. [CrossRef]
45. Yurdakul, O.C.; Subathra, M.; George, S.T. Detection of Parkinson's Disease from gait using Neighborhood Representation Local Binary Patterns. *Biomed. Signal Process. Control* **2020**, *62*, 102070. [CrossRef]
46. Li, B.; Yao, Z.; Wang, J.; Wang, S.; Yang, X.; Sun, Y. Improved Deep Learning Technique to Detect Freezing of Gait in Parkinson's Disease Based on Wearable Sensors. *Electronics* **2020**, *9*, 1919. [CrossRef]
47. Rana, A.; Dumka, A.; Singh, R.; Panda, M.K.; Priyadarshi, N.; Twala, B. Imperative Role of Machine Learning Algorithm for Detection of Parkinson's Disease: Review, Challenges and Recommendations. *Diagnostics* **2022**, *12*, 2003. [CrossRef]
48. Masoudi-Sobhanzadeh, Y.; MotieGhader, H.; Masoudi-Nejad, A. FeatureSelect: A software for feature selection based on machine learning approaches. *BMC Bioinform.* **2019**, *20*, 107. [CrossRef]
49. Rahmaninia, M.; Moradi, P. OSFSMI: Online stream feature selection method based on mutual information. *Appl. Soft Comput.* **2018**, *68*, 733–746. [CrossRef]
50. Pourbahrami, S. Improving PSO global method for feature selection according to iterations global search and chaotic theory. *arXiv* **2018**, *preprint*. arXiv:1811.08701.
51. Yu, L.; Liu, H. Feature selection for high-dimensional data: A fast correlation-based filter solution. In Proceedings of the 20th International Conference on Machine Learning (ICML-03), Washington, DC, USA, 21–24 August 2003; pp. 856–863.
52. Blum, A.L.; Langley, P. Selection of relevant features and examples in machine learning. *Artif. Intell.* **1997**, *97*, 245–271. [CrossRef]
53. Raileanu, L.E.; Stoffel, K. Theoretical Comparison between the Gini Index and Information Gain Criteria. *Ann. Math. Artif. Intell.* **2004**, *41*, 77–93. [CrossRef]

54. Jolliffe, I.T. *Principal Component Analysis*, 2nd ed.; Springer: New York, NY, USA, 2002.
55. Miao, Y.; Lou, X.; Wu, H. The Diagnosis of Parkinson's Disease Based on Gait, Speech Analysis and Machine Learning Techniques. In Proceedings of the 2021 International Conference on Bioinformatics and Intelligent Computing (BIC 2021). Association for Computing Machinery, New York, NY, USA, 22–24 January 2021; pp. 358–371.
56. Little, M.; McSharry, P.; Hunter, E.; Spielman, J.; Ramig, L. Suitability of dysphonia measurements for telemonitoring of Parkinson's disease. *Nat. Preced.* **2008**, 1. [CrossRef]
57. Lichman, M. *UCI Machine Learning Repository*; University of California, School of Information and Computer Science: Irvine, CA, USA; Available online: http://archive.ics.uci.edu/ml (accessed on 25 September 2022).
58. Rewar, S. A systematic review on Parkinson's disease (PD). *Indian J. Res. Pharm. Biotechnol.* **2015**, *3*, 176.
59. Arora, S.; Venkataraman, V.; Zhan, A.; Donohue, S.; Biglan, K.; Dorsey, E.; Little, M. Detecting and monitoring the symptoms of Parkinson's disease using smartphones: A pilot study. *Park. Relat. Disord.* **2015**, *21*, 650–653. [CrossRef] [PubMed]
60. Miljkovic, D.; Aleksovski, D.; Podpečan, V.; Lavrač, N.; Malle, B.; Holzinger, A. Machine Learning and Data Mining Methods for Managing Parkinson's Disease. In *Machine Learning for Health Informatics*; Springer: Cham, Switzerland, 2016; pp. 209–220. [CrossRef]
61. Challa, K.N.R.; Pagolu, V.S.; Panda, G.; Majhi, B. An improved approach for prediction of Parkinson's disease using machine learning techniques. In Proceedings of the 2016 International Conference on Signal Processing, Communication, Power and Embedded System (SCOPES), Odisha, India, 3–5 October 2016; pp. 1446–1451. [CrossRef]
62. Lee, G.S.; Lin, S.H. Changes of rhythm of vocal fundamental frequency in sensorineural hearing loss and in Parkinson's disease. *Chin. J. Physiol.* **2009**, *52*, 446–450. [CrossRef] [PubMed]
63. Asmae, O.; Abdelhadi, R.; Bouchaib, C.; Sara, S.; Tajeddine, K. Parkinson's Disease Identification using KNN and ANN Algorithms based on Voice Disorder. In Proceedings of the 2020 1st International Conference on Innovative Research in Applied Science, Engineering and Technology (IRASET), Meknes, Morocco, 16–19 April 2020; Institute of Electrical and Electronics Engineers (IEEE): New York, NY, USA, 2020; pp. 1–6.
64. Wu, X.; Kumar, V.; Ross Quinlan, J.; Ghosh, J.; Yang, Q.; Motoda, H.; Steinberg, D. Top 10 algorithms in data mining. *Knowl. Inf. Syst.* **2008**, *14*, 1–37. [CrossRef]
65. Ray, P.K.; Mohanty, A.; Panigrahi, T. Power quality analysis in solar PV integrated microgrid using independent component analysis and support vector machine. *Optik* **2019**, *180*, 691–698. [CrossRef]
66. Lahmiri, S.; Shmuel, A. Detection of Parkinson's disease based on voice patterns ranking and optimized support vector machine. *Biomed. Signal Process. Control* **2018**, *49*, 427–433. [CrossRef]
67. Bhatia, A.; Sulekh, R. Predictive Model for Parkinson's disease through Naïve Bayes Classification. *Int. J. Comput. Sci. Commun.* **2017**, *9*, 194–202.
68. Rana, A.; Bahuguna, H.; Bijalwan, A. Artificial Neural Network based Diagnosis System. *Int. J. Comput. Trends Technol.* **2017**, *3*, 189–191.
69. Alzubaidi, M.S.; Shah, U.; DhiaZubaydi, H.; Dolaat, K.; Abd-Alrazaq, A.A.; Ahmed, A.; Househ, M. The Role of Neural Network for the Detection of Parkinson's Disease: A Scoping Review. *Healthcare* **2021**, *9*, 740. [CrossRef]
70. Sakar, B.E.; Isenkul, M.E.; Sakar, C.O.; Sertbas, A.; Gurgen, F.; Delil, S.; Apaydin, H.; Kursun, O. Collection and Analysis of a Parkinson Speech Dataset With Multiple Types of Sound Recordings. *IEEE J. Biomed. Health Inform.* **2013**, *17*, 828–834. [CrossRef]
71. Vadovský, M.; Paralič, J. Parkinson's disease patients classification based on the speech signals. In Proceedings of the 2017 IEEE 15th International Symposium on Applied Machine Intelligence and Informatics (SAMI), Herlany, Slovakia, 26–28 January 2017; pp. 000321–000326.
72. Ouhmida, A.; Raihani, A.; Cherradi, B.; Terrada, O. A Novel Approach for Parkinson's Disease Detection Based on Voice Classification and Features Selection Techniques. *Int. J. Online Biomed. Eng.* **2021**, *17*, 111–130. [CrossRef]
73. Mabrouk, R.; Chikhaoui, B.; Bentabet, L. Machine Learning Based Classification Using Clinical and DaTSCAN SPECT Imaging Features: A Study on Parkinson's Disease and SWEDD. *IEEE Trans. Radiat. Plasma Med. Sci.* **2018**, *3*, 170–177. [CrossRef]
74. Benba, A.; Jilbab, A.; Hammouch, A. Using Human Factor Cepstral Coefficient on Multiple Types of Voice Recordings for Detecting Patients with Parkinson's Disease. *Irbm* **2017**, *38*, 346–351. [CrossRef]

electronics

MDPI

Article

Predicting Alzheimer's Disease Using Deep Neuro-Functional Networks with Resting-State fMRI

Sambath Kumar Sethuraman [1], Nandhini Malaiyappan [2], Rajakumar Ramalingam [3], Shakila Basheer [4], Mamoon Rashid [5,6,*] and Nazir Ahmad [7]

1   Department of Computer Science, Lovely Professional University, Phagwara 140011, India
2   Department of Computer Science, Pondicherry University, Pondicherry 605014, India
3   Department of Computer Science and Technology, Madanapalle Institute of Technology & Science, Madanapalle 517325, India
4   Department of Information Systems, College of computer and Information Science, Princess Nourah Bint Abdulrahman University, P.O. Box 84428, Riyadh 11671, Saudi Arabia
5   Department of Computer Engineering, Faculty of Science and Technology, Vishwakarma University, Pune 411048, India
6   Research Center of Excellence for Health Informatics, Vishwakarma University, Pune 411048, India
7   Department of Information System, College of Applied Sciences, King Khalid University, P.O. Box 61913, Muhayel 63317, Saudi Arabia
*   Correspondence: mamoon.rashid@vupune.ac.in

Abstract: Resting-state functional connectivity has been widely used for the past few years to forecast Alzheimer's disease (AD). However, the conventional correlation calculation does not consider different frequency band features that may hold the brain atrophies' original functional connectivity relationships. Previous works focuses on low-order neurodynamics and precisely manipulates the mono-band frequency span of resting-state functional magnetic imaging (rs-fMRI). They specifically use the mono-band frequency span of rs-fMRI, leaving out the high-order neurodynamics. By creating a high-order neuro-dynamic functional network employing several levels of rs-fMRI time-series data, such as slow4, slow5, and full-band ranges of (0.027 to 0.08 Hz), (0.01 to 0.027 Hz), and (0.01 to 0.08 Hz), we suggest an automated AD diagnosis system to address these challenges. It combines multiple customized deep learning models to provide unbiased evaluation, and a tenfold cross-validation is observed We have determined that to differentiate AD disorders from NC, the entire band ranges and slow4 and slow5, referred to as higher and lower frequency band approaches, are applied. The first method uses the SVM and KNN to deal with AD diseases. The second method uses the customized Alexnet and Inception blocks with rs-fMRI datasets from the ADNI organizations. We also tested the other machine learning and deep learning approaches by modifying various parameters and attained good accuracy levels. Our proposed model achieves good performance using three bands without any external feature selection. The results show that our system performance of accuracy (96.61%)/AUC (0.9663) is achieved in differentiating the AD subjects from normal controls. Furthermore, the good accuracies in classifying multiple stages of AD show the potentiality of our method for the clinical value of AD prediction.

Keywords: rs-fMRI; classifications; high-order neuro-dynamic functional network; deep learning; Alzheimer's disease

Citation: Sethuraman, S.K.; Malaiyappan, N.; Ramalingam, R.; Basheer, S.; Rashid, M.; Ahmad, N. Predicting Alzheimer's Disease Using Deep Neuro-Functional Networks with Resting-State fMRI. *Electronics* **2023**, *12*, 1031. https://doi.org/10.3390/electronics12041031

Academic Editors: Alberto Fernandez Hilario and Gabriella Olmo

Received: 24 December 2022
Revised: 16 February 2023
Accepted: 17 February 2023
Published: 19 February 2023

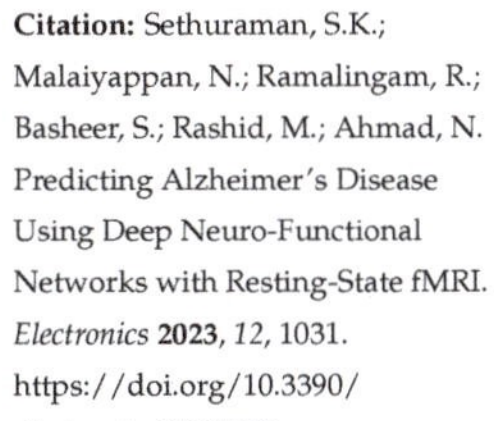

1. Introduction

AD is a chronic, developing, acute abnormality that affects people over 60 years of age [1,2]. Classified as the typical cause of dementia, it comprises memory loss, loss of spatial orientation, lack of time sense, behavioral issues and, at the acute stages, retrograde amnesia and mild cognitive impairment [3,4]. The disease is characterized by the unique "clumps" found in the brains of patients, termed medically as amyloid plaques and tangled fibers called neurofibrillary tangles. As the ailment progresses, the above-listed anomalies in the brain result in the degradation of the neural networks, causing the gradual loss of bodily functions. The progression from the loss of mental capabilities to physical degradation renders AD physically and mentally taxing to the families of the afflicted and the caregivers. Recent advancements in medical care have increased life expectancy, which in turn has increased the aged population [5]. Thus, the fraction of the human population susceptible to AD has also increased. Some researchers have begun using computer techniques such as neural networks, optimization, machine learning, and so on to solve the medical domain issues [6].

In existing ML techniques, a field expert manually extracts and labels features. Especially in the field of computer vision, deep learning (DL), an advanced machine learning (ML) technique, outperforms classical ML in terms of detecting inclusive structures in complex, high-dimensional data. The main benefit of DL algorithms is that they attempt to incrementally learn high-level properties from the brain imaging data, minimizing the need for domain expertise. DL outperforms ML since it can accurately handle enormous volumes of data, while ML algorithms require a specific processing step.

Objectives of the Study

Automated diagnosis systems have gained importance in the field of medical image analysis. The recurring patterns in images have the potential to determine the conformation, function, and activities of the brain. Unlike most popular AD discovery algorithms, the input dataset is extensive. Therefore, an efficient technique is essential.

The main objective of this research work is to propose an automated AD diagnosis system by developing a high-order neuro-dynamic functional network. LFOs (Low-Frequency Oscillations) also refer to slow brain activity fluctuations between 0.01 to 0.08 Hz. To understand brain atrophies, these slow fluctuations are analyzed using different levels, such as slow4, slow5, and full-band ranges. The use of LFOs in brain studies allows for examining slow changes in brain activity that may be relevant to various neurological conditions.

2. Related Works

Many researchers are interested in revisiting this area to identify a treatment for AD because of the relevance of early detection of the disease. Therefore, the most significant studies in this area will be presented in this section. The classified approach of the MCI (mild cognitive impairment) and AD patients using different network approaches with strengths and weaknesses are described. Ting Ma et al. [7] have extracted two important parameters constructed via the pre-processed data of rs-fMRI, such as ALFF and ReHo. In addition, their findings imply that during deterioration, ROIs in the brain may experience various physiological alterations. Evanthia E. Tripoliti et al. [8] created the five phases of the method by including preprocessing fMRI to remove non-task-related variability, modelling the BOLD material resulting in the stimulus, extracting from fMRI image data, features selection, and finally, the random forest algorithm. The methods assist in classifying the disease, with 80.5–87% accuracy. Dachena et al. [9] elaborated on MRI and fMRI shared with the misuse of MMSE to discriminate AD using SVM classification. Additionally, the multimodal approach (MRI, fMRI, MMSE) provides more accuracy of 95.65% and specificity of 97.22% with a sensitivity of 93.39%. Zhe wang Li et al. [10] classified AD, MCI, and NC and proposed a regularized LDA approach that reduces the noise effect by using two required shrinkage methods. Furthermore, they investigated the relationship between LDA and Maximum Likelihood-based classifications. These developed methods can be applied to a limited sample size.

Babajani-Feremi et al. [11] have developed an approach that can discriminate possible MCI-decliners using structural and functional MRI integration for AD identification. A multi-scale time series kernel-based learning model was used to diagnose brain diseases as the foundation for the traditional statistical analysis technique proposed by Fei Guo et al. [12]. They found that this method has advantages for accurately identifying brain diseases. Xia-an Bi et al. [13] classified AD, patients' abnormal brain regions, and HC by proposing a random neural network cluster based on fMRI data and found that a neural network cluster is a suitable approach for identifying AD. His group also integrates other imaging to know brain activity and combines brain and cerebellum activity. In addition, they differentiate AD and HC patients. In this study, the authors examined 138 participants using various accuracy criteria.

The model for early-stage detection from functional alterations in MRI images was created by Modupe Odusami et al. [14] using ResNet18. The accuracy of the ResNet18 is as follows: the attained results are 99.9% for EMCI (Early Mild Cognitive Impairment) against AD, 99.95% for LMCI (Late Mild Cognitive Impairment) against AD, and 99.95% for MCI against EMCI. Accuracy, sensitivity, and specificity were all improved using the created model. Then, a novel three-dimensional two stage-age-network (TSAN) was used to compute brain age using T1-weighted MRI data. The two-stage network design used by TSAN was demonstrated [15]. (i) The first stage network more accurately calculates the approximate brain age from the discretized brain age, and (ii) the first stage network measures brain age. Additionally, some researchers used machine learning methods to categorize AD. Feature extraction from ADNI's fMRI images was used in this work [16], and the performance analysis is based on the confusion matrix. Additionally, the author has developed various techniques for CNN architecture and machine learning classifiers (SVM, KNN, DT, RF, and LDA). The accuracy levels provided by the suggested model are 85.8%, 77.5%, 91.7%, 96.7%, and 79.5%. Finally, the accuracy levels provided by the CNN architecture are 98.1%, 95.2%, 87.5%, and 89.0%, respectively.

Quamzheng Li et al. [17] built the R-fMRI data to calculate the functional connectivity of different brain areas. Additionally, a standard control-targeted autoencoder network was constructed to distinguish between MCI and normal ageing. The technique offers accurate AD classifiers and discriminative brain network characteristics. Deep learning outperforms the more traditional R-fMRI method in categorizing high-dimensional multimedia data, as shown by the accuracy increases of 31.21%.

Unmang Gupta et al. [18] presented an architecture that operates by using the 2D-CNN model to encode each 2D slice of the MRI. It shows that when compared to the most cutting-edge methods, the permutation invariant layers train more quickly and produce better predictions. Additionally, they provide more accurate estimates of healthy participants' brain ages. Cross-validation of the sMRI-fMRI model by Vince D. Calhoun et al. [19] indicates that it performed better than a unimodal prediction analysis. Additionally, some research is based on data from the correlation coefficients between the R-fMRI signal and functional intellectual network creation. Compared to the former method, this method demonstrated an increased diagnostic accuracy of about 25%. The convolutional component of the Spatial-Temporal Net is employed to describe the spatial dependency between the time series segments of various brain areas and to predict the course of AD using rs-fMRI time series data. This method performed better than the most recent methodology in terms of categorization accuracy. Furthermore, it sheds light on the pathogenic chain that underlies AD [20].

Based on an examination of fMRI data, Yifei Zhang et al. [21] explains a unique technique for differentiating AD patients from normal (healthy) individuals: functional connectivity between the brain's activity voxels. The predicted AD patients are significantly influenced by the FC between activity voxels inside the prefrontal lobe and those between the prefrontal and parietal lobes, according to the suggested technique, which demonstrated higher classification accuracy. It also has a high prospective value.

Uttam Khatri et al. [22] examined the dynamic frequency functional networks at frequency response time series, including full-band, slow-4, and slow-5 bands, using the rs-fMRI data amassed by the ADNI. His team also combined four frequency bands with dynamic frequency brain functional network elements to aid in the early identification of AD. In addition, it also offers a fresh perspective on how the brain network functions and offers early Alzheimer's detection. The author also attained a 94.10% classification accuracy level, 96.75% specificity, and 90.95% sensitivity. The High-Order Dynamic Functional Connectivity model's experimental results can improve the classification performance with different levels of evaluation matrices to identify the AD.

The author in [23] implemented two significant approaches—first, normal CNN methods with 2D and 3D structural brain images. Second, transfer learning methods were applied achieved 97%. Deep learning methods reached 95.17% and 93.61% accuracy for 3D and 2D multiclass AD and MCI classifications. The authors in [24] incorporated unsupervised convolutional spiking neural networks trained with the preprocessed ADNI datasets. They achieved three binary categories without the spike of 86.90%, 83.25%, and 76.70%.

The Motivation for Study

- Studies with the literature reviews have revealed conventional/existing approaches, and correlation calculation has not considered different frequency band features, which eventually maintains the accurate functional connectivity relationships of brain atrophies.
- Moreover, existing works focus on low-order neurodynamics, manipulating the mono-band frequency span of rs-fMRI by leaving out high-order neurodynamics.
- Researchers also claim great potential lies in the deep learning and rs-fMRI-enabled classification models.

However, all the existing methods have their bottlenecks and limitations. Therefore, the model aims to establish a novel DL method that can push the classification accuracy boundaries towards the most accurate AD and MCI classification approaches. This research model leads to finding out the limitations of an early diagnosis of AD.

So, with the advancements in rs-fMRI and the deep learning approach, unique ways have been developed to introduce a diagnosis system for high-order neuro-dynamic functional networks using various levels, which motivated us to create such a classification model. Results of the study claimed optimal performance with the D2 model using three bands (slow4, slow5, and full-band) without any external feature selection compared to other models.

3. Methods and Materials

From the literature, it was deduced that low-order neurodynamics precisely manipulate the mono-band frequency span of resting-state functional magnetic imaging (rs-fMRI), leaving out the high-order neurodynamics. These were then hypothesized to be outperformed by DL techniques. Experimentally, we also propose an automated AD diagnosis system by developing a high-order neuro-dynamic functional network using various levels such as slow4, slow5, and full-band ranges (0.027 to 0.08 Hz), (0.01 to 0.027 Hz), and (0.01 to 0.08 Hz) of rs-fMRI time-series data.

3.1. ADNI Dataset

ADNI (https://adni.loni.usc.edu/ accessed on 1 September 2022) comprises multimode neuroimages of people who have Alzheimer's disease and is developed by the National Institute of Aging (NIA). The ADNI database consists of three classes of biological markers: AD, MCI, and NC. A total of 153 baseline subjects were selected. Table 1 shows the demographic details of the selected rs-fMRI subjects.

The MRI protocol for ADNI1 (2004–2009) focused on consistent longitudinal structural imaging with 1.5T scanners using T1- and dual-echo T2-weighted sequences. One-fourth of ADNI1 subjects were scanned using the same protocol on 3T scanners. ADNI-GO/ADNI2 (2010–2016) imaging was performed at 3T with T1-weighted imaging parameters similar to ADNI1. In place of the dual-echo T2-weighted image from ADNI1, 2D FLAIR and T2-weighted imaging were added at all sites. Both fully sampled and accelerated T1-weighted images were acquired in each imaging session.

Table 1. The demographic of the rs-fMRI subjects for our proposed model.

Number of Samples	AD (n = 51)		MCI (n = 51)		NC (n = 51)	
	Average	Standard Deviation (SD)	Average	Standard Deviation (SD)	Average	Standard Deviation (SD)
Age	75.2	7.4	75.3	7.0	75.3	5.2
Education	14.7	3.6	15.9	2.9	15.8	3.2
MMSE	23.8	2.0	27.1	1.7	29.0	1.2
CDR	0.7	0.3	0.5	0.0	0.0	0.0

3.2. Data Pre-Processing

SPM 12 was used to pre-process the ADNI dataset and segment it into grey, white, and cerebrospinal fluid planes. The initial ten volumes were removed to permit dynamic equilibrium in each subject. All the slices were resampled with the slice-time correction to provide uniformity in time variation. Here, the middle slice was taken as a reference. It is followed by the realignment technique based on the reference slice. The individual averaged functional slices were co-registered using the landmark-based registration technique to their corresponding MRI. Later, the segmentation process was performed to extract the brain parts such as White Matter (WM), Gray Matter (GM), and cerebrospinal fluid (CSF). Every fMRI slice was resized/normalized to MNI (Montreal Neurological Institutes) space, and resampling was performed with a $3 \times 3 \times 3$ mm^3 setting.

A Gaussian kernel was used for smoothing. Last, low frequencies are categorized based on their ranges—slow4 (0.027 to 0.08 Hz), slow5(0.01 to 0.027 Hz), and full-band (0.01 to 0.08 Hz). Zhang et al. [25] proposed the new model," hybrid high order fully connected networks", to describe the previously unenclosed intermediary relationship between down and up-order brain networks, getting the highest accuracy. Even the existing model was not able to address the dynamic brain changes. This work proposes a novel method using an automated AD diagnosis system by developing a high-order neuro-dynamic functional network using various frequency levels (slow4, slow5, and full-band) of rs-fMRI time-series data. Another common transformation is the imaging time series, which converts time series into images. One significant benefit of this transformation is the ability to retrieve data for any two time points given a time series. These imaging time series have been classified using deep neural networks [26], particularly convolutional neural networks.

Higher-order functional brain connections across several frequency bands are used in customized deep-learning models to distinguish AD and MCI from normal healthy levels. Thus, the combination of higher-order dynamic and frequency division-based brain networks opens a new window into diagnosing AD. We have used an "ensemble process" to increase the current model's performance by integrating many models into a single robust model. Figure 1 illustrates the complete workflow of the proposed model using deep neural networks.

The SPM12 software [27] and the toolboxes DPARSF (Data Processing Assistant for rsMRI) [28] and REST (Resting-state fMRI Data Analysis Toolkit) [29] were used to process the input scans. The initial ten volumes were removed to permit the dynamic equilibrium in each subject. Normalization for images has been performed, by which they were normalized from 0 to 1. We used the Inception V2 architecture [30] to identify abnormalities in the brain and detect them, leading to better results with less computational effort. The primary aim of this network was to select a particular layer at each level. This Inception V2 network uses a single filter size on the input brain MRI image (1×1) for which max pooling action is involved as a result of this inclusion.

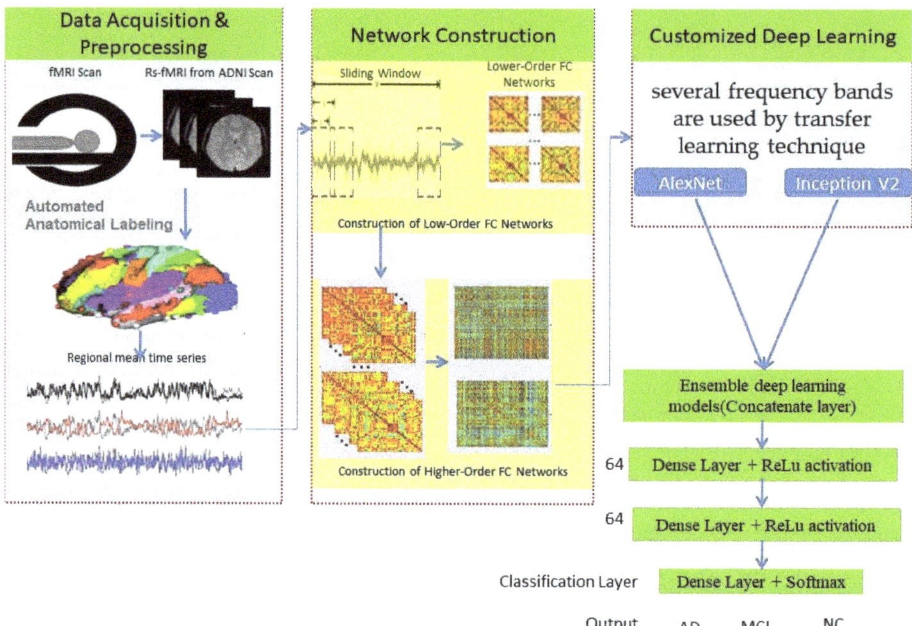

Figure 1. Proposed model using deep networks.

This neuro-dynamic functional network provided better accuracy, sensitivity, and specificity, apart from being speedy and efficient for detecting AD using rs-fMRI images. The ADNI image dataset's performance has been evaluated using various metrics, such as recall, specificity, and overall accuracy.

1. Disease Identification method: This is used to classify AD images collected by a medical expert during screening/monitoring programs.

2. Computer-Assisted Diagnosis: These methods are used to find the chances of disease based on rs-fMRI image changes.

3. Biomarkers: These are used for evaluating AD disease according to its severity.

This paper aimed to instigate an ML and CNN method for classifying AD from normal controls.

3.3. Methods

3.3.1. Customized AlexNet

The final three layers are replaced to solve the issue and achieve maximum accuracy. The last three layers of AlexNet—FC, SoftMax, and classification layer—replace the pre-trained network. These layers with the altered hyperparameters were eventually included by fine-tuning the previous layers and training the new layer of the AlexNet model using the ADNI dataset. The pre-trained model improved classification using the extensive ImageNet database using the feature extraction method. Minor tweaks are needed for the pre-trained parameters to adjust to the new MRI brain images. The modified hyperparameters define a small portion of the freshly transferred network.

Transfer learning is an essential statistical model for developing an efficient DL strategy. The critical regions of the brain can be recognized from MRI images by using newly updated parameters in a pre-trained network. These models have good convergence and are primarily used to extract the features and their classification. For this parameter learning, stochastic gradient descent with momentum optimizer is used.

As an extension to CNNs, the customized AlexNet architecture was developed to be competitive at the object detection task. Our proposed model achieves good performance using three bands without any external feature selection, reducing the task of exhaustive search using a set of heuristic approaches.

3.3.2. Customized Inception V2

In this section, the Inception V2 CNN architecture is presented. For AD detection, TI-weighted MR images and non-invasive methods are used.

The Inception V2 architecture being shown may identify abnormalities in the brain and detect them, leading to better results with less computational effort. The main goal of this network is to select a particular layer at each level. This Inception V2 network utilizes a single filter size on the input brain MRI image (a1 $\times$ a1). A max-pooling action is included as a result of the inclusion. The inception V2's four pipes operate simultaneously. The architecture employs (1 $\times$ 1 $\times$ 1) filters in the first block to decrease the network by reducing the dimensions. The network begins with three convolutions dimension (a1 $\times$ a1 $\times$ a1, b3 $\times$ b3 $\times$ b3) matrix, which is comparable to the traditional network approaches (c5 $\times$ c5 $\times$ c5), (a1 $\times$ b3), and (c3 $\times$ 1). The filter size in the conventional network modal is (b3 $\times$ b3), which is divided into (a1 $\times$ b3) and (b3 $\times$ a1) convolutions. As an example, convolutions in the form of (b3 $\times$ b3) or (c5 $\times$ c5) are comparable to convolutions in the form of (a1 $\times$ b3) or (a1 $\times$ c5) and need less computation than (b3 $\times$ b3) dimension convolutions. In addition, the network comprises only two convolutions (a1 $\times$ a1 $\times$ a1); in the third part, it has only the pooling layer, and the fourth has only (a1 $\times$ a1 $\times$ a1) filters of convolution.

Similarly, all bands—aside from the max-pooling operation—begin with (a1 $\times$ a1 $\times$ a1) convolution filters. All conventional network layers are subjected to batch normalization to expedite training and reduce the risk of overfitting. After batch normalization is implemented, the convolution network improves and paves the way for data regularization in each network's hidden layers; finally, a leaky ReLU function, the activation function, is implemented. The slope of this function is marginally negative at (0.01). The function's slope is somewhat negative (0.01, or so on). The process is as follows in Equation (1):

$$f(x) = 1(x < 0)(x\alpha) + 1(x \geq 0)(x) \tag{1}$$

where α denoted as a negligible constant. The suggested Inception V2 has n properties, as shown in Figure 2, based on the input. In Figure 2, $*$ symbol indicates the multiplication operation. The primary advantage of the proposed model is the drastically reduced number of network parameters. The network's primary goal is to transfer specific data from the origin to the target feature space. The primary notion of this network has changed the feature space of the spatial data from source to destination. The CONV 3D (s.m) represents the three-dimensional convolution with size (S) and filters (m), whereas the max pool three-dimensional (p.q) represents the three-dimensional max pool layer for down-sampling with the stride IQ and size of pool P. The convolution's filter size nxn has been divided into ixn and nxi convolutions. It is demonstrated that (a1 $\times$ b3) or (a1 $\times$ c5) convolutions, which perform (b3 $\times$ a1) or (c5 $\times$ a1) convolutions and are an output of the last layer, are comparable to (b3 $\times$ b3) or (c5 $\times$ c5) dimension convolutions. Finally, (b3 $\times$ b3), which is less expensive than other convolutions, is the concentration of two convolutions.

The customized model still has its limitations. Pre-trained models such as VGG/inception often produce valuable features. The big difference is the formation of the problem, especially the VGG/inception, which was designed for multiclass classification, which means learning a lot of irrelevant information. All the issues can be solved by fine-tuning a pre-trained VGG argument with a few layers augmented with a few layers for binary classification, thus changing the intra-network AD vs. MCI, MCI vs. NL, and NL vs. AD. The weight stored internally can also be too much, requiring additional regularization. During the design of the network, we incorporated an Adam optimizer to require fewer parameters for tuning and implement a faster computation time.

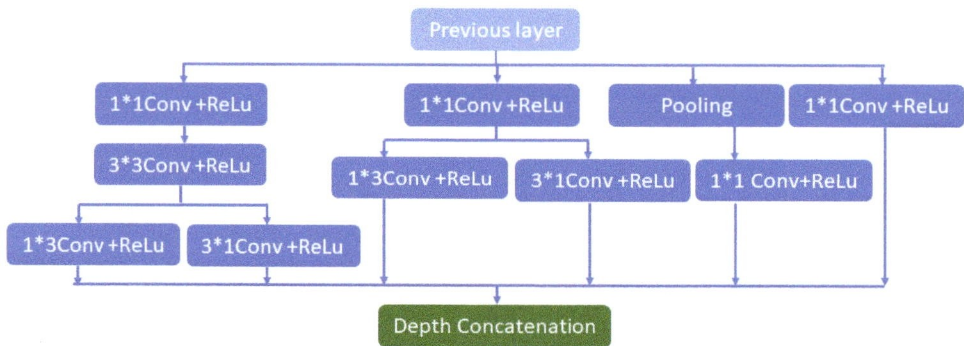

Figure 2. Inception V2 network architecture.

In the traditional training method, the learning rate always remains the same. Recently, it has been suggested that the learning rate should be gradually changed, but this method has not been used in migration learning. Frequent changes in the learning rate not only accelerate the network convergence but also solve the problem that the loss value oscillates and is challenging to converge, with the learning weight also being gradually reduced. In top-level network training, better weight parameters are learned. A set of experiments are carried out to ensure the enhanced convergence rate of the model with improved recognition accuracy. The hyperparameter value is shown in Table 2.

Table 2. Training Parameters for Customized AlexNet and Inception V2.

Training Parameter	Value
Batch size	4
Epochs	50
L1 (learning rate)	1×10^{-5}
Op (optimizer)	Adam optimizer
β_1	0.9
β_2	0.999
ε (epsilon)	1×10^{-7}

3.3.3. Ensemble Deep Learning Model (D2)

The customized models of Alexnet and Inception V2 described in the previous sections are separately tested on the dataset. The ensemble output is then created by adding the probabilities of each participant's output.

The ensemble process merges various learning algorithms to gain their collective performance or to enhance the performance of current models by mixing different models to produce one trustworthy model. An ensemble framework works best when the participating systems are statistically varied because ensemble learning attempts to assemble complementing information from its numerous contributing models. Information fusion for improving classification performance is the primary justification for employing an ensemble learning model. To acquire a more reliable result, models trained using various data distributions related to the same set of classes are used while making predictions. The primary sources of error in learning models are noise, variation, and bias. Deep learning (DL) algorithms are accurate and stable due to the ensemble methods' capacity to reduce these error-causing elements. SVM and KNN are two different learning methods. SVM makes the quite restricted assumption that a hyperplane separates the data points. In contrast, KNN attempts to approximate the underlying distribution of the data in a non-parametric way.

4. Experimentation Setup and Results Analysis

A typical Windows 10 system with 8 GB ram is used to develop an automated tool in MATLAB for the comparison of the results. The pre-processing steps of ADNI subjects are given in Section 2. The same input images are provided to all the transformations to get an unbiased estimation of their performance. One hundred fifty-three brain subjects are chosen and utilized for this purpose. Tenfold cross-validation is used for calculating the accuracies. The number of cross-validation sets is created from the entire dataset, and the result averaged precision from all those sets. The input array used for the deep learning models varies for different low frequencies. The slow4 features are resized to 50×50, slow5 features are resized to 60×60, and full-band features are resized to 110×110. This size is fixed based on the number of corresponding features. Figure 3 illustrates the entire execution process model.

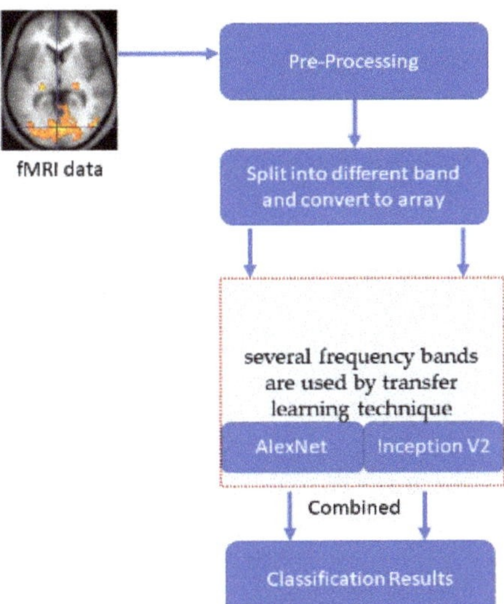

Figure 3. The overall process of the execution model.

4.1. Performance Analysis

Performance analysis is necessary to evaluate the performance of any classification system. The tests are conducted in the following ways: three external bands, such as slow4, slow5, and full-band, are used to extract the dataset's features.

Specificity: It correctly identifies negatively labelled classes using Equation (2).

$$S = \frac{True\ Negative}{True\ Negative + False\ Positive} \tag{2}$$

Recall/Sensitivity: It identifies correctly positive labelled classes by using Equation (3)

$$R = \frac{True\ Positive}{True\ positive + False\ Negative} \tag{3}$$

Accuracy: It is an overall accuracy of true positive and negative out of the total number of observations that are examined by using Equation (4),

$$Accuracy = \frac{TP + TN}{TP + FP + FN + TN} \tag{4}$$

4.2. Result Analysis

To provide an unbiased evaluation, repeated tenfold cross-validation is observed, and the mean results are presented in this section. This study demonstrates the customized AlexNet, InceptionNet, and D2 stacked models. The images were normalized from 0 to 1. The Training Parameters for AlexNet and InceptionNet models are given in Table 2.

Table 3 lists the performance of customized AlexNet, InceptionNet, and D2 models using slow4, slow5, and full-band features. Their corresponding area under the curve (AUC) is plotted in Figure 4. With AlexNet, slow5 parts give the highest performance. For the AD again NL dataset, the majority baseline classification performance is 93.34%, and the AUC is 0.9488. For the MCI vs. NL dataset, the majority baseline classification accuracy is 76.56%, and for the AUC is 0.76.03. For the AD vs. MCI dataset, the majority baseline classification accuracy is 76.34%, and the AUC is 0.7780.

Table 3. Summarizes the performance of customized AlexNet, InceptionNet, and D2 models using slow4, slow5, and full-band features.

Group	Frequency Band	Classifers	Acc (%)	SPE (%)	SEN (%)	AUC
	Full-band		58.67	65.37	68.45	0.6221
AD/MCI	Slow4	AlexNet	75.47	74.56	70.46	0.7603
	Slow5		76.34	76.88	77.45	0.778
	Full-band		63.23	60.23	70.34	0.6444
AD/NL	Slow4	AlexNet	91.34	90.45	92.42	0.9163
	Slow5		93.43	92.34	92.45	0.9488
	Full-band		74.29	72.34	70.67	0.7573
MCI/NL	Slow4	AlexNet	75.23	77.45	77.95	0.7675
	Slow5		76.56	77.56	78.34	0.7754
	Full-band		78.89	77.37	75.56	0.7967
AD/MCI	Slow4	InceptionNet	79.79	74.37	75.56	0.8021
	Slow5		80.45	82.45	81.45	0.8196
	Full-band		65.9	62.38	72.56	0.6878
AD/NL	Slow4	InceptionNet	92.38	92.36	91.46	0.9390
	Slow5		94.47	93.87	93.56	0.9518
	Full-band		62.67	64.38	62.47	0.6320
MCI/NL	Slow4	InceptionNet	75.59	78.59	72.57	0.7668
	Slow5		79.67	78.56	76.34	0.7989
	Full-band		76.67	77.37	75.88	0.7782
AD/MCI	Slow4	D2	78.58	75.31	75.96	0.7959
	Slow5		82.67	81.15	80.16	0.8445
	Full-band		64.67	61.2	73.67	0.6540
AD/NL	Slow4	D2	94.56	92.12	94.26	0.9546
	Slow5		96.61	94.34	94.96	0.9663
	Full-band		66.99	67.65	62.4	0.6758
MCI/NL	Slow4	D2	74.19	77.12	71.25	0.7514
	Slow5		81.87	79.86	75.47	0.8221

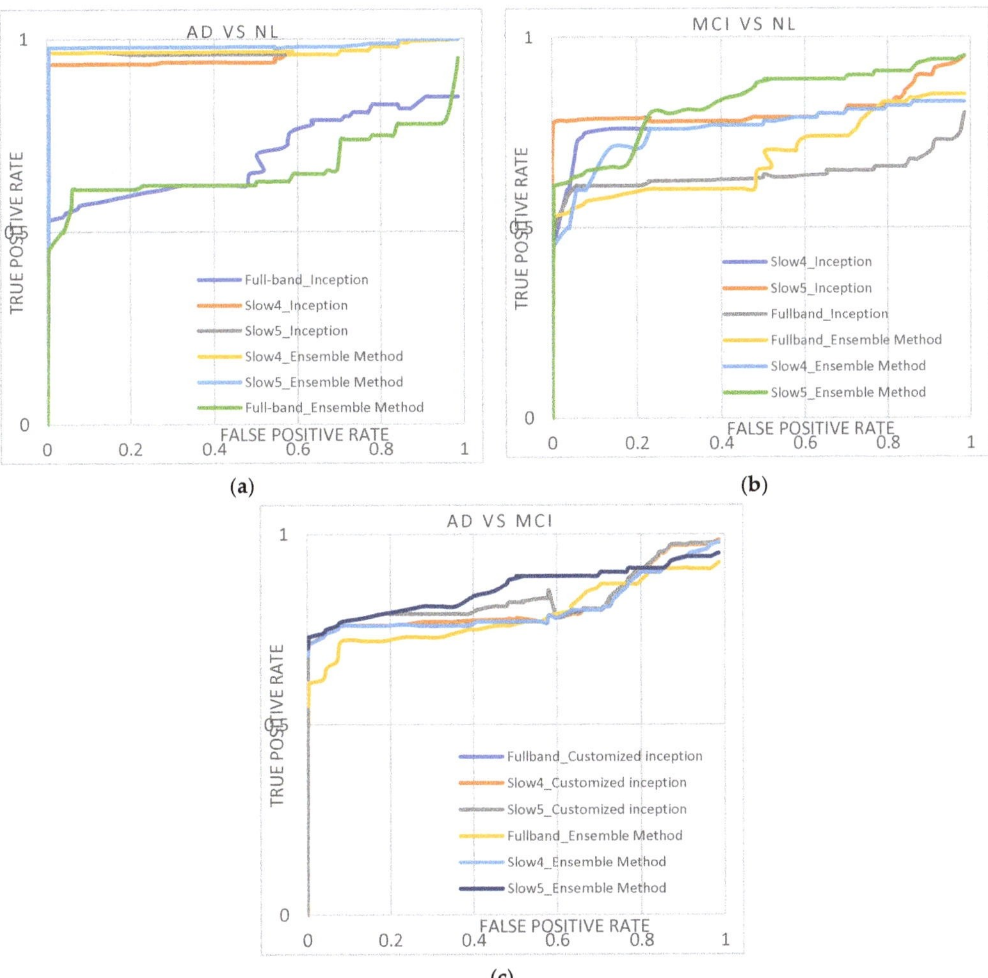

Figure 4. (**a**–**c**) shows the ROC curve obtained from the AD/NL, MCI/NL, and AD/MCI.

InceptionNet trained with slow5 frequencies reaches the highest performance. For the AD vs. NL dataset, the baseline classification accuracy/AUC is 94.47%/0.9518, and for the MCI vs. NL dataset, the majority baseline classification accuracy/AUC is 79.67.47%/0.7989. For the AD vs. MCI dataset, the majority baseline classification accuracy/AUC is 80.45%/0.8145.

The D2 model with slow5 frequencies achieves the highest performance compared to all the other features. For the AD vs. NL dataset, the baseline classification accuracy/AUC is 96.61%/0.9663; for the MCI vs. NL dataset, the majority baseline classification accuracy/AUC is 81.87%/0.8221, and for the AD vs. MCI dataset, the majority baseline classification accuracy/AUC is 82.67%/0.8445. Among the three features, the slow4 and full-band perform lower than the slow5 frequencies for all the classification tasks. Additionally, it is noted that the D2 model outperforms better than other conventional models. It shows the diversity of the results produced using the AlexNet and InceptionNet models.

Next, the results are compared with conventional ML algorithms and are furnished in Table 4. The performance of our ensemble model outperforms the machine learning algorithms by 5–9%. It is also highlighted that other research also indicates that slow5 features perform better than using slow4 or full-band frequencies. The hyperparameter value is shown in Table 5.

Table 4. Compares the performance of KNN and SVM model using slow4, slow5, and full-band features with traditional machine learning models.

Group	Frequency Band	Classifers	Acc (%)	SPE (%)	SEN (%)	AUC
AD/MCI	Full-band	KNN	58.1	60.45	56.89	0.5782
	Slow4		74.61	71.25	68.59	0.7539
	Slow5		73.01	77.46	78.11	0.7503
AD/NL	Full-band	KNN	61.49	55.93	67.15	0.6215
	Slow4		89.56	89.87	92.42	0.8986
	Slow5		91.06	91.46	90.57	0.9166
MCI/NL	Full-band	KNN	65.16	69.45	68.15	0.6780
	Slow4		71.13	74.15	69.45	0.7225
	Slow5		71.06	71.89	74.98	0.7279
AD/MCI	Full-band	SVM	56.9	60.24	69.67	0.6215
	Slow4		73.67	72.04	67.57	0.7492
	Slow5		74.56	77.57	79.59	0.7501
AD/NL	Full-band	SVM	62.79	57.13	68.45	0.6312
	Slow4		90.06	89.87	92.42	0.9199
	Slow5		91.76	93.46	91.97	0.9226
MCI/NL	Full-band	SVM	68.46	70.45	69.45	0.6951
	Slow4		72.33	76.45	71.67	0.7386
	Slow5		73.76	71.23	74.34	0.7418

Table 5. Training parameters for SVM and KNN.

Classifiers	Hyper Parameter	Optimized Value
SVM	Kernel type	RBF
	Cost	0.1
	Gamma	0.001
	Kernel degree	2
	Coefficient	9
KNN	K-neighbors	500
	Weighting	Similarity

5. Discussion

In this study, dynamic neuro-functional deep ensemble networks are proposed and implemented to predict several AD periods, including the MCI (prodromal stage), using different frequency signals of rs-fMRI. In the three features (slow4, slow5, and full-band frequencies), slow5 achieves the top performance using the D2 model, with an accuracy/AUC of 96.61% for differentiating AD from NL subjects. In the D2 model trained for the MCI vs. NL task, the accuracy/AUC is 81.87%/0.8221, and for the AD vs. MCI task, the accuracy/AUC is 82.67%/0.8445 with the D2 model. The results in Figure 5 and Table 6 illustrate that the full-band and slow4 frequencies showed no substantial increase in accuracy. However, the slow5 features yield better performance when compared to the other two components. According to [20], slow5 features capture more discriminatory atrophies in different AD classifications. Our work is wholly automated. There is no need for feature selection.

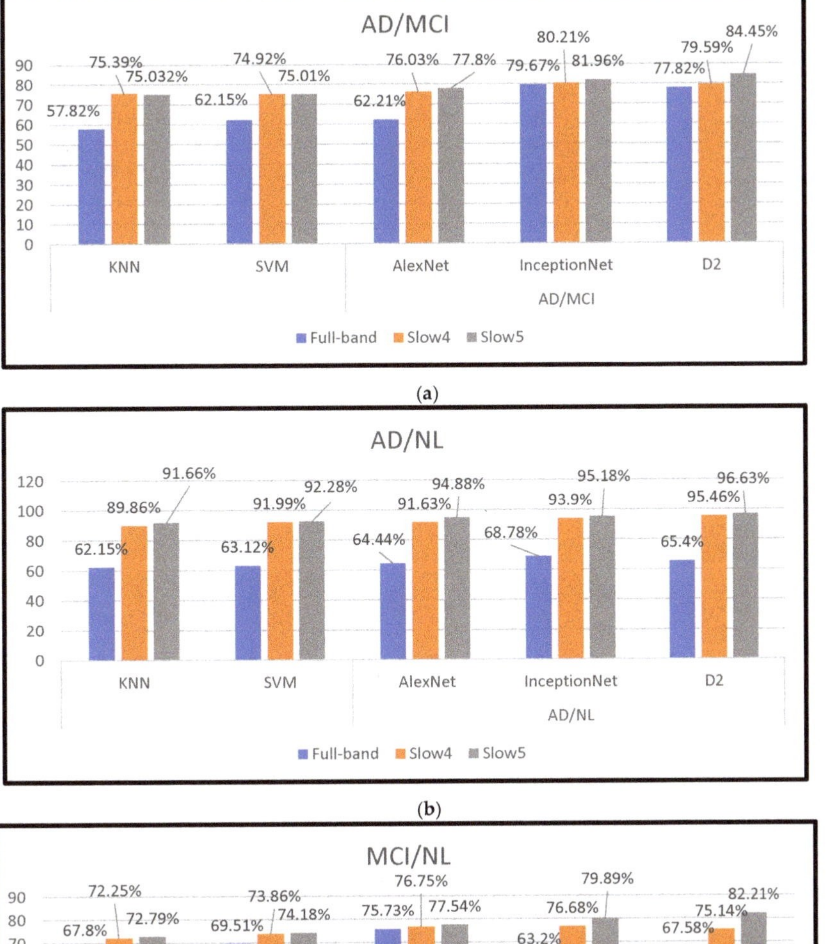

(a)

(b)

(c)

Figure 5. (**a–c**) shows the different frequency bands obtained from the AD vs. MCI, AD vs. NL, and MCI vs. NL.

Furthermore, other works investigated fMRI neuroimaging modalities to identify regions with different levels of atrophy for the various AD classifications. It is noted that the additional training and testing of data used in other research make it difficult to compare our proposed method directly. The current methods use different features/feature selection in exploring various binary classifications in AD/MCI; the performance of multiple spans is illustrated in Table 6. The results from Table 6 indicate that dynamic neuro-functional deep learning ensemble networks with slow5 frequencies achieve better classification performance than other machine learning methods, including those performed in [31,32]. Gaussian/Regression models are used in most of the previous brain network models. Additionally, KNN and SVM classifiers use Fisher score feature selection to pick relevant features [15,33], and SVM with two kernels (radio basis functions and polynomial) are used along with the Fisher score features reaching 90% accuracy, whereas the linear kernel yield 100% performance. For the same Fisher characteristics, the KNN provides an accuracy of 87.5%. KNN and SVM classification algorithm models are used to analyze and classify AD diseases.

Table 6. Compares the performance of fMRI-based recent methods with our proposed D2 model.

Authors	Subjects	Task	ACC	SENS	SPE
Challis et al. [31]	20 NL/50 MCI/27 AD	AD again MCI	80.12	70.97	90.24
		MCI again NC	75.15	100	50.52
Frankde Vos et al. [32]	173 NL/27 AD	AD again NL	76.54	71.75	82.47
Khatri et al. [26]	35 NL/61 MCI/35 AD	AD again NL	94.10	90.95	96.75
		AD again MCI	87.14	91.05	86.91
		NC again MCI	85.85	93.89	90.01
Ramzan et al. [34]	25 NL/25 AD	AD again NL	97.88	-	-
Duc et al. [35]	198NL/133 AD	AD again NL	85.27		
Parmar et al. [36]	-	AD again NL	96.55		
Al-Khuzaie et al. [37]	-	AD again NL	99.30	-	-
Bhaskaran et al. [38]	-	AD again NL	97.54	-	-
Luo et al. [39]	33 NL/27 MCI/24 AD	AD again NL	95.67	-	-
Emily et al. [40]		AD again NL	79.97		
		AD again MCI	73.94		
		NC again MCI	70.42		
Han et al. [41]		AD again NL	94.99		
		AD again MCI	83.88		
		NC again MCI	79.52		
Huang et al. [42]	174 NL/99 MCI/116 AD	AD again NL	95.12		
		AD again MCI	82.32		
		NC again MCI	78.88		
Li et al.-1 [43]		AD again NL	96.47		
		AD again MCI	88.47		
		NC again MCI	81.17		
Li et al.-2 [43]		AD again NL	97.37		
		AD again MCI	92.11	-	-
		NC again MCI	88.12		
Proposed D2 model	51 NL/51 MCI/51 AD	AD again NL	96.61	94.34	94.96
		AD again MCI	82.67	81.15	80.16
		NC again MCI	81.87	79.86	75.47

ACC: accuracy; SENS: sensitivity or recall; SPE: specificity.

In machine learning, the supervised model k-Nearest Neighbors (KNN) is used. Supervised learning is the process through which a model learns from data that has been labelled. A set of input items and output values are fed into a supervised learning model. The method is then trained using the data to figure out how to translate the inputs into the required outputs, enabling it to forecast data that has not yet been observed. We must configure several basic settings for KNN.

5.1. SVM Parameter Setup

The parameter setup for SVM with a fixed number of neighbors is five. Gamma and c are SVM parameters for the Radial Basis Function (RBF) kernel. With low values signifying "far" and big values suggesting "near," the γ parameter indicates the range of a particular training example's influence. The model's support vector samples' radius of influence can be compared to the inverse of the γ parameters. The C parameter compromises good training sample classification for an increase in the margin of the decision function. Greater values of C can tolerate a smaller margin if the decision function is more accurate at correctly detecting all training points. A lower C reduces the training accuracy at the expense of a more significant margin and, hence, a more straightforward decision function. Our suggested dynamic neurofunctional deep ensemble networks demonstrate equal good performance in the prediction of numerous stages of AD when compared to other cutting-edge techniques.

The Brain-Connectivity networks, VGG19 [31], ResNet50 [33], Densenet121 [34], C3d [35], and C3d-LSTM [35], show good performance for the classifications. However, it is noted that the training set consists of both baseline and longitudinal images, and optimization parameters such as L1 and L2 regularization are needed to set up these networks. When compared with these models, it can be said that our work shows good compatible accuracy for all the classifications. This work reveals that there exist unique discriminatory frequency values in different bands, which is a major factor in determining the performance of our proposed research work. We have discussed the accuracies of different conventional ML and recent DL models on fMRI-based datasets in Table 6. In addition, our technique has an inherent feature selection, a deciding factor for improved accuracy. It is highlighted that our method requires less parameter optimization and is fully automated. This makes our method distinct from all the other methods.

From the last ten years, the different network approaches were used. In that, most authors use fewer datasets with low-frequency time series data and achieve more than 90% accuracy with varying band levels.

Our ensemble model also required fewer epochs during training; however, because of the large number of kernels at the first and second layers, the number of parameters is very high, increasing the model's time complexity. Comparing this model to other feature extraction and classification models, the rs-fMRI slice technique effectively reduces the complexity of pre-processing. The drawbacks found in the low-order neurodynamics precisely manipulate the mono-band frequency span of rs-fMRI, leaving out the high-order neurodynamics. We propose an automated AD system to overcome these issues by developing a high-order neuro-dynamic functional network using various bands. The confusion matrix of AD is obtained as a $2 * 2$ matrix from the experiments performed with the segmentation and classification.

Table 7 presents the confusion matrix of our proposed model. From this, out of twenty subjects, the classifier predicted fourteen subjects as one correctly, and six subjects were misclassified as zero.

Table 7. Confusion matrix of AD/NL classification.

Actual/Predicted (n = 20)	Predicted as 1	Predicted as 0
Actual 1 (AD)	8	2
Actual 0 (NL)	6	4

Table 8 presents the confusion matrix of our proposed model. From this, out of twenty subjects, the classifier predicted fourteen subjects as one correctly, and six subjects were misclassified as zero.

Table 8. Confusion matrix of MCI/NL classification with the hippocampal region.

Actual/Predicted (n = 20)	Predicted as 1	Predicted as 0
Actual 1 (MCI)	9	1
Actual 0 (NL)	5	5

Table 9 presents the confusion matrix of our proposed model. From this, out of twenty subjects, the classifier predicted fifteen subjects as one correctly, and five subjects were misclassified as zero. The classifier's efficiency is evaluated from the $2 * 2$ matrix for each region through TP, FP, TN, and FN. The results obtained give desired performances with the testing subjects. The observations show no significant differences over "AD vs. NL, MCI vs. NL and AD vs. MCI". Hence it portrays that the proposed model outperforms the multiclass classification problems.

Table 9. Confusion matrix of AD/MCI classification with the hippocampal region.

Actual/Predicted (n = 20)	Predicted as 1	Predicted as 0
Actual 1 (AD)	7	3
Actual 0 (MCI)	8	2

5.2. Limitations of the Work

The classification of medical images is a fundamental and significant issue in computer innovation, which has undergone much research over the past few decades. Even though the reliability of various medical image classification methods has significantly increased, these methods may not offer correct AD because of their non-universality, vulnerability to illumination and spoofing effects, and insufficient accuracy via the poor data quality. Therefore, in many real-world applications, standard medical picture categorization may not be able to deliver the needed performance. In this study, we solely used the ADNI dataset to categorize the three frequency ranges of the various phases of AD. The dataset we used here is small for the entire experimentation. Additionally, this work solely uses traditional methods for AD classification, such as SVM and KNN, instead of alternative techniques.

6. Conclusions

In this work, the dynamic neuro-functional deep ensemble networks use various frequencies in resting-state fMRI to diagnose different stages of AD from real-time ADNI datasets. The excellent performance is achieved with our proposed D2 model using three bands (slow4, slow5, and full-band) without any external feature selection, and it is a combination of two deep learning models. Among the three bands evaluated, the results show that the slow5 features, when trained with various customized Alex and Inception networks, perform better for AD/MCI classifications. It is also mentioned that additional studies are required to develop these networks to increase the precision of AD classifications. We have contrasted our networks against established machine learning techniques and more contemporary deep learning techniques. It demonstrates that rs-fMRI multi-band characteristics have a higher potential for being AD biomarkers than single-band features. It is also noted that more research is needed to optimize these networks to improve the accuracy of AD classifications. We have compared our networks with traditional machine learning methods and current deep learning methods. Our study shows that the multi-band features of rs-fMRI have more potential to be AD biomarkers than single-band features. Additionally, the performance of the proposed ensemble model outperforms the conventional ML algorithms by 5–9%. The proposed model is less complex to train and requires fewer hardware resources. Furthermore, the proposed model had surpassed in terms of accuracy the various existing models. In the future, we aim to test and apply this model on a more extensive and richer dataset. Moreover, we hope to implement single-cell transcriptome data using variational neighborhood preserving quantum embeddings and deep learning. In the future, the use of image augmentation for AD classification may be added with different image augmentation methods such as flipping, padding, etc.

Author Contributions: Conceptualization, S.K.S.; methodology, S.K.S. and N.M.; validation, M.R. and R.R.; formal analysis, N.M., S.B. and N.A.; writing—original draft preparation, S.K.S.; writing—review and editing, R.R., M.R. and S.B., supervision, N.M. Funding Acquisition, S.B. All authors have read and agreed to the published version of the manuscript.

Funding: This research is supported by Princess Nourah bint Abdulrahman University Researchers Supporting Project number (PNURSP2023R195) Princess Nourah bint Abdulrahman University, Riyadh, Saudi Arabia.

Data Availability Statement: Data in this research paper will be shared upon request with the corresponding author. Data used in the preparation of this article were obtained from the Alzheimer's Disease Neuroimaging Initiative (ADNI)database (https://ida.loni.usc.edu/home/projectPage.jsp?project=ADNI, accessed on 1 September 2022). As such, the investigators within the ADNI contributed to the design and implementation of ADNI and/or provided data but did not participate in analysis or writing of this report. A complete listing of ADNI investigators is available at: https://adni.loni.usc.edu/wp-content/uploads/how_to_apply/ADNI_Acknowledgement_List.pdf (accessed on 1 September 2022).

Acknowledgments: This research is supported by Princess Nourah bint Abdulrahman University Researchers Supporting Project number (PNURSP2023R195) Princess Nourah bint Abdulrahman University, Riyadh, Saudi Arabia.

Conflicts of Interest: The authors declare no conflict of interest.

References

1. Burns, A.; Iliffe, S. Alzheimer's disease. *BMJ* **2009**, *338*, b158. [CrossRef] [PubMed]
2. McKhann, G.; Drachman, D.; Folstein, M.; Katzman, R.; Price, D.; Stadlan, E.M. Clinical diagnosis of Alzheimer's disease: Report of the NINCDS-ADRDA Work Group under the auspices of Department of Health and Human Services Task Force on Alzheimer's Disease. *Neurology* **1984**, *34*, 939–944. [CrossRef] [PubMed]
3. Querfurth, H.W.; Laferla, F.M. Alzheimer's Disease. *J. Med.* **2011**, *362*, 329–344. [CrossRef]
4. Li, X.; Li, Y.; Li, X. Predicting Clinical Outcomes of Alzheimer's Disease from Complex Brain Networks. *Int. Conf. Adv. Data Min. Appl.* **2017**, *3*, 519–525.
5. National Institute for Health and Care Excellence. Dementia: Assessment, management and support for people living with dementia and their carers. *Grants Regist.* **2018**, 1–43. Available online: www.nice.org.uk/guidance/ng97 (accessed on 1 September 2022).
6. Li, T.; Li, J.; Liu, J.; Huang, M.; Chen, Y.-W.; Bhatti, U.A. Robust watermarking algorithm for medical images based on log-polar transform. *EURASIP J. Wirel. Commun. Netw.* **2022**, *2022*, 1–11. [CrossRef]
7. Mao, S.; Zhang, C.; Gao, N.; Wang, Y.; Yang, Y.; Guo, X.; Ma, T. A study of feature extraction for Alzheimer's disease based on resting-state fMRI. In Proceedings of the 2017 39th Annual International Conference of the IEEE Engineering in Medicine and Biology Society (EMBC), Jeju, Republic of Korea, 11–15 July 2017; pp. 517–520. [CrossRef]
8. Tripoliti, E.E.; Fotiadis, D.I.; Argyropoulou, M. A supervised method to assist the diagnosis and monitor progression of Alzheimer's disease using data from an fMRI experiment. *Artif. Intell. Med.* **2011**, *53*, 35–45. [CrossRef] [PubMed]
9. Dachena, C.; Casu, S.; Lodi, M.B.; Fanti, A.; Mazzarella, G. Application of MRI, fMRI and Cognitive Data for Alzheimer's Disease detection. In Proceedings of the 2020 14th European Conference on Antennas and Propagation (EuCAP), Copenhagen, Denmark, 15–20 March 2020. [CrossRef]
10. Wang, Z.; Zheng, Y.; Zhu, D.C.; Bozoki, A.C.; Li, T. Classification of Alzheimer's Disease, Mild Cognitive Impairment and Normal Control Subjects Using Resting-State fMRI Based Network Connectivity Analysis. *IEEE J. Transl. Eng. Health Med.* **2018**, *6*, 1–9. [CrossRef]
11. Hojjati, S.H.; Ebrahimzadeh, A.; Babajani-Feremi, A. Identification of the Early Stage of Alzheimer's Disease Using Structural MRI and Resting-State fMRI. *Front. Neurol.* **2019**, *10*, 904. [CrossRef]
12. Zhang, Z.; Ding, J.; Xu, J.; Tang, J.; Guo, F. Multi-Scale Time-Series Kernel-Based Learning Method for Brain Disease Diagnosis. *IEEE J. Biomed. Health Inform.* **2020**, *25*, 209–217. [CrossRef] [PubMed]
13. Bi, X.-A.; Jiang, Q.; Sun, Q.; Shu, Q.; Liu, Y. Analysis of Alzheimer's Disease Based on the Random Neural Network Cluster in fMRI. *Front. Neuroinformatics* **2018**, *12*, 60. [CrossRef] [PubMed]

14. Odusami, M.; Maskeliūnas, R.; Damaševičius, R.; Krilavičius, T. Analysis of Features of Alzheimer's Disease: Detection of Early Stage from Functional Brain Changes in Magnetic Resonance Images Using a Finetuned ResNet18 Network. *Diagnostics* **2021**, *11*, 1071. [CrossRef]

15. Cheng, J.; Liu, Z.; Guan, H.; Wu, Z.; Zhu, H.; Jiang, J.; Wen, W.; Tao, D.; Liu, T. Brain Age Estimation From MRI Using Cascade Networks With Ranking Loss. *IEEE Trans. Med. Imaging* **2021**, *40*, 3400–3412. [CrossRef]

16. Amini, M.; Pedram, M.; Moradi, A.; Ouchani, M. Diagnosis of Alzheimer's Disease Severity with fMRI Images Using Robust Multitask Feature Extraction Method and Convolutional Neural Network (CNN). *Comput. Math. Methods Med.* **2021**, *2021*, 1–15. [CrossRef]

17. Ju, R.; Hu, C.; Zhou, P.; Li, Q. Early Diagnosis of Alzheimer's Disease Based on Resting-State Brain Networks and Deep Learning. *IEEE/ACM Trans. Comput. Biol. Bioinform.* **2017**, *16*, 244–257. [CrossRef] [PubMed]

18. Gupta, U.; Lam, P.K.; Steeg, G.V.; Thompson, P.M. Improved Brain Age Estimation With Slice-Based Set Networks. In Proceedings of the 2021 IEEE 18th International Symposium on Biomedical Imaging (ISBI), Nice, France, 13–16 April 2021; pp. 840–844. [CrossRef]

19. Abrol, A.; Fu, Z.; Du, Y.; Calhoun, V.D. Multimodal Data Fusion of Deep Learning and Dynamic Functional Connectivity Features to Predict Alzheimer's Disease Progression. In Proceedings of the 2019 41st Annual International Conference of the IEEE Engineering in Medicine and Biology Society (EMBC), Berlin, Germany, 23–27 July 2019; Volume 2019, pp. 4409–4413. [CrossRef]

20. Wang, M.; Lian, C.; Yao, D.; Zhang, D.; Liu, M.; Shen, D. Spatial-Temporal Dependency Modeling and Network Hub Detection for Functional MRI Analysis via Convolutional-Recurrent Network. *IEEE Trans. Biomed. Eng.* **2019**, *67*, 2241–2252. [CrossRef] [PubMed]

21. Shi, Y.; Zeng, W.; Deng, J.; Nie, W.; Zhang, Y. The Identification of Alzheimer's Disease Using Functional Connectivity Between Activity Voxels in Resting-State fMRI Data. *Adv. Intell. Technol. Dement.* **2020**, *8*, 1400211. [CrossRef]

22. Khatri, U.; Kwon, G.-R. Classification of Alzheimer's Disease and Mild-Cognitive Impairment Base on High-Order Dynamic Functional Connectivity at Different Frequency Band. *Mathematics* **2022**, *10*, 805. [CrossRef]

23. Helaly, H.A.; Badawy, M.; Haikal, A.Y. Deep Learning Approach for Early Detection of Alzheimer's Disease. *Cogn. Comput.* **2021**, *14*, 1711–1727. [CrossRef]

24. Turkson, R.E.; Qu, H.; Mawuli, C.B.; Eghan, M.J. Classification of Alzheimer's Disease Using Deep Convolutional Spiking Neural Network. *Neural Process. Lett.* **2021**, *53*, 2649–2663. [CrossRef]

25. Zhang, Y.; Zhang, H.; Chen, X.; Lee, S.-W.; Shen, D. Hybrid High-order Functional Connectivity Networks Using Resting-state Functional MRI for Mild Cognitive Impairment Diagnosis. *Sci. Rep.* **2017**, *7*, 6530. [CrossRef] [PubMed]

26. Arafa, D.A.; Moustafa, H.E.-D.; Ali-Eldin, A.M.T.; Ali, H.A. Early detection of Alzheimer's disease based on the state-of-the-art deep learning approach: A comprehensive survey. *Multimed. Tools Appl.* **2022**, *81*, 23735–23776. [CrossRef]

27. Neuroimaging B Members & Collaborations of the WCFH. SPM, Statistical Parametric Mapping n.d. Available online: http://www.fil.ion.ucl.ac.uk/spm (accessed on 1 September 2022).

28. Yan, C.; Zang, Y. DPARSF: A MATLAB toolbox for "pipeline" data analysis of resting-state fMRI. *Front. Syst. Neurosci.* **2010**, *4*, 13. [CrossRef] [PubMed]

29. Song, X.-W.; Dong, Z.-Y.; Long, X.-Y.; Li, S.-F.; Zuo, X.-N.; Zhu, C.-Z.; He, Y.; Yan, C.-G.; Zang, Y.-F. REST: A Toolkit for Resting-State Functional Magnetic Resonance Imaging Data Processing. *PLoS ONE* **2011**, *6*, e25031. [CrossRef] [PubMed]

30. Bose, S.R.; Kumar, V.S. Efficient inception V2 based deep convolutional neural network for real-time hand action recognition. *IET Image Process.* **2020**, *14*, 688–696. [CrossRef]

31. Challis, E.; Hurley, P.; Serra, L.; Bozzali, M.; Oliver, S.; Cercignani, M. Gaussian process classification of Alzheimer's disease and mild cognitive impairment from resting-state fMRI. *Neuroimage* **2015**, *112*, 232–243. [CrossRef]

32. De Vos, F.; Koini, M.; Schouten, T.M.; Seiler, S.; van der Grond, J.; Lechner, A.; Schmidt, R.; de Rooij, M.; Rombouts, S.A.R.B. A comprehensive analysis of resting state fMRI measures to classify individ-ual patients with Alzheimer's disease. *Neuroimage* **2018**, *167*, 62–72. [CrossRef]

33. Vaithinathan, K.; Parthiban, L. A Novel Texture Extraction Technique with T1 Weighted MRI for the Classification of Alzheimer's Disease. *J. Neurosci. Methods* **2019**, *318*, 84–99. [CrossRef]

34. Ramzan, F.; Khan, M.U.G.; Rehmat, A.; Iqbal, S.; Saba, T.; Rehman, A.; Mehmood, Z. A Deep Learning Approach for Automated Diagnosis and Multi-Class Classification of Alzheimer's Disease Stages Using Resting-State fMRI and Residual Neural Networks. *J. Med. Syst.* **2019**, *44*, 37. [CrossRef]

35. Duc, N.T.; Ryu, S.; Qureshi, M.N.I.; Choi, M.; Lee, K.H.; Lee, B. 3D-Deep Learning Based Automatic Diagnosis of Alzheimer's Disease with Joint MMSE Prediction Using Resting-State fMRI. *Neuroinformatics* **2019**, *18*, 71–86. [CrossRef]

36. Parmar, H.S.; Nutter, B.; Long, R.; Antani, S.; Mitra, S. Deep learning of volumetric 3D CNN for fMRI in Alzheimer's disease classification. In Proceedings of the Medical Imaging 2020: Biomedical Applications in Molecular, Structural, and Functional Imaging, Houston, TX, USA, 15–20 February 2020; Volume 11317, p. 113170C. [CrossRef]

37. Al-Khuzaie, F.E.K.; Bayat, O.; Duru, A.D. Diagnosis of Alzheimer Disease Using 2D MRI Slices by Convolutional Neural Network. *Appl. Bionics Biomech.* **2021**, *2021*, 1–9. [CrossRef] [PubMed]

38. Bhaskaran, B.; Anandan, K. Assessment of Graph Metrics and Lateralization of Brain Connectivity in Progression of Alzheimer's Disease Using fMRI. *Int. J. Softw. Sci. Comput. Intell.* **2017**, *9*, 46–66. [CrossRef]

39. Luo, Y.; Sun, T.; Ma, C.; Zhang, X.; Ji, Y.; Fu, X.; Ni, H. Alterations of Brain Networks in Alzheimer's Disease and Mild Cognitive Impairment: A Resting State fMRI Study Based on a Population-specific Brain Template. *Neuroscience* **2020**, *452*, 192–207. [CrossRef]
40. Dennis, E.L.; Thompson, P.M. Functional Brain Connectivity Using fMRI in Aging and Alzheimer's Disease. *Neuropsychol. Rev.* **2014**, *24*, 49–62. [CrossRef] [PubMed]
41. He, K.; Zhang, X.; Ren, S.; Sun, J. Deep residual learning for image recognition. In Proceedings of the IEEE Computer Society Conference on Computer Vision and Pattern Recognition (CVPR), Las Vegas, NV, USA, 27–30 June 2016; pp. 770–778. [CrossRef]
42. Huang, G.; Liu, Z.; Van Der Maaten, L.; Weinberger, K.Q. Densely connected convolutional networks. In Proceedings of the 30th IEEE Conference on Computer Vision and Pattern Recognition, CVPR 2017, Honolulu, HI, USA, 21–26 July 2017; pp. 2261–2269.
43. Li, W.; Lin, X.; Chen, X. Detecting Alzheimer's disease Based on 4D fMRI: An exploration under deep learning framework. *Neurocomputing* **2020**, *388*, 280–287. [CrossRef]

 *electronics*

Article

Classical FE Analysis to Classify Parkinson's Disease Patients

Nestor Rafael Calvo-Ariza [1,*,†] , Luis Felipe Gómez-Gómez [1,2,†] and Juan Rafael Orozco-Arroyave [1,3,*]

1 GITA Lab, Electronics and Telecommunications Department, Faculty of Engineering, Universidad de Antioquia, Medellín 050010, Colombia
2 School of Engineering, Universidad Autónoma de Madrid, 28049 Madrid, Spain
3 Pattern Recognition Lab, Friedrich-Alexander-Universität Erlangen-Nürnberg, 91054 Erlangen, Germany
* Correspondence: nestor.calvo@udea.edu.co (N.R.C.-A.); rafael.orozco@udea.edu.co (J.R.O.-A.)
† These authors contributed equally to this work.

Abstract: Parkinson's disease (PD) is a neurodegenerative condition that affects the correct functioning of the motor system in the human body. Patients exhibit a reduced capability to produce facial expressions (FEs) among different symptoms, namely hypomimia. Being a disease so hard to be detected in its early stages, automatic systems can be created to help physicians in assessing and screening patients using basic bio-markers. In this paper, we present several experiments where features are extracted from images of FEs produced by PD patients and healthy controls. Classical machine learning methods such as local binary patterns and histograms of oriented gradients are used to model the images. Similarly, a well-known classification method, namely support vector machine is used for the discrimination between PD patients and healthy subjects. The most informative regions of the faces are found with a principal component analysis algorithm. Three different FEs were modeled: angry, happy, and surprise. Good results were obtained in most of the cases; however, happiness was the one that yielded better results, with accuracies of up to 80.4%. The methods used in this paper are classical and well-known by the research community; however, their main advantage is that they provide clear interpretability, which is valuable for many researchers and especially for clinicians. This work can be considered as a good baseline such that motivates other researchers to propose new methodologies that yield better results while keep the characteristic of providing interpretability.

Keywords: Parkinson's Disease; image processing; hypomimia; FE; classic techniques; machine learning

Citation: Calvo-Ariza, N.R.; Gómez-Gómez, L.F.; Orozco-Arroyave, J.R. Classical FE Analyses to Classify Parkinson's Disease Patients. *Electronics* **2022**, *11*, 3533. https://doi.org/10.3390/electronics11213533

Academic Editor: Enzo Pasquale Scilingo

Received: 12 September 2022
Accepted: 26 October 2022
Published: 29 October 2022

Publisher's Note: MDPI stays neutral with regard to jurisdictional claims in published maps and institutional affiliations.

1. Introduction

Parkinson's disease (PD) is a neurodegenerative condition that affects the basal ganglia, and it is responsible for the correct functioning of the cortical and sub-cortical motor systems. PD patients often exhibit reduced facial expressivity and develop difficulties producing facial expressions (FEs). The cortical motor system modulates expressions that are executed with consciousness, while the sub-cortical one is related to genuine expressional expressions which cannot be consciously moderated [1]. Studies suggest that PD patients show significantly less overall facial movement than healthy controls (HC) [2]. Reduced facial activity derives from impaired production of smiles and other expressions due to partial or permanent disabilities to move certain muscle groups, i.e., bradykinesia [3].

In the last decade, technological innovations have motivated the inclusion of machine learning (ML) techniques in different fields, including a diverse spectrum of topics within medicine [4–6]. ML contributes to this field by helping in medical assessments with predictive models that have been demonstrated to be accurate and reliable in a wide variety of applications [7,8]. Similarly, with the growth of ML methods, deep learning (DL) algorithms have been widely used thanks to the possibility to automatically extract features from raw data, perform the prepossessing and give a decision based on the data [9].

In [10], the authors showed that classical methods typically used to extract information and/or classify subjects, have been less used over years, while neural networks (NN) structures have increased their popularity. Although this is a global trend in many research areas, classical approaches can still yield good results. Classical models have a good performance with fewer computational requirements and offer the possibility to interpret the result, which is not possible in most of the cases where DL is used.

This paper intends to set a baseline for the automatic classification of PD patients and HC subjects. To this end, different FEs produced by the participants are modeled with classical and well-known feature extraction techniques including local binary patterns (LBP) and histograms of oriented gradients (HOG). Different expressions were studied including anger, happiness, and surprise. A support vector machine (SVM) was used as a classifier because it has been extensively used in other studies where FEs are modeled. Finally, an analysis of the regions that provide the most discriminant information was also performed.

2. Related Work

The interest in analyzing FEs in PD is growing in the ML community. One of the main challenges is to automatically detect the expression, which has been a hot topic in the past decade and multiple contributions have been done recently. For instance, In [11] proposed a method called Discriminative Kernel Facial Emotion Recognition (DKFER) which focuses on the integration of information from static facial features and motion-dependent features, the first set of features is extracted from a single image, the authors extracted landmarks of the faces of the Japanese Female FE (JAFFE) database [12] to obtain geometrical information; meanwhile, the motion dependent features are based on the Euclidean distance of the landmarks between the static state and the peak of the emotion, the result of this work was the definition of a new technique to merge both static and dynamic information.

One year later in [13], the authors extracted features using LBP from near-infrared (NIR) video sequences to classify different FEs, the authors used SVM and sparse representation classifier (SRC) and found that NIR videos help to reduce indoor light that changes depending on multiple factors and can affect the quality of the classification but has a drawback, which is that the working distance of NIR is limited. Later, the authors in [14] validated that there are a few facial muscles that are essential to discriminate different FEs. This result was achieved by extracting features from the Cohn–Kanade Database (CK+) [15] using LBP, which previously showed to be a powerful descriptor in FEs recognition [16].

In 2014, refs. [17,18] worked in understanding the contribution of different facial muscles in the performance of a FEs, and how this can be used to obtain a better classification of the FEs. The authors in [17] use landmarks to detect the main parts of the face such as the eyes, eyebrows corners, nose, and lip corners. The face is detected and later extracted using the Viola–Jones technique of Haar-like features [19]. The experiments were performed in both JAFFE and CK+ databases. The authors used an SVM as the classifier due to its simplicity and success in related works. On the other hand, the authors in [18] used only the CK+ database and proposed a new set of features called Muscle force-based features, which uses prior knowledge of facial anatomy to estimate the different activation levels of the muscles depending on the FEs. A wireframe model of the face, called HIgh polygon GEneric face Model (HI-GEM), originally introduced in [20] is used by the authors to extract information on key points of the face. Information on the involved muscles, forces, and direction are also extracted. Three classifiers are evaluated in [18], a Naive Bayes, an SVM, and a k-Nearest Neighbors (kNN). Better results were obtained with the SVM and KNN classification methods. Later, in [21] the authors extracted LBP and HOG features from both CK+ and JAFFE databases and found that different subjects have different ways to produce FEs, and such differences can be observed in LBP features, which makes this method a good feature extractor to model FEs. Other works such as [22–28] have used combinations of features, classifiers, and techniques to detect FEs; however, few works have addressed the problem of modeling FEs in PD patients. For instance, Bandini et al. [29] classified 17 PD patients and 17 HC subjects using landmarks extracted using information from the Microsoft Kinect Sensor. A Multi-Class SVM with a Gaussian kernel was trained for each expression: neutral, happiness, anger, disgust, and sadness. The classifiers were trained with the CK+ database and the Radboud Faces Database (RaFD) [30]. A 10-Fold Cross-Validation (CV) strategy was performed to optimize the meta-parameters of the classifiers, and the authors reported an average accuracy (ACC) of 88%. Specific results per expression indicate accuracies in test of 98% for happiness, 90% for disgust, 88% for anger, 84% for neutral, and 74% for sadness. In 2018 Rajnoha et al. [31] considered 50 PD patients and 50 HC subjects to automatically identify hypomimia through conventional classifiers such as random forest, XGBoost, and decision trees. The authors reported an average ACC of 67.33% in the classification between PD and HC subjects. One year later, Grammatikopoulou et al. [32] used similar features as the ones used by Bandini et al. to model FEs produced by 23 PD patients and 11 HC. The authors tried to classify three groups of subjects according to the FEs score in the MDS-UPDRS-III scale [33]. The Google Face API and Microsoft Face API were used to extract 8 facial landmarks and 27 facial landmarks, respectively. Two individual models were trained to estimate two Hypomimima Severity indexes (HSi1 for Google and HSi2 for Microsoft features). The authors reported a sensitivity of 0.79 and a specificity of 0.82 for the HSi1, while for HSi2 the results were 0.89 and 0.73, respectively. More recently, Jin et al. [34] used Face++ to automatically locate facial landmarks from an image, providing 106 landmark points. The work focused on the analysis of the tremor caused by movement disorders, which makes the key points tremble while trying to maintain the expression. The classification was performed with a long short-term memory (LSTM) and the authors reported accuracies of 86.76%.

Another recent contribution includes the one made by Gomez et al. [35]. In that work, the authors used the FacePark-GITA database, which includes a total of 54 participants (30 PD patients and 24 HC subjects). The authors implemented a multimodal study based on static and dynamic features. A set with 17 dynamic features was combined with 2048 static ones. They reported accuracies of 77.36% for static features and 71.15% for the dynamic ones. When the combination was considered the ACC improved to 88.76%. More recently, in 2021 the authors in [36] analyzed action units activation variance from Open-Face predictions. The three most relevant action units per expression were used to discriminate between PD patients and HC subjects by using an SVM classifier. The analysis was performed on 61 PD patients and 534 HC subjects of the PARK dataset [37]. The reported precision and recall were 95.8%, and 94.3%, respectively.

From the studies mentioned above, we observed that the use of landmarks and geometric features as well as classical classifiers are the most popular approaches and provide interpretable results. For this reason, in this paper we proposed to use LBP and HOG features to model FE produced by PD patients and HC subjects. Both methods are based on transformations over the images and return feature vectors widely used in FE recognition. Further analysis in classification stages can indicate which regions of the images may have influenced the decisions made. We are aware that there exist more sophisticated methods to perform FE analysis; however, we want to present this work as a rationale baseline for future studies. We expect other researchers to motivate to evaluate other methods, hopefully with better results but keeping a high level of interpretability, which is the strongest argument in favor of classical approaches.

3. Contributions of This Work

Two classical feature extraction techniques, LBP and HOG are used to extract features from video frames of PD patients and HC subjects who produced three different FEs, namely happiness, surprise, and angriness. An SVM classifier is considered to perform the classification between PD patients and HC subjects. The three FEs are modeled with the two feature extraction methods for comparison purposes. Furthermore, the information extracted from the features was analyzed to find those areas of the face that are more informative depending on the FEs and the feature extractor. This work can be considered as a baseline for the topic of considering FEs to discriminate between PD and HC subjects.

4. Methods

4.1. Methodology

Video recordings from both PD and HC groups will be separated into frames, from which only five frames will be used according to the findings reported in [35]. The sequence consisted of five images: Normal, Onset, Apex, Offset, and Normal, as shown in Figure 1.

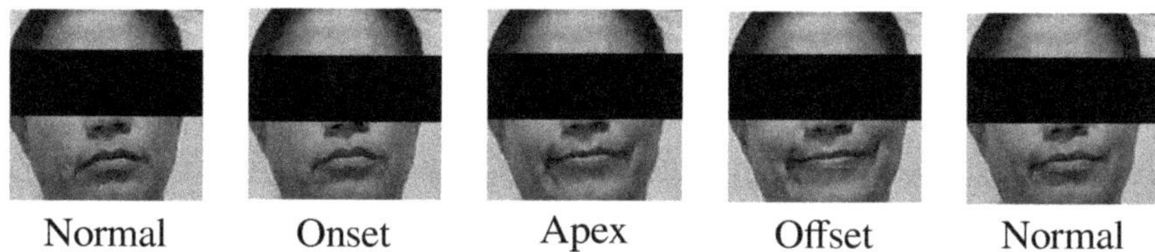

Figure 1. Sequence of frames selected for each subject.

The face was extracted from each frame using the multi-task cascaded convolutional networks (MTCNN) algorithm, which removes the background noise to avoid unnecessary variability. The resulting image is resized to 80×80 pixels, gray-scaled, and normalized using facial landmarks as shown in Figure 2.

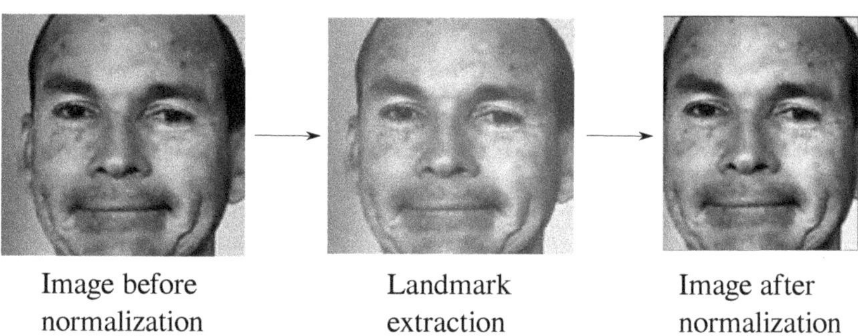

Figure 2. Normalization process using Landmarks.

LBP and HOG features were extracted. For LBP, the image is transformed and divided into 20×20 sectors, for each sector the histogram is calculated and concatenated to form a 4096-dimensional feature vector (16 sectors $\times$ 256 values of the histograms). This process is illustrated in Figure 3.

In the case of HOG, the algorithm requires the number of pixels per cell in one of the parameters, which is set to 20×20 to use the same separation grid as the one used in LBP. For each block, eight orientations are extracted, then a principal component analysis (PCA) transformation with 95% of the variance is performed to select the most relevant features and to perform an analysis that allows the identification of areas in the face that are relevant for each expression. To achieve this, we check the coefficients a_{ji} found in the PCA transformation, the magnitudes of these coefficients give an idea of the relevance of each original feature in the new set of features.

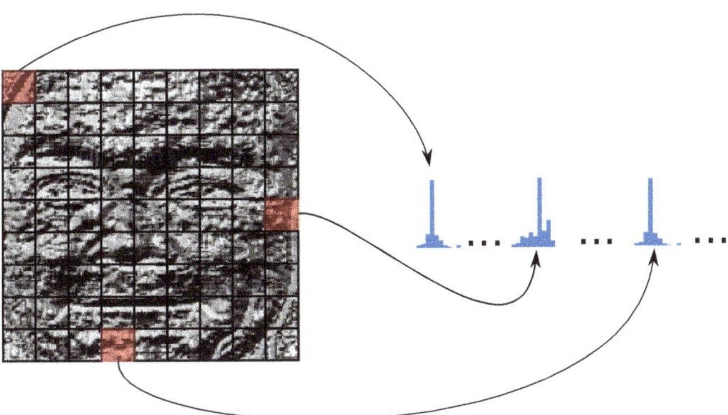

Figure 3. LBP feature extraction.

The resulting features are then mapped back to the original image to identify the portion within the image from which it was extracted, a histogram with the contribution of each feature to the PCA transformation is created and reshaped to a 4×4 matrix, and later resized into an 80×80 image to show the areas of the faces that are selected the most.

4.2. Participants and Data Collection

The corpus considered for this work includes 31 PD patients and 23 HC subjects. All participants gave informed consent to participate in the study. Patients performed a variety of tasks including speech production, handwriting, gait, and the posed FEs exercises. Only the tasks about producing FEs are considered in this work. After completing those tasks, each patient visited the neurologist, who administered the MDS-UPDRS-III scale and provided the resulting scores. In the FEs tasks, patients were asked to imitate a specific expression presented by an avatar on a screen. For this study videos of angriness, surprise and happiness were considered. Figures 4 and 5 show the distribution of demographic and clinical information of the participants, more detailed information can be found in Table 1. Possible biases due to age and gender were discarded according to a Welch's t-test ($p = 0.16$) and a chi-square test ($p = 0.57$), respectively. All patients were recorded in ON-state, i.e., no more than 3 h after the medication intake.

Table 1. Demographic information of the patients and healthy controls considered in this study.

	PD Patients		HC Subjects	
	Men	**Women**	**Men**	**Women**
# of participants	19	12	12	12
Age	70.1 ± 10.4	67.4 ± 10.9	65.3 ± 10.2	65.2 ± 8.7
Age range	52–90	53–87	49–80	49–83
Time since diagnosis	8.7 ± 5.4	15.6 ± 17.3	–	–
Range time since diagnosis	2–20	1–45	–	–
MDS-UPDRS-III	35.4 ± 13.9	29.7 ± 12.3	–	–

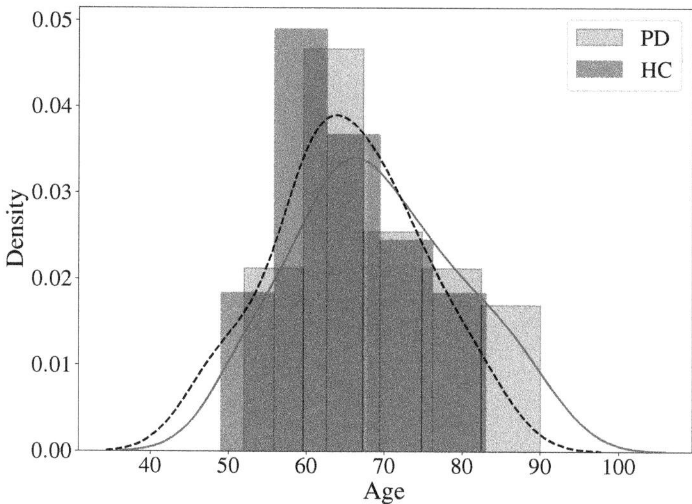

Figure 4. Age distribution for both HC and PD.

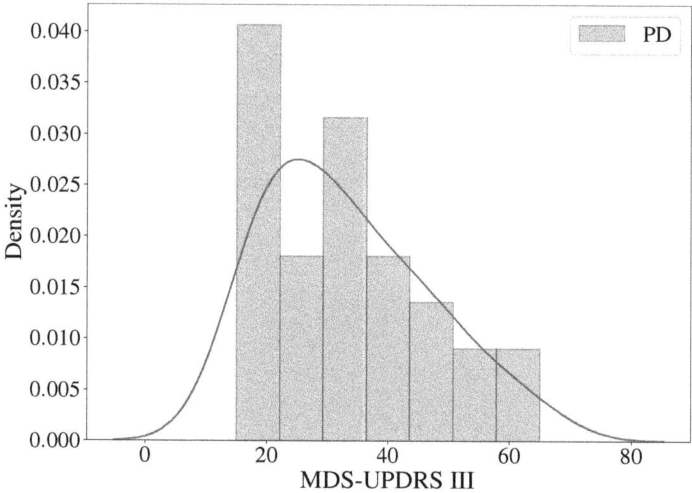

Figure 5. Label count for the MDS-UPDRS-III scale.

4.3. Multi-Task Cascaded Convolutional Networks

Face detection is the first step before removing environmental noise, allowing the system to focus on the subjects' faces. Cascade classifiers are commonly used for this aim. These methods consider features based on pixel intensities on images. For example, weighted classifiers detect contrasting face parts, such as the nose bridge and eyes. The algorithms work through small classifiers that ensemble a more robust one by detecting a face multiple times using different filters; the lower the complexity of the aforementioned small classifiers the more efficient the resulting system.

Multi-Task convolutional networks implement a novel and efficient approach to detect faces in images. The image is first resized multiple times in what is called an image pyramid, these resized images are then passed through a three-stage cascade network P-Net, R-Net, and O-Net [38]. P-Net is used to localize possible windows where a face can be found, the other two networks focus on the refinement and final decision of the window and its bounding boxes, as shown in Figure 6.

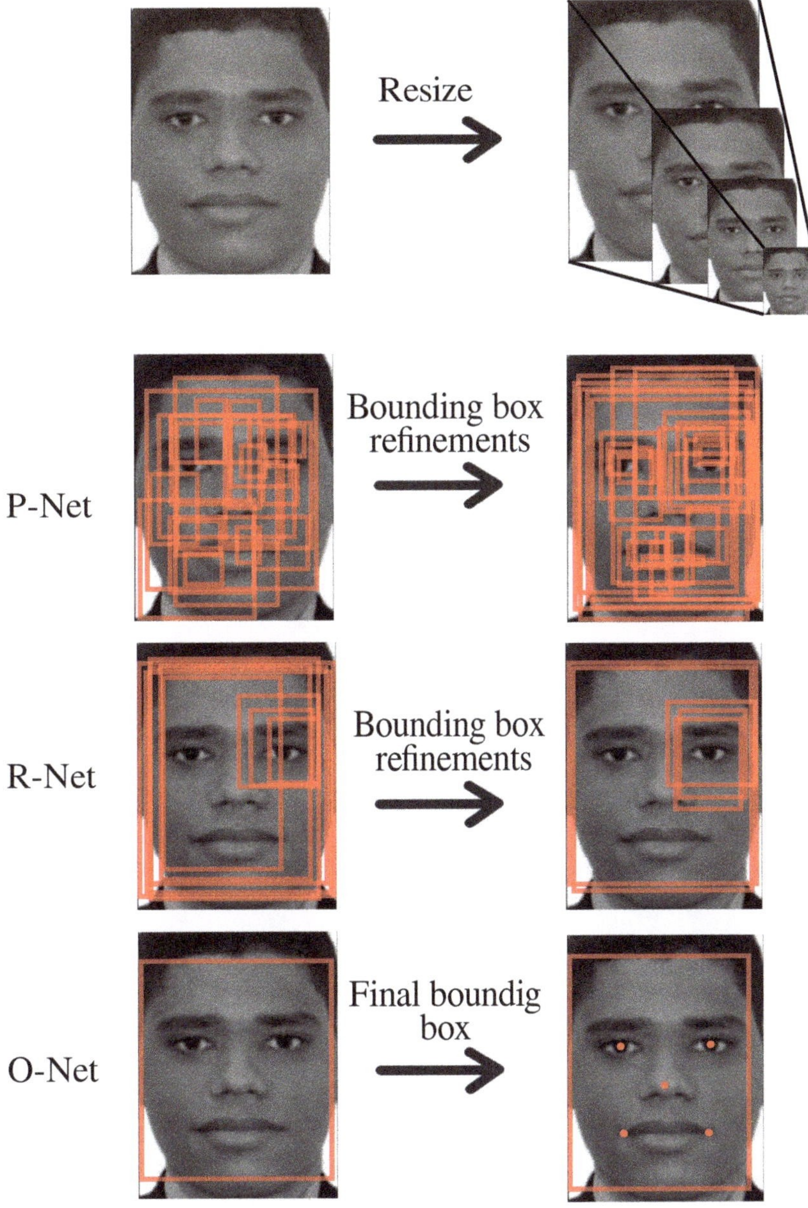

Figure 6. MTCNN sequence to find a face in an image.

4.4. Local Binary Patterns

LBP is a visual descriptor-based method that considers differences between pixel neighborhoods to recognize features in images. The workflow of this method is as follows:

1. The image color space is set to gray-scale.

2. A radius hyper-parameter is chosen and the image is divided into cells.
3. The central pixel of each cell is compared against its N neighbors. If the intensity of the center pixel is greater than or equal then a value of 1 is set in the neighbors' position, otherwise, the value is set to 0.
4. Starting clockwise from the top-right (Figure 7) a binary number is formed with the 1 s and 0 s from the previous step, this binary representation is then converted into decimal and stored in the central pixel position.
5. With this new representation a 2^N feature histogram is formed.
6. The process is repeated for each region and the histograms are concatenated forming the feature vector.

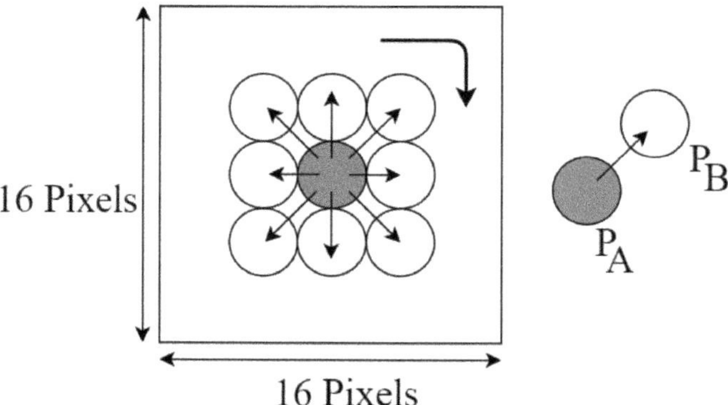

Figure 7. Neighbors operation in LBP.

4.5. Histogram of Oriented Gradients

HOG considers shapes, objects, and textures by computing the intensity and direction of gradients [39]. The flow of this algorithm is as follows:

1. The image color space is set to gray-scale.
2. For each pixel in the image, the gradient is calculated in the x and y axes, generating G_x and G_y
3. The magnitude and angle are calculated as shown in Equation (1):

$$|G| = \sqrt{G_x^2 + G_y^2}$$
$$\theta = \arctan G_y / G_x \tag{1}$$

4. The gradient matrix is divided into $N \times N$ cells where the histogram is calculated.
5. Each histogram is normalized across local groups of cells using the $L1$ normalization. This step is necessary to compensate for different changes in illumination and contrast between neighboring cells.
6. An x-dimensional feature vector is computed across the resulting histograms.

The resulting vector is the one that will be used as a feature vector for the classification process, but we can also see how the magnitudes and angles change depending on the parameters, Figure 8 shows an example of an image where HOG was applied.

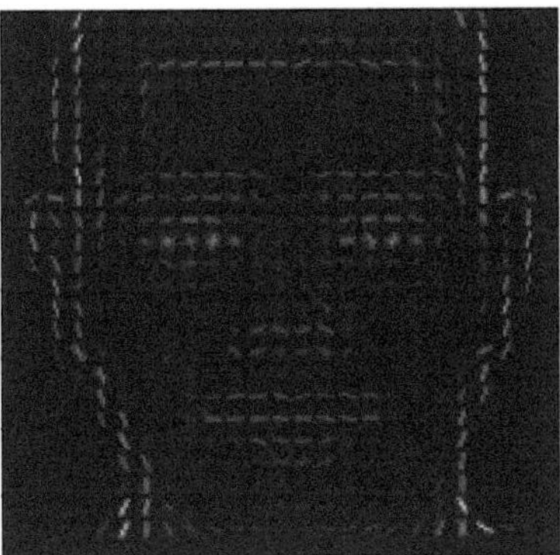

Figure 8. Gradients obtained by the HOG algorithm.

4.6. Landmarks

A landmark is a point of correspondence that "matches between and within populations" [40]. This set of points raised the interest of researchers due to its successful use in face and FEs recognition [41]. The commonly used landmarks are focused in areas around the eyes, nose tip, nostrils, mouth, ears, and chin. A total of 68 landmarks are typically used to improve the representation of the face. External landmarks are also used to normalize the image. In this work, the pre-trained facial landmark detector introduced in [42] is used to estimate the 68 landmarks.

4.7. Principal Component Analysis (PCA)

PCA is a well-known transformation method commonly used to reduce the dimensionality of a large dataset. It can also be used to perform a feature selection that allows identifying which are the most relevant features in a given problem. PCA intends to capture the data with the most variance [43], and components that give less information are removed. The data are transformed through a linear combination of the variables, as shown in Equation (2):

$$\mathbf{Y}_1 = a_{11}\mathbf{X}_1 + \ldots + a_{1p}\mathbf{X}_p$$
$$\mathbf{Y}_2 = a_{21}\mathbf{X}_1 + \ldots + a_{2p}\mathbf{X}_p$$
$$\ldots$$
$$\mathbf{Y}_d = a_{d1}\mathbf{X}_1 + \ldots + a_{dp}\mathbf{X}_p \tag{2}$$

where $\mathbf{X}_i$ the original set of features and $\mathbf{Y}_j$ is the new set of features created with the linearly combination between $\mathbf{X}_i$ and the constant values a_{ji}, where $i \in \{1, 2, \ldots, p\}$ and $j \in \{1, 2, \ldots, d\}$ with p and d are the original dimensionality and the new dimensionality respectively. After the transformation, there will be a linear combination of the original feature set multiplied by a constant for each principal component. Such a constant value can be considered as the "weight" for each feature.

4.8. Support Vector Machine (SVM)

SVM is a supervised machine learning method that focuses on finding the best margin or hyperplane that separates the data into two classes as shown in Figure 9. To find the best margin, an optimization process is performed by looking at the largest distance between the hyperplane and the data [44].

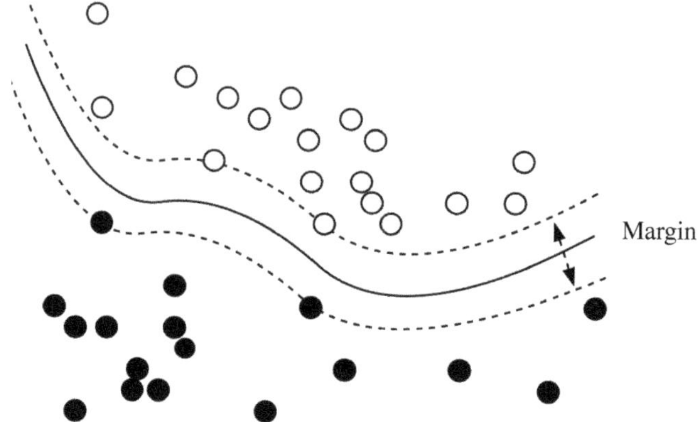

Figure 9. Support vector machine.

The SVM intends to find the optimal hyperplane but the data are not always linearly separable. Therefore, kernels are implemented, and the main aim of a kernel is to transform the data into another space where the separability of the classes is linear. The kernel trick allows operating in the new space without mapping the data [45]. Some kernels are ($\mathbf{x}$ represents the feature matrix in all cases):

1. Linear: $\mathbf{x} \cdot \mathbf{x}'$
2. Polynomial: $(\mathbf{x} \cdot \mathbf{x}' + 1)^d$, where d is the degree of the polynomial.
3. Gaussian: $e^{\gamma ||\mathbf{x}-\mathbf{x}'||/2}$, where γ is the kernel bandwidth.
4. Sigmoid: $tanh(\alpha(\mathbf{x} \cdot \mathbf{x}') + r)$, where r is a shifting parameter that controls the threshold of the mapping and α is a scaling parameter for the input data [46].

SVM was selected as the classifier for this work given the fact that it has been extensively used in similar works where FEs are intended to be modeled. In fact, according to our literature revision, SVM is the classifier that most of the times yields better results. Another advantage of this classification method is its direct interpretability regarding the distance of the samples to the separating hyperplane.

5. Experiments and Results

5.1. Classification

The hyperparameters of the SVM are optimized following a subject-independent nested cross-validation strategy. This helps in reducing the bias when combining the hyperparameter tuning and model selection. Given that each subject has five images for the outer loop of the nested cross-validation, stratified cross-validation is applied to split the dataset into train and test. These cross-validation method returns fold balanced in classes and with non-overlapping groups. Notice that in this case, the groups are each subject. A grid-search up to powers of ten where $C \in \{10^{-5}, 10^{-4}, \ldots 10^2\}$ and $\gamma \in \{10^{-5}, 10^{-4}, \ldots 10^2\}$ is performed for the inner loop to obtain optimal parameters. shows the range of the hyper-parameter considered in the grid search. Optimal parameters used for the test are selected according to the mode along the training process. Notice that two different kernels are used, namely linear and Gaussian (also known as radial basis function—RBF).

5.2. Results

Tables 2–4 show both the parameters used in the classifier and the performance obtained, for the tree expressions the best feature extractor was LBP. The score distribution, as well as the ROC curve, are shown in Figures 10 and 11.

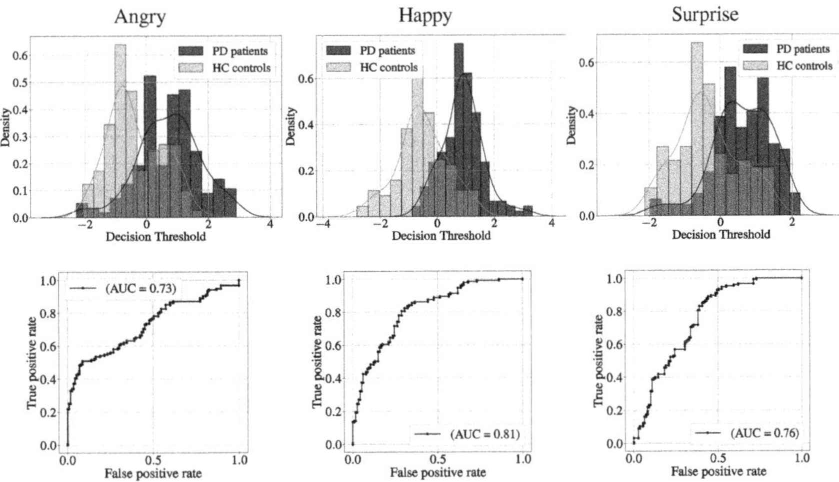

Figure 10. Score distribution (**Top**) and ROC curve (**Bottom**).

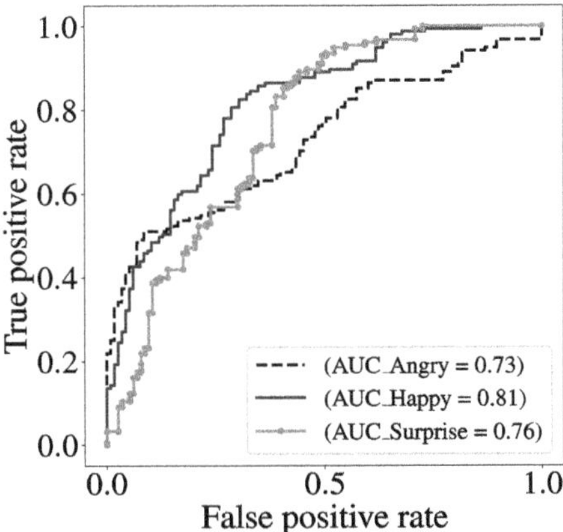

Figure 11. ROC curve with the three expressions.

Table 2. PD classification results and optimal parameters analyzing angry expression.

Feature	Classifier	ACC	SEN	SPE
LBP	SVM: C = 0.001, kernel = linear	**72.8**	**75.8**	**68.3**
HOG	SVM: C = 0.1, kernel = linear	66.1	88.0	37.8

Table 3. PD classification results and optimal parameters in analyzing happiness expression.

Feature	Classifier	ACC	SEN	SPE
LBP	SVM: C = 0.001, kernel = linear	**80.4**	**84.6**	**74.6**
HOG	SVM: C = 0.1, kernel = linear	62.1	79.6	41.0

Table 4. PD classification results and optimal parameters analyzing surprise expression.

Feature	Classifier	ACC	SEN	SPE
LBP	SVM: SVM: C = 0.001, kernel = linear	**75.8**	**80.1**	**70.6**
HOG	SVM: SVM: C = 0.1, kernel = linear	64.9	75.6	51.4

Apart from the classification experiments, statistical tests were performed to evaluate the distribution of the scores obtained with the classifier for each facial expression. First, a Shapiro–Wilk test was used to know whether the data follow a normal distribution. This test was performed for each class and facial expression, a total of six p-values were calculated (PD and HC in each emotion) and all of them were smaller than 0.05 which was the threshold to either reject or accept the null hypothesis. In this case, the test showed that none of the distributions are normal, so tests such as the *t*-test could not be applied. Nevertheless, another tests can be used e.g the Mann–Whitney U test. Table 5 shows the results of the Mann–Whitney U test and also whether the null hypothesis was rejected or not.

Table 5. Mann–Whitney U tests to compare the scores obtained from the classifier per emotion.

Emotion	*p*-Value (Mann-Whitney U Test)	H_0
Angry	3.81×10^{-14}	Rejected
Happiness	3.23×10^{-29}	Rejected
Surprise	1.35×10^{-22}	Rejected

After classifying the subjects, it is important to discuss what are the most relevant zones for both feature extractors in all expressions, this can be later compared with the classifier performance to understand which information is being considered by the classifiers to make the final decision. For both anger and surprise, we can notice that LBP shows more emphasis on the upper part of the face, while HOG focuses more on the eyes for both cases, as observed in Figures 12 and 13. This behavior is somewhat expected because when the patient is performing the facial expression, these zones that change due to the facial expression are considered the zones with more variance and hence, the ones that stand out when the PCA algorithm is applied.

Figure 12. Most important zones of the faces for angry.

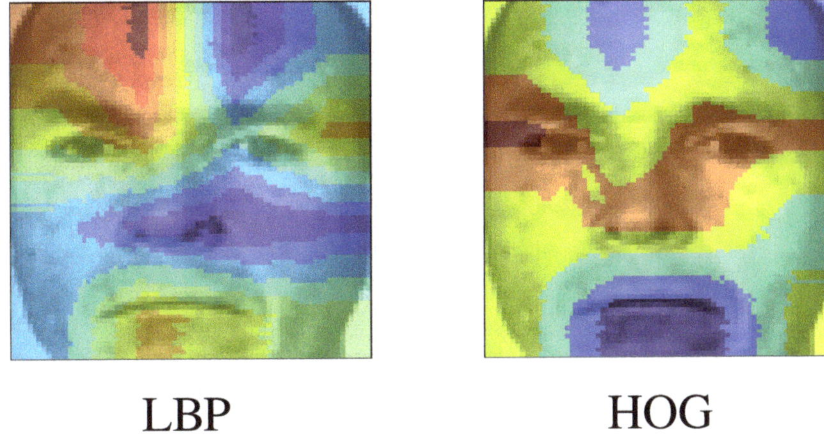

Figure 13. Most important zones of the faces for surprise.

For the case of happiness, both feature extractors focus on the lower part of the face, LBP focuses more on the mouth region, while HOG focuses more on the cheek regions as seen in Figure 14. Both regions are the main regions involved in happiness expression. This show that both extractors when a PCA is applied can focus on different facial parts depending on the expression.

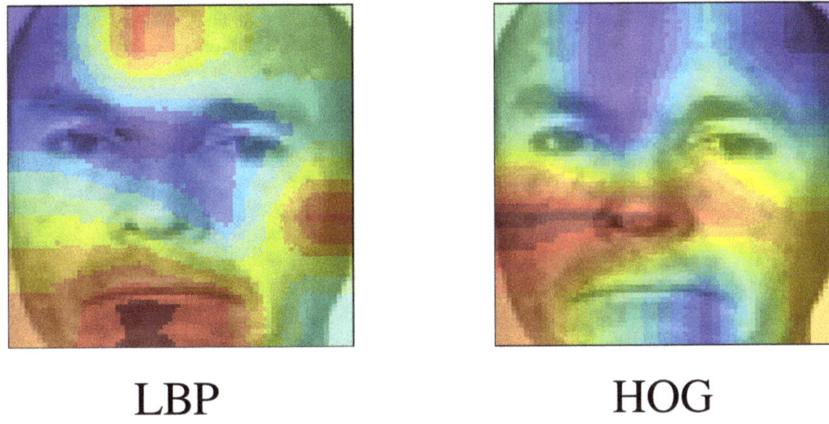

Figure 14. Most important zones of the faces for happy.

6. Discusion

This study covers the analysis of FEs in PD using HOG and LBP features. The objective was to classify PD patients vs. HC subjects. 31 PD patients and 24 HC subjects were considered for the classification task. 55 videos were used for each expression. For each patient, five images are used, following the pattern: Normal, Onset, Apex, Offset, and Normal. A PCA reduction with a variance ratio of 95% is applied to remove redundant features and analyze these features to identify where the most important features are placed. Each feature set was considered separately for the classification tasks. An SVM classifier is considered. Nested cross-validation was used to optimize hyper-parameters and divide the dataset into train and test. The best result for anger has an ACC = 72.8%, for surprise, the classifier had an accuracy of 75.8%, and happiness an accuracy of 80.4%, making it the best facial expression for the classification of PD vs. HC.

The results in this work could relate to [29], a similar behavior was experimented with these expressions, although the validation scheme and features are different. This is an exploratory analysis of FEs in PD using classical approaches. The performance of the models proved to be adequate and robust to classify impaired expressions (i.e., models with ACC = 80.0%) despite the PDs with low UPDRS values.

The focus of this work is not only to be able to classify PD patients vs. HC subjects, but to perform a more detailed of the results and to understand what the classifier is seeking in order to separate the classes.

Further research is needed with more expressions to find out which is the most suitable for this task, as well as extracting features using deep learning architecture such as convolutional neural network (CNN), which has been widely used to automatically extract information from an image, this information can be compared with the hand-crafted features extracted in this work and also a combination of this features can be performed in order to improve the performance of the models. There is also a need for experiments related to clinical personnel to find which features are more suitable for clinical evaluations and find possible clinical interpretations of the results obtained with these models.

7. Limitations and Constraints

The main limitation of this work is the use of classical techniques both for feature extraction and for classification. We are aware that there exist more sophisticated methods in the literature such as those based on deep learning; however, we believe that classical methods have been overshadowed in the last few years, mostly because deep learning models require much less knowledge about the application (in the particular case of this paper, Parkinson's disease and hypomimia), and achieve better results. The main point in favor of classical approaches is their interpretability (at the features and classification levels), which makes them more attractive for clinical applications in the real world. Another limitation of this study is the small size of the corpus. Although according to the literature revision the number of patients considered in this study is within the average, we are aware that more data would help in finding more conclusive results. Notice that this limitation also supports the fact of not using DL methods, because these approaches require much more data than the classical ones.

8. Conclusions

The production of FEs is a sensible bio-marker for the classification between PD patients and HC subjects. Given the fact that different muscles are activated depending on which FE to be produced, accurate and interpretable models able to extract information from different FEs are necessary. This work presents a study where classical yet interpretable techniques are used to create models that allow the automatic discrimination between PD patients and HC subjects. The classifier used in this study showed high sensitivity in most of the cases. However, the specificity decreased when the HOG features were considered. This is possibly due to similarities between the facial abilities of some HC subjects and PD patients who were in a low to intermediate state of the disease. The normalization performed with Landmarks reduced the variability of the background which helped in reducing errors when the models were focusing on the important zones. Regarding the different FEs produced by the patients, happiness has yielded the highest accuracy; however, the results obtained with the other two FEs suggest that the three of them can be used to perform the classification and obtain better results. LBP shows the best results for the three FEs, although the zones highlighted with HOG features are also interesting to look at. The main contribution of this work is to set a baseline with classical and interpretable methods such that motivate other researchers to study other approaches that likely yield also interpretable results with higher accuracies. Although the results in this work cannot be directly compared to those in the state-of-the-art because the datasets are different, we believe that in terms of classical approaches, the results presented here are competitive and result in a good baseline model. Future studies should focus on developing more sophisticated methodologies that provide better classification results while keeping a clear interpretability for clinicians, patients, and caregivers.

Author Contributions: Conceptualization, N.R.C.-A., L.F.G.-G. and J.R.O.-A.; methodology, N.R.C.-A. and L.F.G.-G.; validation, N.R.C.-A. and L.F.G.-G.; formal analysis, N.R.C.-A., L.F.G.-G. and J.R.O.-A.; investigation, L.F.G.-G.; resources, L.F.G.-G. and J.R.O.-A.; data curation, N.R.C.-A. and L.F.G.-G.; writing—original draft preparation, N.R.C.-A.; writing—review and editing, N.R.C.-A., L.F.G.-G. and J.R.O.-A.; visualization, N.R.C.-A.; supervision, L.F.G.-G. and J.R.O.-A.; project administration, J.R.O.-A.; funding acquisition, J.R.O.-A. All authors have read and agreed to the published version of the manuscript.

Funding: The study was partially funded by CODI at Universidad de Antioquia grant # PRG2017-15530 and by the planning direction and institutional development at Universidad de Antioquia, project # ES92210001.

Institutional Review Board Statement: This study was approved by the Ethical Research Committee of the University of Antioquia and according to the Helsinki declaration (1964) and its later amendments.

Informed Consent Statement: Written informed consent has been obtained from the patient(s) to publish this paper.

Data Availability Statement: Not applicable.

Acknowledgments: The authors would like to thank the patients of the Parkinson's Foundation in Medellín, Colombia (Fundalianza https://www.fundalianzaparkinson.org/) for their cooperation during the development of this study. Without their contribution, it would have been impossible to address this work.

Conflicts of Interest: The authors declare no conflict of interest. The funders had no role in the design of the study; in the collection, analyses, or interpretation of data; in the writing of the manuscript; or in the decision to publish the results.

References

1. Rinn, W. The neuropsychology of facial expression: A review of the neurological and psychological mechanisms for producing facial expressions. *Psychol. Bull.* **1984**, *95*, 52–77. [CrossRef] [PubMed]
2. Smith, M.C.; Smith, M.K.; Ellgring, H. Spontaneous and posed facial expression in Parkinson's Disease. *J. Int. Neuropsychol. Soc.* **1996**, *2*, 383–391. [CrossRef] [PubMed]
3. Bologna, M.; Fabbrini, G.; Marsili, L.; Defazio, G.; Thompson, P.D.; Berardelli, A. Facial bradykinesia. *J. Neurol. Neurosurg. Psychiatry* **2013**, *84*, 681–685. [CrossRef] [PubMed]
4. Dargan, S.; Kumar, M.; Ayyagari, M.R.; Kumar, G. A Survey of Deep Learning and Its Applications: A New Paradigm to Machine Learning. *Arch. Comput. Methods Eng.* **2020**, *27*, 1071–1092. [CrossRef]
5. Prakash, A.J.; Ari, S. A system for automatic cardiac arrhythmia recognition using electrocardiogram signal. *Bioelectron. Med. Devices* **2019**, *2019*, 891–911.
6. Safi, K.; Fouad Aly, W.H.; AlAkkoumi, M.; Kanj, H.; Ghedira, M.; Hutin, E. EMD-Based Method for Supervised Classification of Parkinson's Disease Patients Using Balance Control Data. *Bioengineering* **2022**, *9*, 283. [CrossRef]
7. Schwartz, W.; Patil, R.; Szolovits, P. Artificial intelligence in medicine. Where do we stand? *N. Engl. J. Med.* **1987**, *316*, 685–688. [CrossRef]
8. Deo, R.C. Machine Learning in Medicine. *Circulation* **2015**, *132*, 1920–1930. [CrossRef]
9. LeCun, Y.; Bengio, Y.; Hinton, G. Deep learning. *Nature* **2015**, *521*, 436–444. [CrossRef]
10. Canal, F.Z.; Müller, T.R.; Matias, J.C.; Scotton, G.G.; de Sa Junior, A.R.; Pozzebon, E.; Sobieranski, A.C. A survey on facial emotion recognition techniques: A state-of-the-art literature review. *Inf. Sci.* **2022**, *582*, 593–617. [CrossRef]
11. Wang, H.; Huang, H.; Hu, Y.; Anderson, M.; Rollins, P.; Makedon, F. Emotion detection via discriminative kernel method. In Proceedings of the 3rd International Conference On Pervasive Technologies Related To Assistive Environments, Samos, Greece, 23–25 June 2010; pp. 1–7.
12. Lyons, M.; Akamatsu, S.; Kamachi, M.; Gyoba, J. Coding facial expressions with gabor wavelets. In Proceedings of the Third IEEE International Conference on Automatic Face And Gesture Recognition, Nara, Japan, 14–16 April 1998; pp. 200–205.
13. Zhao, G.; Huang, X.; Taini, M.; Li, S.Z.; Pietikälnen, M. Facial expression recognition from near-infrared videos. *Image Vis. Comput.* **2011**, *29*, 607–619. [CrossRef]
14. Zhong, L.; Liu, Q.; Yang, P.; Liu, B.; Huang, J.; Metaxas, D.N. Learning active facial patches for expression analysis. In Proceedings of the 2012 IEEE Conference on Computer Vision and Pattern Recognition, Providence, RI, USA, 16–21 June 2012; pp. 2562–2569.

15. Lucey, P.; Cohn, J.; Kanade, T.; Saragih, J.; Ambadar, Z.; Matthews, I. The Extended Cohn-Kanade Dataset (CK+): A complete dataset for action unit and emotion-specified expression. In Proceedings of the 2010 IEEE Computer Society Conference on Computer Vision and Pattern Recognition-Workshops, San Francisco, CA, USA, 13–18 June 2010; pp. 94–101. [CrossRef]
16. Shan, C.; Gong, S.; McOwan, P.W. Facial expression recognition based on local binary patterns: A comprehensive study. *Image Vis. Comput.* **2009**, *27*, 803–816. [CrossRef]
17. Happy, S.; Routray, A. Automatic facial expression recognition using features of salient facial patches. *IEEE Trans. Affect. Comput.* **2014**, *6*, 1–12. [CrossRef]
18. Eskil, M.T.; Benli, K.S. Facial expression recognition based on anatomy. *Comput. Vis. Image Underst.* **2014**, *119*, 1–14. [CrossRef]
19. Viola, P.; Jones, M. Rapid object detection using a boosted cascade of simple features. In Proceedings of the 2001 IEEE Computer Society Conference On Computer Vision And Pattern Recognition, CVPR 2001, Kauai, HI, USA, 8–14 December 2001; Volume 1, p. I.
20. Abeysundera, H.P.; Benli, K.S.; Eskil, M.T. Nearest neighbor weighted average customization for modeling faces. *Mach. Vis. Appl.* **2013**, *24*, 1525–1537. [CrossRef]
21. Liu, Y.; Li, Y.; Ma, X.; Song, R. Facial expression recognition with fusion features extracted from salient facial areas. *Sensors* **2017**, *17*, 712. [CrossRef]
22. Varma, S.; Shinde, M.; Chavan, S.S. Analysis of PCA and LDA features for facial expression recognition using SVM and HMM classifiers. In *Techno-Societal 2018*; Springer: Berlin/Heidelberg, Germany, 2020; pp. 109–119.
23. Reddy, C.V.R.; Reddy, U.S.; Kishore, K.V.K. Facial emotion recognition using NLPCA and SVM. *Trait. Signal* **2019**, *36*, 13–22. [CrossRef]
24. Nazir, M.; Jan, Z.; Sajjad, M. Facial expression recognition using histogram of oriented gradients based transformed features. *Clust. Comput.* **2018**, *21*, 539–548. [CrossRef]
25. Acevedo, D.; Negri, P.; Buemi, M.E.; Fernández, F.G.; Mejail, M. A simple geometric-based descriptor for facial expression recognition. In Proceedings of the 2017 12th IEEE International Conference on Automatic Face & Gesture Recognition (FG 2017), Washington, DC, USA, 30 May–3 June 2017; pp. 802–808.
26. Chaabane, S.B.; Hijji, M.; Harrabi, R.; Seddik, H. Face recognition based on statistical features and SVM classifier. *Multimed. Tools Appl.* **2022**, *81*, 8767–8784. [CrossRef]
27. Bhangale, K.B.; Jadhav, K.M.; Shirke, Y.R. Robust pose invariant face recognition using DCP and LBP. *Int. J. Manag. Technol. Eng.* **2018**, *8*, 1026–1034.
28. Eng, S.; Ali, H.; Cheah, A.; Chong, Y. Facial expression recognition in JAFFE and KDEF Datasets using histogram of oriented gradients and support vector machine. In *Proceedings of the IOP Conference Series: Materials Science And Engineering*; IOP Publishing: Bristol, UK, 2019; Volume 705, p. 012031.
29. Bandini, A.; Orlandi, S.; Escalante, H.J.; Giovannelli, F.; Cincotta, M.; Reyes-Garcia, C.A.; Vanni, P.; Zaccara, G.; Manfredi, C. Analysis of facial expressions in parkinson's disease through video-based automatic methods. *J. Neurosci. Methods* **2017**, *281*, 7–20. [CrossRef] [PubMed]
30. Langner, O.; Dotsch, R.; Bijlstra, G.; Wigboldus, D.H.J.; Hawk, S.T.; van Knippenberg, A. Presentation and validation of the Radboud Faces Database. *Cogn. Emot.* **2010**, *24*, 1377–1388. [CrossRef]
31. Rajnoha, M.; Mekyska, J.; Burget, R.; Eliasova, I.; Kostalova, M.; Rektorova, I. Towards Identification of Hypomimia in Parkinson's Disease Based on Face Recognition Methods. In Proceedings of the 2018 10th International Congress on Ultra Modern Telecommunications and Control Systems and Workshops (ICUMT), Moscow, Russia, 5–9 November 2018; pp. 1–4. [CrossRef]
32. Grammatikopoulou, A.; Grammalidis, N.; Bostantjopoulou, S.; Katsarou, Z. Detecting hypomimia symptoms by selfie photo analysis: For early Parkinson disease detection. In Proceedings of the 12th ACM International Conference on PErvasive Technologies Related to Assistive Environments, Island of Rhodes, Greece, 5–7 June 2019; pp. 517–522.
33. Goetz, C.G.; Tilley, B.C.; Shaftman, S.R.; Stebbins, G.T.; Fahn, S.; Martinez-Martin, P.; Poewe, W.; Sampaio, C.; Stern, M.B.; Dodel, R.; et al. Movement Disorder Society-sponsored revision of the Unified Parkinson's Disease Rating Scale (MDS-UPDRS): Scale presentation and clinimetric testing results. *Mov. Disord. Off. J. Mov. Disord. Soc.* **2008**, *23*, 2129–2170. [CrossRef] [PubMed]
34. Jin, B.; Qu, Y.; Zhang, L.; Gao, Z. Diagnosing Parkinson disease through facial expression recognition: Video analysis. *J. Med. Internet Res.* **2020**, *22*, e18697. [CrossRef] [PubMed]
35. Gómez-Gómez, L.F.; Morales, A.; Fierrez, J.; Orozco-Arroyave, J.R. Exploring Facial Expressions and Action Unit Domains for Parkinson Detection. *SSRN* **2022**. https://dx.doi.org/10.2139/ssrn.4069648. [CrossRef]
36. Ali, M.R.; Myers, T.; Wagner, E.; Ratnu, H.; Dorsey, E.; Hoque, E. Facial expressions can detect Parkinson's disease: Preliminary evidence from videos collected online. *NPJ Digit. Med.* **2021**, *4*, 1–4. [CrossRef]
37. Langevin, R.; Ali, M.R.; Sen, T.; Snyder, C.; Myers, T.; Dorsey, E.R.; Hoque, M.E. The PARK framework for automated analysis of Parkinson's disease characteristics. *Proc. ACM Interact. Mob. Wearable Ubiquitous Technol.* **2019**, *3*, 1–22. [CrossRef]
38. Zhang, K.; Zhang, Z.; Li, Z.; Qiao, Y. Joint face detection and alignment using multitask cascaded convolutional networks. *IEEE Signal Process. Lett.* **2016**, *23*, 1499–1503. [CrossRef]
39. Dalal, N.; Triggs, B. Histograms of oriented gradients for human detection. In Proceedings of the 2005 IEEE Computer Society Conference On Computer Vision And Pattern Recognition (CVPR'05), San Diego, VA, USA, 20–25 June 2005; Volume 1, pp. 886–893.

40. Dryden, I.L.; Mardia, K.V. *Statistical Shape Analysis: With Applications in R*; John Wiley & Sons: Hoboken, NJ, USA, 2016; Volume 995.
41. Shi, J.; Samal, A.; Marx, D. How effective are landmarks and their geometry for face recognition? *Comput. Vis. Image Underst.* **2006**, *102*, 117–133. [CrossRef]
42. Kazemi, V.; Sullivan, J. One millisecond face alignment with an ensemble of regression trees. In Proceedings of the IEEE Conference On Computer Vision And Pattern Recognition, Washington, DC, USA, 23–28 June 2014; pp. 1867–1874.
43. Xu, Y.; Zhang, D.; Yang, J.Y. A feature extraction method for use with bimodal biometrics. *Pattern Recognit.* **2010**, *43*, 1106–1115. [CrossRef]
44. Schölkopf, B.; Smola, A.J.; Bach, F. *Learning with Kernels: Support Vector Machines, Regularization, Optimization, And Beyond*; MIT Press: Cambridge, MA, USA, 2002.
45. Soentpiet, R. *Advances in Kernel Methods: Support Vector Learning*; MIT Press: Cambridge, MA, USA, 1999.
46. Lin, H.T.; Lin, C.J. A study on sigmoid kernels for SVM and the training of non-PSD kernels by SMO-type methods. *Neural Comput.* **2003**, *3*, 16.

MDPI

Article

EdgeTrust: A Lightweight Data-Centric Trust Management Approach for IoT-Based Healthcare 4.0

Kamran Ahmad Awan [1], Ikram Ud Din [1,*], Ahmad Almogren [2,*], Hasan Ali Khattak [3] and Joel J. P. C. Rodrigues [4,5]

1. Department of Information Technology, The University of Haripur, Haripur 22620, Pakistan
2. Chair of Cyber Security, Department of Computer Science, College of Computer and Information Sciences, King Saud University, Riyadh 11633, Saudi Arabia
3. School of Electrical Engineering & Computer Science (SEECS), National University of Sciences & Technology (NUST), H12, Islamabad 44000, Pakistan
4. College of Computer Science and Technology, China University of Petroleum (East China), Qingdao 266555, China
5. Instituto de Telecomunicações, 6201-001 Covilhã, Portugal
* Correspondence: ikramuddin205@yahoo.com (I.U.D.); ahalmogren@ksu.edu.sa (A.A.)

Abstract: Internet of Things (IoT) is bringing a revolution in today's world where devices in our surroundings become smart and perform daily-life activities and operations with more precision. The architecture of IoT is heterogeneous, providing autonomy to nodes so that they can communicate with other nodes and exchange information at any time. IoT and healthcare together provide notable facilities for patient monitoring. However, one of the most critical challenges is the identification of malicious and compromised nodes. In this article, we propose a machine learning-based trust management approach for edge nodes to identify nodes with malicious behavior. The proposed mechanism utilizes knowledge and experience components of trust, where knowledge is further based on several parameters. To prevent the successful execution of good and bad-mouthing attacks, the proposed approach utilizes edge clouds, i.e., local data centers, to collect recommendations to evaluate indirect and aggregated trust. The trustworthiness of nodes is ranked between a certain limit, and only those nodes that satisfy the threshold value can participate in the network. To validate the performance of the proposed approach, we have performed extensive simulations in comparison with existing approaches. The results show the effectiveness of the proposed approach against several potential attacks.

Keywords: Internet of Things; trust management; healthcare; digital revolution; edge clouds; security; privacy preservation

Citation: Awan, K.A.; Ud Din, I.; Almogren, A.; Khattak, H.A.; Rodrigues, J.J.P.C. EdgeTrust: A Lightweight Data-Centric Trust Management Approach for IoT-Based Healthcare 4.0. *Electronics* **2023**, *12*, 140. https://doi.org/10.3390/electronics12010140

Academic Editor: Cheng-Chi Lee

Received: 15 December 2022
Revised: 20 December 2022
Accepted: 24 December 2022
Published: 28 December 2022

1. Introduction

Internet of Things (IoT) [1] consists of diverse standards of nodes in a heterogeneous environment connected with the Internet to communicate and exchange information in the network [2]. The classification of these nodes can be created based on their processing power wherein edge devices, such as sensors, contain the least processing power causing vulnerabilities [3]. The generic architecture of IoT consists of multiple layers, i.e., business, application, middleware, and perception layers [4], which are illustrated in Figure 1. The business layer contains system management solutions that may be varied according to the requirements [5]. The middleware layer is the most critical layer that consists of information processing [6], ubiquitous computing [7], services management [8], databases [9], and decision units [10]. The network layer consists of transmission networks that provide a source by which IoT participating nodes can transmit information among them [11]. These transmission connections will be 4G, 5G, etc. [12]. The perception layer consists of edge nodes that can be RFID [13], sensors [14], or any physical object [15]. In [4], a generic

IoT trust architecture is proposed that integrates trust into all these layers as an integral component to manage security. IoT faces several security challenges [16], e.g., authentication [17,18], access control [19], trust management in cross-domain along with smart edge nodes [20], security management in IoT equipped with VANET nodes, policy enforcement, secure middleware, and confidentiality.

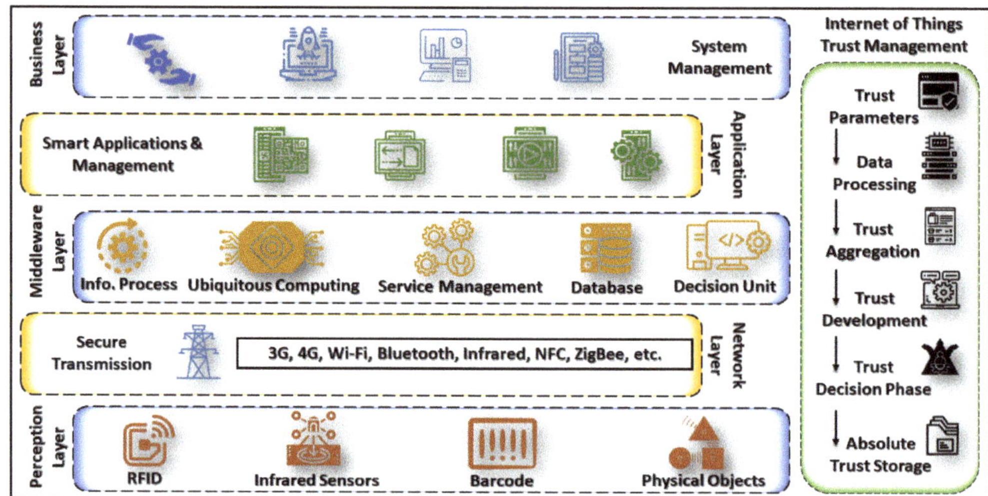

Figure 1. IoT architecture with the integration of trust management.

Due to the heterogeneous environment of IoT, it is inevitable to implement robust approaches that maintain a secure environment by eliminating malicious nodes and are also robust enough to keep resilience towards several potential attacks [21–23]. The maintenance of security is a significant challenge due to wireless technologies that have been extensively deployed in the IoT environment [24]. Healthcare 4.0 [25] is the term used to describe the next generation of healthcare technology, which is focused on harnessing data, and analytic and digital tools to improve patient care and outcomes. The cost of healthcare is a major issue for many people, and patients are looking for ways to obtain the care they need without breaking the bank. The trend toward more affordable healthcare [26] has led to an incredibly promising new technology, i.e., IoT, which allows us to connect devices with sensors so that we can track our health in real-time [27]. With IoT, doctors can use AI [28] to analyze your health data and make predictions about your future health prospects. Healthcare monitoring contains patients' electronic health records [29] that are transmitted to doctors for monitoring. The transmitted data become vulnerable to potential IoT attacks. The most prominent way to maintain a trustworthy environment is to identify and eliminate such nodes. Trust is proposed as the most prominent lightweight mechanism that helps to maintain a secure environment by utilizing parameters.

In this article, we have proposed a trust management approach (EdgeTrust) for those nodes which are not capable to perform complex computations. The proposed approach is a combination of centralized and distributed trust management architectures. The EdgeTrust working consists of two major components, i.e., distributed edge devices and centralized data centers/edge clouds. The proposed mechanism utilizes the direct and indirect trust evaluation mechanism where the pre-observations required to evaluate the trust are provided by a central authority. The absolute direct trust evaluation consists of observations provided by central authority along with the observations stored on nodes' local storage. For indirect trust evaluation, nodes also do not require generating the request to neighboring nodes as the recommendation is to gather by a central authority. The advantage of utilizing recommendations of the centralized authority reduces the time required to evaluate the trust. The trust is further compared with the threshold value for decision-making.

The structure of the rest of the article is as follows: Section 2 discusses and elaborates on the existing trust management approaches. Section 3 explains the working of the proposed mechanism such as trust parameters, computations, trust aggregation, and threshold comparison of trust. Section 4 elaborates and discusses the simulation outcomes and performance comparison of EdgeTrust with existing approaches. Finally, Section 5 concludes the paper.

2. Literature Review

There are several trust management approaches proposed for IoT-based Healthcare, but significant research attention is required to address the computational challenges associated with IoT edge devices that are not capable of performing complex computations. This section will elaborate on the existing approaches along with their contribution and limitation to identify the research gaps, also illustrated in Table 1.

A trust management mechanism is proposed for the Social IoT that maintains trust by self-enforcing in a decentralized manner [30]. The proposed mechanism architecture consists of multiple IoT devices owned by numerous users who interact with others at particular time intervals. After the interaction, these nodes submit user ratings to the IoT decentralized database shared among nodes. These ratings consist of feedback and zero knowledge. The major contribution of the proposed mechanism is the integration of a database that contains the feedback of the nodes. However, the decentralized database can cause data integrity challenges as it is shared and stored without utilizing any central authority.

A game theory-based decentralized trust management mechanism is proposed for IoT to maintain robustness among nodes [31]. The proposed mechanism applies the game theory to identify nodes that are executing good or bad-mouthing attacks by sending mendacious trust degrees. For updating the trust degrees of nodes, the proposed approach utilizes the Dempster–Shafer theory that collects the scores for updating process by excluding disparate scores. To perform a trust computation, the approach utilizes Fuzzy theory to classify trust into none, low, high, and definitely. The major contribution of the proposed mechanism is the utilization of the Fuzzy rule to classify trust. However, the performance of the proposed mechanism needs to be evaluated against potential IoT attacks such as on-off, whitewashing, etc.

In 2017, a study was proposed to design an architecture and protocol for eHealth monitoring with the integration of 5G [32]. The study focuses on the continuous monitoring of patient's health and concludes no notable difference between 4G and 5G. The architecture of the proposed scheme consists of a user, a 5G network-enabled antenna, and a database server on the hospital side. Users/patients are monitored using Bluetooth wearable sensors and gadgets, whereas the monitored data are forwarded to the hospital using a 5G network. The monitored data are received by the database server, which acts as the central authority between the hospital and its users. The database also receives medical analytical data from hospitals and forwards alarms to patients in case of emergency.

In 2020, a blockchain-based trust protocol was proposed for IoT, which maintains trust in a decentralized manner [33]. The study stated that an IoT object can communicate and exchange information, which makes the environment highly dynamic and raise security challenges. The proposed mechanism is a hierarchical blockchain protocol that also supports mobility where the architecture of the proposed mechanism consists of a fog layer, a private blockchain layer, and an IoT layer with different clusters/zones.

In 2018, an energy-efficient trust management mechanism (EET-IoT) [34] is proposed to protect the IoT network and primarily focus on smart cities [35,36]. The proposed mechanism utilizes the IEEE 802.14 protocol to perform computations. The purpose of using the IEEE 802.14 protocol is to sustain the efficiency of the IEET-IoT. The proposed mechanism further uses Jasang's Subjective Logic (JSL) to examine the ambiguity of an entity. The EET-IoT uses a triple variable concept, i.e., *b*, *d*, and *u*. Variable *b* expresses the belief, *d* represents the disbelief, and *u* denotes the uncertainty. The evaluation of EET-IoT shows a significant decrease in energy utilization. The energy consumption evaluation of the proposed algorithms shows that LT consumes maximum energy followed by LDE and NDLF. However, optimization at the MAC Layer is required to overcome adequate energy consumption.

A smart middle-ware mechanism (Smart-TM) [37] is proposed to detect on-off attacks in IoT. The focus of the proposed mechanism is to automatically assess the resources of IoT trust by evaluating the attributes of service providers. The Smart-TM utilizes an approach of machine learning based on the One-Class Support Vector Machine (OneClass-SVM) method. The degree of trust is estimated by examining the distance from a function of the Hyper-plane model. Moreover, the middleware implements the decision function to estimate the trust, and nodes with a higher degree of trust are listed as trusted nodes, while nodes with a lower degree of trust are classified as untrusted ones or specified as attackers. The performance evaluation of Smart-TM represents that the proposed approach successfully distinguishes the behavior to recognize on-off attacks. However, the proposed mechanism is unable to specify the framework of information gathering, trust dissemination, updating, and maintenance.

A scheme of trust management (Tm-SecPro) [38] is proposed that adopts two methods, i.e., maximum ratios combining and selection combining. In Tm-SecPro, service providers and seekers communicate with each other directly, and the mechanism preserves trust between them. The proposed mechanism estimates and concludes the results in three phases. In the first phase, the information about trust control is transmitted to the lower layer. In the second phase, the specified model is used to calculate the trust values. While in the last phase, all relations related to these phases are extracted from each layer. The considerable aspect of this scheme is a fusion of MRC and SC that will help to maintain the reliability of Tm-SecPro.

Table 1. The comparative analysis of the existing approaches.

Ref.	Contribution	Limitation
[30]	Integration of database that contains feedback of the nodes.	Decentralized databases can cause integrity challenges.
[31]	Utilization of Fuzzy rule to classify trustworthy and malicious nodes.	Performance needed to be evaluated in the IoT Environment.
[33]	Hierarchical blockchain protocol that also supports mobility.	Not suitable for nodes with less computational capabilities due to complexity.
[34]	The utilization of Jasang's Subjective Logic (JSL) to examine the ambiguity of an entity.	Optimization at MAC Layer is required to overcome adequate energy consumption.
[37]	Hyper-plane model along with middleware implements the decision function.	Unable to specify the framework of trust management.
[38]	Fusion of MRC and SC that will help to maintain reliability.	Transmission of trust computation between multiple layers may raise integrity challenges.

3. Proposed EdgeTrust Approach

The identification of malicious and compromised is one of the important challenges in Healthcare 4.0 that can affect the network security and privacy of users. In this article, we have proposed EdgeTrust to address the challenges caused by these malicious nodes. The architecture of the proposed approach consists of three major layers which are data center/edge clouds, trust management, and edge nodes as illustrated in Figure 2. The data center contains the data center and edge cloud that have the capability of Naive Bayes [39,40] for the identification and classification and behavior prediction of malicious and compromised nodes by utilizing the stored direct observation collected by the network nodes. These observations are utilized further to formulate direct trust for edge nodes. Indirect trust at the data center layer can be formulated with the help of recommendations collected by the edge nodes. The trust management evaluation is a combination of events and time-driven under a different scenario. The direct trust degree is evaluated-based on the knowledge and experience component, which also involves the trust aggregation, threshold comparison, and decision-making phase. The edge nodes in IoT can be classified concerning their computational power and internal capabilities.

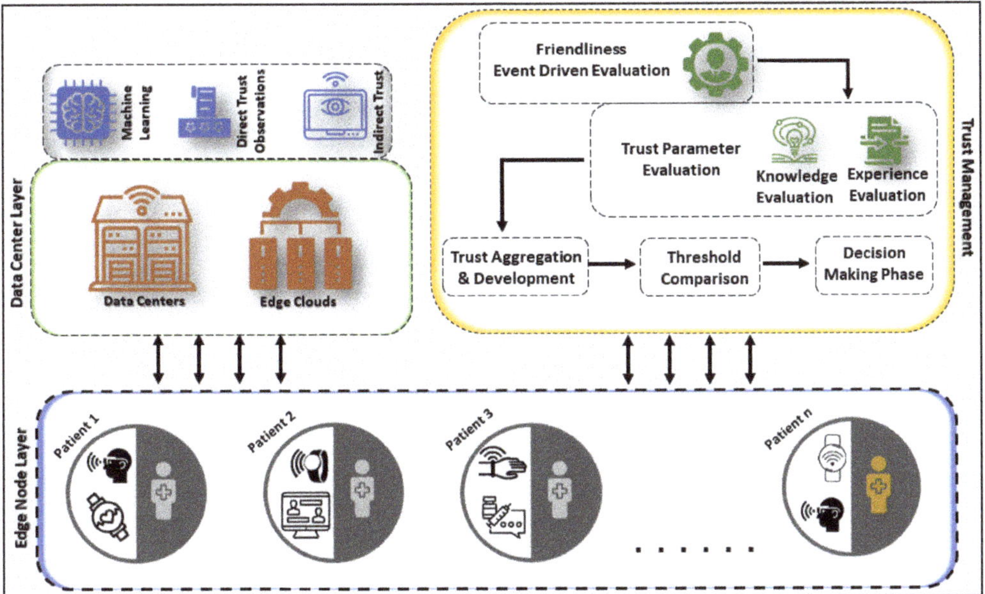

Figure 2. The proposed EdgeTrust architecture.

In the proposed approach, these edge nodes are classified based on their categories, i.e., sensors, home appliances, and smart mobile devices among others. The training phase of the proposed mechanism includes five distinct phases which are features selection, feature scaling, classifier implementation, dataset training, and classification of malicious and compromised nodes. The features of trust parameters used are reliability, cooperativeness along with experience, and the computation depends on sessions created between nodes which are denoted as friendliness. If the friendliness of nodes is higher, then the computations are computed in a time-driven manner while, in the case of low friendliness, trust computation is performed based on events. The next phase is to scale the features in which all the features involved in computations are scaled between 0.0–1, where 0.0 represents the lowest trust and 1 represents a higher trust degree. The complete workflow of trust computation and decision-making is illustrated by Algorithm 1.

Algorithm 1 EdgeTrust trust computation process

1: **procedure** TRUST DEGREE EVALUATION(d_t^{ob})

2:      Central authority computation: $\overset{ob}{t} = \sum \left[ob_{n-id}^1 + ob_{n-id}^2 + ob_{n-id}^3 + ... + ob_{n-id}^n \right]$

3:      Recommendation-based evaluation: $r_{c-id}^{itrust} = \sum \left[rec_{r_1}^{e_i \to e_j} + rec_{r_2}^{e_i \to e_j} + ... + rec_{r_n}^{e_i \to e_j} \right]$

4: **procedure** EXPERIENCE GATHERING AND FORMULATION($ab_{e_i \to e_j}^{t \to aggregate}$)

5:      Previous computation: $ep_{e_{absolute}}^{node-id} = \sum \left[ep_{e_i \to e_j}^{p_1} + ep_{e_i \to e_j}^{p_2} + ... + ep_{e_i \to e_j}^{p_n} \right]$

6:      Aggregated trust: $ab_{e_i \to e_j}^{t \to aggregate} = d_t^{ob} + ep_{e_{absolute}}^{node-id}$

7: **procedure** TRUST COMPUTATION

8:      Friendliness computations: $fr_{n_{id}}^{tr}$

9:      Reliability observation: $obp_{e_i \to e_j}^{rt} = ob_{n-id}^{t(e_i \to e_j)_1} + ob_{n-id}^{t(e_i \to e_j)_2} + ... + ob_{n-id}^{t(e_i \to e_j)_n}$

10:      Reliability formulation: $rt_{e_i \to e_j}^{dt}$

11:      Cooperativeness computation: $obp_{e_i \to e_j}^{cpt} = ob_{n-id}^{cpt_1} + ob_{n-id}^{cpt_2} + ob_{n-id}^{cpt_3} + ... + ob_{n-id}^{cpt_n}$

12: **procedure** TRUST AGGREGATION($f_{te_i \to e_j}^{dt_{exp}}$)

13:      Absolute computation: $ct_{de_i \to e_j}^{ag} = rt_{e_i \to e_j}^{dt} + cpt_{e_i \to e_j}^{dt}$

14:      Experience formulation: $t_{exp_{e_i \to e_j}}^{pt} = \sum_{i=0}^{n} \left[et_{e_i \to e_j}^{o_1} + et_{e_i \to e_j}^{o_2} + ... + et_{e_i \to e_j}^{o_n} \right]$

15:      Absolute trust degree: $f_{te_i \to e_j}^{dt_{exp}} = ct_{de_i \to e_j}^{ag} + t_{exp_{e_i \to e_j}}^{pt}$

16: **procedure** DECISION MAKING

17:      Decision making: $\theta = t_{exp_{e_i \to e_j}}^{pt}$

18: Exit

To perform classification and prediction, we adopt the Naive Bayes classifier due to its accuracy and low energy consumption for classification. After selecting the classifier, the training phase begins, using a dataset of 120,766 trust values per feature for the classifier to learn from. After training, the classifier calculates the error difference between computed and actual trust values to increase precision.

3.1. Data Centers and Edge Clouds

In the proposed approach, the data center layer is responsible for performing three major operations: machine learning-based prediction and direct and indirect trust observation evaluation. The data centers and edge clouds are able to make predictions based on direct observations transmitted by the nodes. These transmitted values are first stored by the central authorities and later used to predict the behavior of edge nodes by applying the Naive Bayes Classifier. The direct trust evaluation at the data center layer is a time-driven process, evaluated after 90 minutes. When an edge node requests data from the data center layer, the central authorities share the already stored observations for further processing. After receiving the request, the central authorities formulate the direct trust degree using Equation (1), where d_t^{ob} represents the available direct trust observation and ob_{n-id}^1 is the number of observations transmitted by a particular node.

$$d_t^{ob} = \sum \left[ob_{n-id}^1 + ob_{n-id}^2 + ob_{n-id}^3 + ... + ob_{n-id}^n \right] \tag{1}$$

The coverage area of central authorities is larger compared to edge trust, so they also provide recommendations that have been computed over a specific time interval. These recommendations help nodes to compute indirect trust. The recommendation-based indirect trust is formulated using Equation (2), where r_{c-id}^{itrust} represents the recommendation-based trust evaluation, and c-id represents the unique identity of a central authority that computed indirect trust.

$$r_{c-id}^{itrust} = \sum \left[rec_{r_1}^{e_i \rightarrow e_j} + rec_{r_2}^{e_i \rightarrow e_j} + \ldots + rec_{r_n}^{e_i \rightarrow e_j} \right] \tag{2}$$

3.2. IoT Edge Nodes

The edge nodes are those that cannot perform complex computations but are crucial to lightening the burden from them to increase the scalability and security of a network. In the proposed EdgeTrust approach, central authorities compute the direct trust and transmit it to the requested node while the edge nodes just have to aggregate that value with the pre-stored experience. The experience component of trust represents the previous experience of a particular node regarding other nodes that provide services. To evaluate the aggregate value, the edge nodes apply the summation function to the previous experience available as represented by Equation (3), where $ep_{e_{absolute}}^{node-id}$ shows the absolute experience formulation of a node with a unique identity. The $ep_{e_i \rightarrow e_j}^{p_1}$ represents the number of previous experiences stored on the internal memory of edge nodes.

$$ep_{e_{absolute}}^{node-id} = \sum \left[ep_{e_i \rightarrow e_j}^{p_1} + ep_{e_i \rightarrow e_j}^{p_2} + \ldots + ep_{e_i \rightarrow e_j}^{p_n} \right] \tag{3}$$

After the formulation of absolute experience, the edge node computes aggregate trust by using the direct trust and experience trust degree computation as represented by Equation (4), where $ab_{e_i \rightarrow e_j}$ represents the absolute trust evaluation of edge node i towards j, d_t^{ob} and $ep_{e_{absolute}}^{node-id}$ represent the direct trust evaluated based on observation and experience evaluation of a node with unique identifiers, respectively.

$$ab_{e_i \rightarrow e_j}^{t \rightarrow aggregate} = d_t^{ob} + ep_{e_{absolute}}^{node-id} \tag{4}$$

3.3. Trust Management Computations

The trust computation in the proposed mechanism consists of multiple features that are computed by the central authorities along with edges to formulate an absolute trust value for decision-making. When edge nodes want to compute the trust value of a particular node, the node transmits a trust computation request to the nearest central authority. The request generated by a particular node consists of the trustee's identification, the trustor's identification, and the previous experience trust degree computed by the edge nodes. The trust computation process begins by first observing the friendliness of the nodes, which represents the number of sessions created over a specific interval of time. If the friendliness of the nodes is high, then trust is computed as time-driven, which reduces the energy consumption of computation. The time-driven trust computation in the case of higher friendliness is 60 min, which means that nodes are not required to compute trust based on events and can use the same trust degree for a pre-defined time. The friendliness is computed based on the sessions created between particular nodes, as represented in Equation (5).

$$fr_{n_{id}}^{tr} = \begin{cases} timedriven & \text{if } fr \geq 50 \\ Eventdriven & \text{if } fr \leq 49 \\ indirecttrust & \text{if } p_{ob} = Yes \end{cases} \tag{5}$$

In Equation (5), fr, $n_i d$, and tr represent friendliness, nodes' unique identify, and trust degree, respectively. For direct trust, if $fr \geq 50$, the trust is computed as time-driven. When the number of sessions formulated between two nodes becomes event-driven and $fr \leq 49$, the trust is also computed using a time-driven approach. In case of no previous observations p_{ob}, the trust is computed by gathering recommendations from central authorities. After evaluating friendliness, the next phase is to compute the trust parameters, i.e., knowledge and experience. TThe knowledge component of trust consists of reliability and cooperativeness, which are computed by central authorities when a particular node generates a request. In the knowledge parameters, the evaluation is initiated by evaluating the reliability by gathering the pre-stored observations received by the central authorities for a while from network nodes. The process of observation gathering is shown in Equation (6), where the reliability trust degree is formulated by applying summation to these pre-available observations:

$$obp_{e_i \to e_j}^{rt} = ob_{n-id}^{t(e_i \to e_j)_1} + ob_{n-id}^{t(e_i \to e_j)_2} + \dots + ob_{n-id}^{t(e_i \to e_j)_n} \tag{6}$$

In Equation (6), *obp* represents previous observations, *rt* shows reliability trust evaluation, and $e_i \to e_j$ is the trust evaluation of edge node i towards j, where $ob_{n-id}^{t(e_i \to e_j)_1}$ represents the pre-stored previous observations. After reliability observation gathering, the proposed mechanism applies a limit to formulate the absolute trust value of the reliability parameter as shown in Equation (7):

$$rt_{e_j}^{dt} = ob_{n-id}^{t(e_i \to e_j)_1} + ob_{n-id}^{t(e_i \to e_j)_2} + \dots + ob_{n-id}^{t(e_i \to e_j)_n} \tag{7a}$$

$$rt_{e_i \to e_j}^{dt} = \sum_{i=0}^{n} [obp_{e_i \to e_j}^{rt} * rt_{e_j}^{dt}] \tag{7b}$$

In Equation (7), $rt_{e_i \to e_j}^{dt}$ represent the evaluation of reliability evaluation based on a direct trust approach, where $\sum_{0.0}^{1}$ is the summation function that applies on the previous trust observation to formulate absolute reliability trust degree with a limit of 0.0–1. The completion of reliability evaluation leads the computation phase to cooperativeness estimation. The cooperativeness evaluation is evaluated with the same process as reliability computation and represented by Equation (8). In Equation (8a), $obp_{e_i \to e_j}^{cpt}$ represents the cooperativeness trust evaluation of edge node i towards j where $ob_{n-id}^{cpt(1...n)}$ represents the available observations utilized for the cooperativeness trust evaluation. In Equation (8b), $cpt_{e_i \to e_j}^{dt}$ represents the formulation of absolute cooperativeness trust degree, while dt shows the direct trust evaluation. After the trust parameter estimation, the central authority will proceed further for the trust formulation along with experience as explained in Section 3.4:

$$obp_{e_i \to e_j}^{cpt} = ob_{n-id}^{cpt_1} + ob_{n-id}^{cpt_2} + ob_{n-id}^{cpt_3} + \dots + ob_{n-id}^{cpt_n} \tag{8a}$$

$$cpt_{e_i \to e_j}^{dt} = \sum_{i=0}^{n} \left[obp_{e_i \to e_j}^{cpt} (ob_{n-id}^{cpt_1} + ob_{n-id}^{cpt_2} + \dots + ob_{n-id}^{cpt_n}) \right] \tag{8b}$$

3.4. Trust Aggregation and Development

The trust aggregation process is the procedure in which the previous trust value has been utilized with the current trust to develop an absolute trust value that is used during the phase of decision-making. In the proposed approach, the aggregation and development process is initiated by developing the trust degree of the parameter. Furthermore, it uses that value to compute the aggregated value of trust with the previous experience trust degree of a node. At that phase, the proposed mechanism formulates the absolute trust degree of knowledge component that consists of reliability, and cooperativeness as illustrated in Equation (9):

$$ct_{d_{e_i \to e_j}}^{ag} = rt_{e_i \to e_j}^{dt} + cpt_{e_i \to e_j}^{dt} \tag{9}$$

In Equation (9), the $ct_{d_{e_i \to e_j}}^{ag}$ represents the direct current trust evaluation of edge node i towards j, where $rt_{e_i \to e_j}^{dt}$ and $cpt_{e_i \to e_j}^{dt}$ illustrate the reliability and cooperativeness trust evaluation. After developing the parameter trust evaluation, the central authorities transmit the trust degree of a particular node towards the edge node for the aggregation of experience with current trust. After receiving the parameter trust degree, the edge node aggregates the experience with current trust by first formulating the previous experience observations using Equation (10):

$$t^{pt}_{exp_{e_i \to e_j}} = \sum_{i=0}^{n} \left[et^{o_1}_{e_i \to e_j} + et^{o_2}_{e_i \to e_j} + \dots + et^{o_n}_{e_i \to e_j} \right] \tag{10a}$$

$$f^{dt_{exp}}_{te_i \to e_j} = ct^{ag}_{de_i \to e_j} + t^{pt}_{exp_{e_i \to e_j}} \tag{10b}$$

In Equation (10a), $t^{pt}_{exp_{e_i \to e_j}}$ represents the absolute experience trust formulation process of edge node i towards j, where $et^{o_{1\dots n}}_{e_i \to e_j}$ illustrates the number of previous experience evaluation available at local storage of edge nodes. In Equation (10b), $f^{dt_{exp}}_{te_i \to e_j}$ represents the formulation process of final trust degree, where $ct^{ag}_{de_i \to e_j}$ is the current trust parameter evaluation and $t^{pt}_{exp_{e_i \to e_j}}$ illustrates the absolute experience trust evaluation. After the formulation of the final trust degree, the edge node can compare it with the threshold value for decision-making as discussed in Section 3.5.

3.5. Trust-Based Decision-Making

The decision-making phase is the final phase that utilizes the absolute final trust degree to compare it with a threshold value to determine if the node is trustworthy or malicious. In the proposed mechanism, the range of trust degree is 0.0 to 1. Newly joined edge nodes have a default trust degree of 0.6. A trust degree of 0.7 to 1 is considered trustworthy, while a trust degree of 0.0 to 0.6 is considered flunk/no trust for old edge nodes, as illustrated in Equation (11).

$$\theta = t^{pt}_{exp_{e_i \to e_j}} \tag{11a}$$

$$\theta = \begin{cases} FlunkTrust & \text{if } \theta \leq 0.6 \\ Trustworthy & \text{if } \theta \geq 0.7 \end{cases} \tag{11b}$$

If a node satisfies the threshold value, it is allowed to communicate and transmit monitoring details to hospitals/doctors. If the trust degree of a particular node is less than the minimum requirement, the node cannot communicate and is not allowed to exchange or share information. Furthermore, at the end of communication, the edge node will evaluate the friendliness to determine whether the process of trust degree evaluation should be time-driven or event-driven in the future. This classification is evaluated in Section 3.3.

3.6. Recommendation-Based Indirect Trust

Recommendation-based trust evaluation is an important factor when a node wants to communicate or take services. Furthermore, there are several nodes that do not have previous observations or experience to evaluate trustworthiness. Recommendation-based trust evaluation provides a way to evaluate trust degree by requesting input from neighboring nodes.

EdgeTrust utilizes recommendations when no previous observations are available. To gather recommendations, the node broadcasts requests to surrounding nodes with the node's unique ID to share stored observations. After receiving the recommendations, EdgeTrust develops trust by applying a summation function and then comparing the result with a threshold for decision-making. In the case of indirect trust, the threshold is different from the threshold used for direct trust evaluation. In recommendation-based evaluation, nodes are required to maintain a minimum trust degree of 0.9 to be considered trustworthy. The conditions for decision-making are illustrated by Equation (12):

$$\theta = t^{rt}_{exp_{e_i \to e_j}} \tag{12a}$$

$$\theta = \begin{cases} FlunkTrust & \text{if } \theta \leq 0.8 \\ Trustworthy & \text{if } \theta \geq 0.9 \end{cases} \tag{12b}$$

4. Results and Discussion

In this section, we elaborate on the performance evaluation of the proposed model in comparison with existing schemes. We used an open-source library (Zetta [41,42]) to create a central authority and the IoTivity library [43] to enable inter-object connectivity. Wireless communication is performed using Zigbee (IEEE 802.15) [44]. The complete simulation setup is given in Table 2. We performed comparative analysis using several existing mechanisms: TMEI [45], RobustD [31], and SGSQ-TM [46].

The simulation was performed under different scenarios and attacks by varying the number of network nodes. During the simulation, the number of varying nodes was 50 to 400, and the percentage of malicious and compromised nodes was 35 to 45. The simulation time (t) was also varied between 600 to 1100 minutes (m), with time-based friendliness being performed when the number of sessions created between nodes was 50 or more. For newly joined nodes, the default trust degree was 0.6, while for old nodes, the flunk/no trust was 0.0 to 0.6. A trust degree of 0.7 to 1 was considered trustworthy.

Table 2. Parameters and simulation setup.

Parameters	Value
Area of Network	300 (m^2)
No. of Nodes	400~600
Simulation Time	600~1100
Trust Degree	0.0-1
MAC	IEEE 802.11
Transmission Rate	3~5 Mbps
Size of Packet	20~30
Peak Transmission Range	323 (m)
Node Placement	Uniform
Maximum Connection	11

4.1. Aggregated Trust Evaluation

Trust aggregation is a process in which certain nodes evaluate the trust degree by using the previous trust and current trust to formulate an absolute trust degree for decision-making. In the proposed mechanism, nodes rank the performance of a particular node after obtaining the services, known as experience, and use that for aggregation purposes in future trust evaluation. We evaluated the impact of experience trust aggregation under two different scenarios in which trust computation is performed by nodes with or without experience aggregation, as illustrated in Figure 3. The figure shows the comparative analysis of trustworthy TWP (Trust with Previous) and trustworthy TNP (Trust with no Previous) observations. The trust evaluation of the trustworthy node with aggregation formulates a stable result and enhances accuracy, while the trust without aggregation illustrates a wavered trust degree over a time interval (t). In the second scenario, we performed an identical evaluation on the trust degree of malicious or compromised nodes, and the result showed similar outcomes in which Flunk TWP (Trust with Previous) represented a uniform trust degree and Flunk TNO (Trust with no Previous) showed notable inconstancy in the trust degree and also assigned a higher trust degree, highlighting the significance of employing previous experience in the proposed approach.

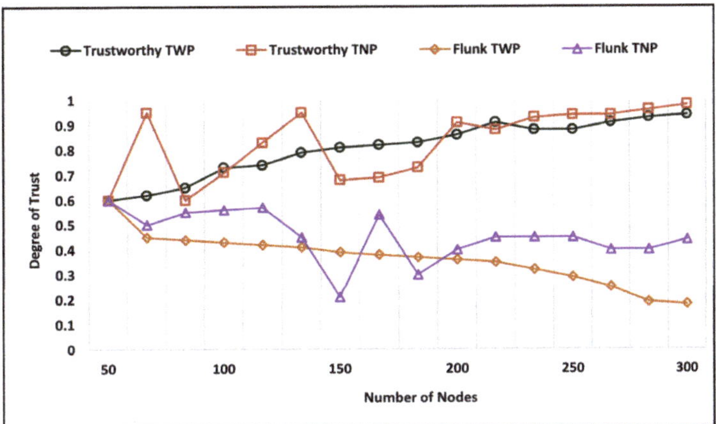

Figure 3. The impact of aggregated trust computation.

4.2. Honest and Dishonest Trust Accuracy

The accuracy of the honest and dishonest trust evaluation is determined by comparing the outcomes of the actual and computed trust degree by the model after the training phase. The simulation was performed to evaluate the trust degree of honest and dishonest nodes, with the comparative analysis illustrated by Figures 4 and 5. The simulation time for the honest and dishonest accuracy evaluation was 300 seconds, with the minimum trust being 0.0 and the maximum trust being 1. The comparative analysis of the computed and actual trust degree of honesty is represented by Figure 4, which shows that the model took 147 seconds to evaluate the actual trust. During the evaluation of the dishonest trust degree, it took 162.5 seconds to remove the difference between computed and actual trust for accurate computations, as illustrated in Figure 5.

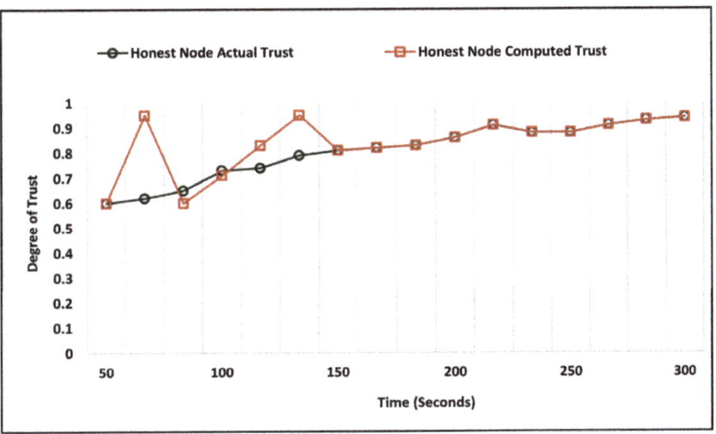

Figure 4. Honest node's trust degree accuracy.

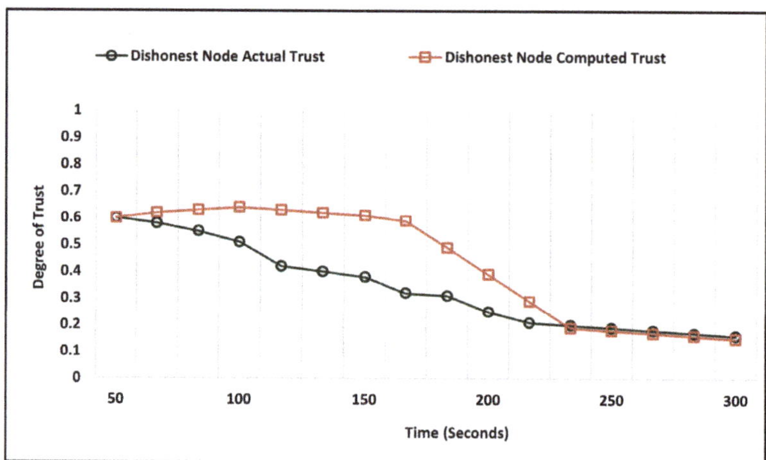

Figure 5. Dishonest node's trust degree accuracy.

4.3. On-Off Attack

The on-off attack is one of the most serious attacks in the IoT heterogeneous environment, where good nodes may become malicious or compromised at any time. It is important to distinguish such nodes that maintain a higher trust degree and whose neighboring nodes also assign a higher rank as an experience, but become malicious after a certain period of time. These nodes may also be compromised by different attacks, making it crucial to recognize these nodes in order to maintain security and privacy. We evaluated the performance of existing approaches under two distinct scenarios by varying the percentage of malicious nodes and time (t).

In the first scenario of an on-off attack, the number of nodes varied from 50 to 400, with a percentage of malicious and compromised nodes at 35%. The simulation time was 600 minutes. Figure 6 shows the simulation outcomes of on-off attack scenario-1, illustrating the performance comparison in which the proposed mechanism successfully recognized the execution and assigned a lower/flunk trust degree as the nodes became malicious after a certain time interval. Initially, the proposed mechanism assigned the default trust degree to nodes with no past experience, and assigned an increasing trust degree at different points that reached 0.64 at point-5, before dropping to 0.55 and then to the lowest trust of 0.01. In the second scenario (Figure 7), the number of nodes was the same as in the previous scenario, and the percentage of malicious nodes increased to 45%. The simulation time was 1100 minutes, with a threshold of 0.0 to 1, and trust was computed with aggregated past experience. The increase in malicious and compromised nodes clearly had an impact on the simulation, and the trust computation assigned to these nodes was lower from the beginning and reached a minimum of 0.25 at the end. In both scenarios, the proposed EdgeTrust mechanism assigned a lower trust degree, indicating the effectiveness of the trust parameters along with the experience component of trust. Therefore, it successfully recognized the on-off attack.

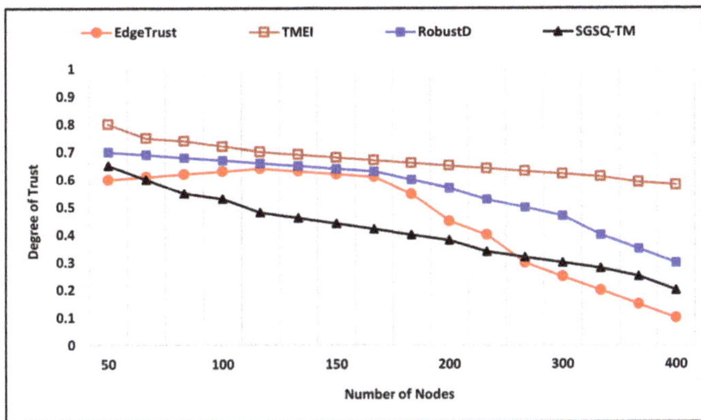

Figure 6. On-off attacks (Scenario-1).

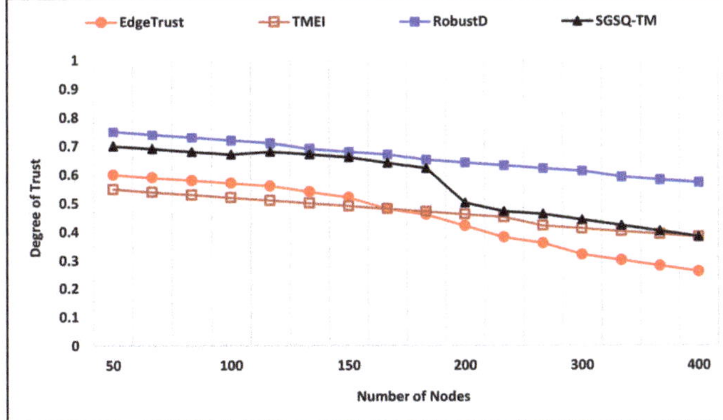

Figure 7. On-off attacks (Scenario-2).

4.4. Self Promoting Attack

It is a kind of attack in which nodes try to promote themselves either alone or in groups to provide the services. The successful execution of a self-promoting attack can have severe consequences that may compromise privacy by gaining access to private and sensitive information. To evaluate the performance of the proposed approach with existing approaches, we have considered two different scenarios in which nodes try to execute a self-promoting attack in different ways. In the first scenario of a self-promoting attack, nodes try to promote themselves alone with any support from the surrounding where the number of nodes is 400 along with varying self-promoting nodes, and the simulation time is 600 (m).

In the first scenario, the total number of nodes is 400 with the percentage of self-promoting nodes being 35%. These nodes self-promote themselves alone and do not have any supporting nodes, where the simulation time consists of 600 (m) with default trust being 0.6 for new nodes, flunk trust is 0.0–0.6, and supreme trust is 0.7–1. Figure 8 illustrates the simulation outcomes of the self-promoting attack in scenario 1, wherein the proposed mechanism assigns the trust degree of 0.86, and the trust degree decreases to reach 0.2, which shows the successful identification of self-promoting nodes. Furthermore, the SGSQ-TM [46] also shows effective performance and assigns a low trust degree, i.e, 0.5. In the second scenario, the total number of nodes is 400 with 45% self-promoting nodes where the simulation time is 600 (m). In this scenario, the self-promoting attack executes in a group, which means a bundle of nodes works in parallel to promote a particular node by assigning a higher fake trust degree. Figure 9 illustrates the simulation outcomes in comparison with the existing approaches, and the results show that the proposed mechanism successfully identifies the malicious nodes and assigns the flunk trust degree of 0.18. Whereas the existing approaches also identify and assign low trust degrees, such as TMEI assigning a lower trust degree of 0.6, RobustD and SGSQ-TM assign a lower trust degree of 0.4 and 0.23, respectively.

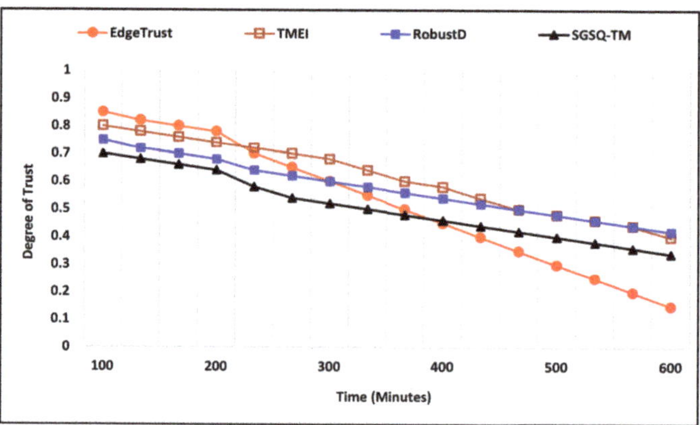

Figure 8. Self-promoting attacks (Scenario-1).

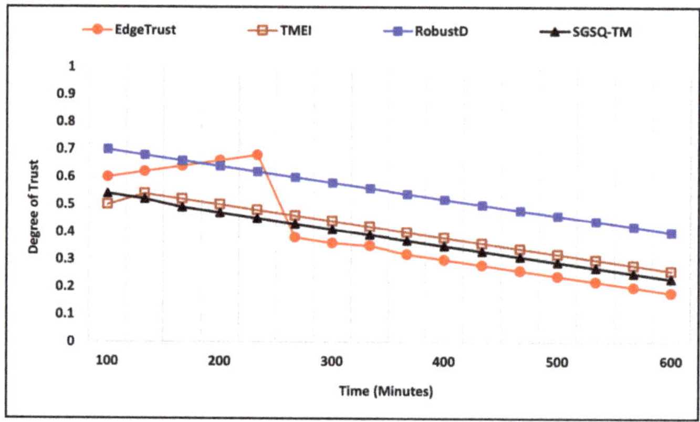

Figure 9. Self-promoting attacks (Scenario-2).

4.5. Good and Bad Mouthing Attacks

Good and Bad mouthing attacks are similar to self-promoting attacks, but in these attacks, nodes do not work together to promote themselves. The good and bad-mouthing attacks are executed by malicious nodes to assign a lower trust degree to the trustworthy nodes called bad-mouthing, while they can assign a higher trust degree to malicious nodes known as a good-mouthing attack. The chances of successful execution of this attack increase when nodes rely on recommendation-based trust evaluation. In the proposed mechanism, the utilization of recommendations is minimal, whereas the central authorities provide the recommendation that has been evaluated based on direct observation. To evaluate the effectiveness of utilizing direct trust-based evaluation as a recommendation, we have performed extensive simulations against good and bad-mouthing attacks under different scenarios. The performance of the proposed approach in comparison to the existing ones is evaluated under two different scenarios for each good and bad-mouthing attack by applying the variation to the number of trustworthy and malicious nodes.

In Figures 10 and 11, the X-axis of the graph shows the simulation time, whereas the Y-axis represents trust, which is computed and assigned at a particular time. In the first scenario of the good mouthing attack, the number of nodes is 600 where the percentage of malicious nodes is 35. Figure 10 illustrates the performance of the proposed mechanism, which shows the trust degree to reach 0.9. After the identification of a good mouthing attack, the trust degree declines to 0.7, and later on, the trust degree assigned by the EdgeTrust declines to flunk trust of 0.4. In comparison, the TMEI and RobustD also show a notable performance and assign a lower trust degree, i.e., 0.4, and 0.5, respectively. In the second scenario of good mouthing evaluation, the number of nodes increases to 800 where the percentage of malicious and compromised nodes is 45, and the simulation time is 600 (m). Figure 11 illustrates the simulation outcomes of the second scenario. In comparison with the first scenario, the result is more fluctuated than what happened due to the percentage ratio of malicious or compromised nodes. When the number of nodes increases and numerous nodes try to execute an attack, then the trust fluctuates between higher and lower degrees. In the second scenario of good mouthing evaluation, the proposed mechanism initially assigns a higher trust degree up to 3 points and then it falls to 0.2 at point 4. Looking at both scenarios, the EdgeTrust assigns the lowest trust to malicious nodes and detects the trustworthy nodes.

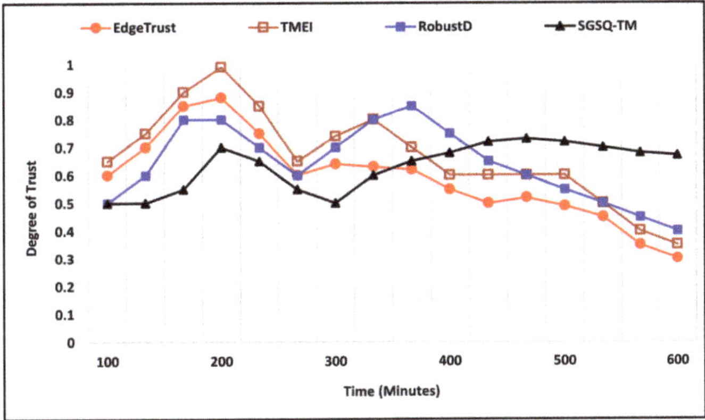

Figure 10. Good mouthing attacks with varying nodes (Scenario-1).

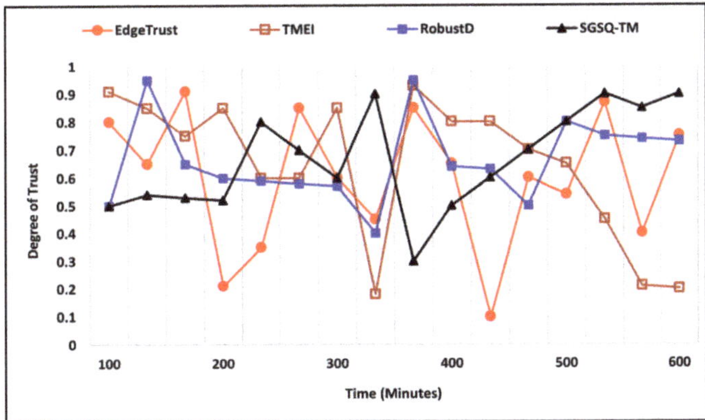

Figure 11. Good mouthing attacks with varying nodes (Scenario-2).

The bad-mouthing attack is also evaluated under two different scenarios by applying variation to the number of total nodes along with the percentage ratio of malicious and compromised nodes. In the first scenario of a bad-mouthing attack, the number of nodes is 400, where the percentage ratio of malicious nodes is 35%, and the simulation time is 600 (m). Figure 12 shows the simulation outcome in which malicious nodes try bad-mouthing trustworthy nodes by assigning a low trust degree while the increasing trust graph of the proposed approach clearly shows that it successfully recognizes the attack and assigns a higher trust degree to the nodes. The proposed EdgeTrust approach initially assigns a lower degree of trust, i.e., 0.3, but later it reaches 0.9, which is the highest trust degree. Furthermore, the existing approaches also show a notable performance against the attack and assign a higher trust degree to the trustworthy nodes. In the second scenario, the total number of nodes is 400, where the malicious and compromised nodes that execute the attack are 45%, and the simulation time is 600 (m). Figure 13 illustrates the comparative performance analysis of the proposed mechanism along with existing approaches. The EdgeTrust approach begins by assigning a default trust degree that increases with time and reaches 0.9, which is the highest trust degree. In comparison, the SGSQ-TM approach also manifests an effective performance and keeps the trust degree of trustworthiness higher, which is 0.4 in the beginning and reaches 0.7. The performance of TMEI is stable and assigns a higher trust degree, whereas the performance of RobustD assigns a lower trust degree, i.e., 0.2, but begins by assigning a higher trust degree after 450 (m) that reaches 0.5.

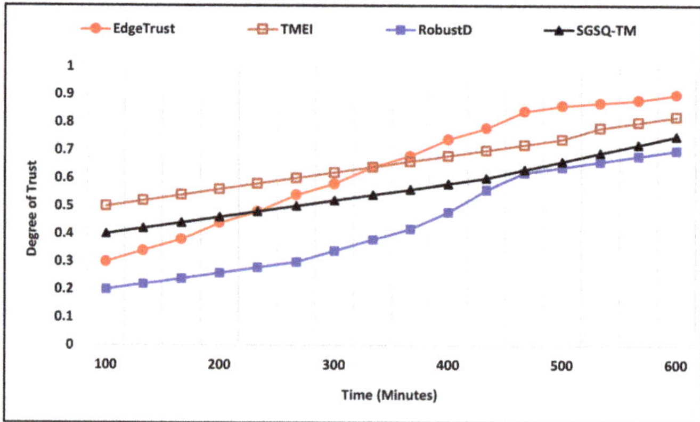

Figure 12. Bad mouthing attacks in Scenario-1.

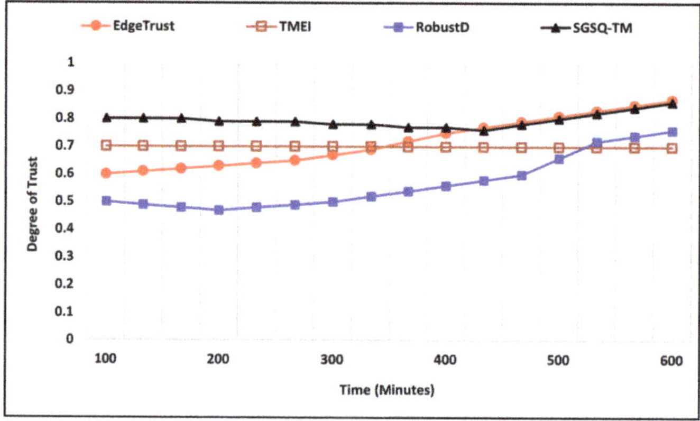

Figure 13. Bad mouthing attacks in Scenario-2.

4.6. Energy Consumption Evaluation

Communication and computation consume a notable amount of energy in IoT, and it is important to propose such approaches that consume less energy to make the implementation of green IoT possible in a real-world scenario. We have evaluated the energy consumption of proposed approaches with existing approaches by applying the variation to the total number of nodes, and the energy consumption is measured in Joules (J). We evaluated the energy consumption of the proposed mechanism with a fixed number of nodes by applying variations to the total time (t). Figure 14 illustrates the simulation which has been performed with 100, 200, up to 600 nodes, where the maximum energy consumed by the proposed approach at 1100 (m) is 240 (J) with 400 nodes, 270 (J) with 500 nodes, and 300 (J) with 600 nodes. The average energy consumption has also been evaluated with varying total numbers of nodes where the simulation time is 1100 (m). Figure 15 illustrates the energy consumption of the approaches that show that the proposed approach has utilized less energy to perform trust computation, whereas, in comparison, RobustD and TMEI use average consumption while SGSQ-TM approaches use a higher amount of energy to perform their computations. The maximum energy consumption of approaches with 600 nodes at 1100 (m) is 360 (J) of EdgeTrust, 450 (J) of TMEI, 400 (J) of RobustD, and 520 (J) of SGSQ-TM. The simulation outcomes of average consumption make the proposed approaches a better way to maintain security among IoT nodes.

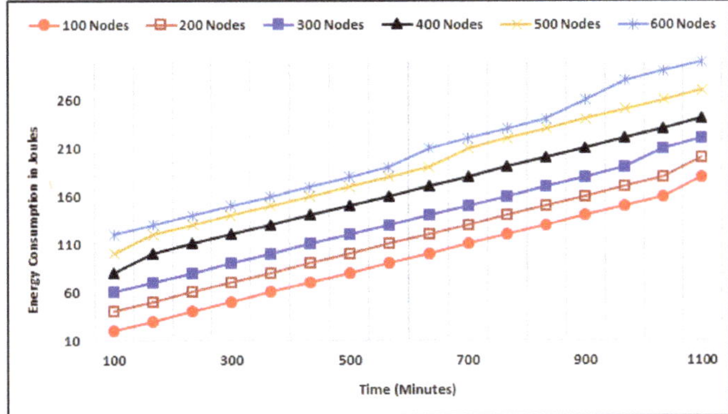

Figure 14. Energy consumption with varying nodes.

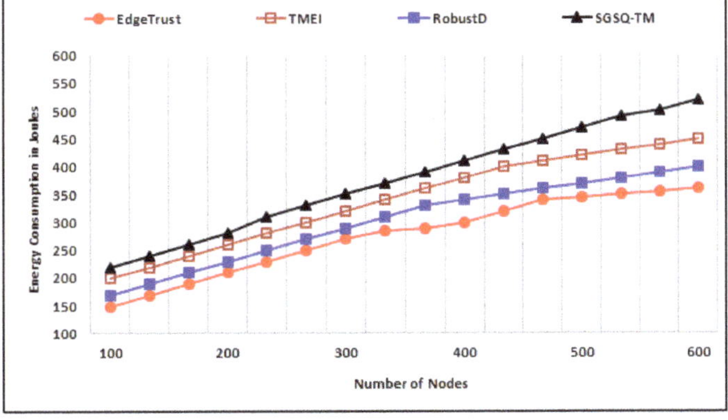

Figure 15. Average energy consumption comparison.

5. Conclusions

Internet of Things (IoT) provides diverse opportunities to the real world to improve daily life by making autonomic devices, which are intelligent and can perform the required operations and given tasks. Healthcare 4.0 and IoT can enhance the facilities provided to patients in remote areas. The monitoring of patients may help to save them in a critical situation. In healthcare 4.0, patients' details are transmitted to the hospital needed to maintain integrity and security. The proposed mechanism addresses the requirements of a lightweight approach to maintain security among nodes. The proposed mechanism utilizes trust parameters and central authority to manage and provide trust observations. The proposed mechanism combines the concept of distributed and centralized trust management along with time-driven and event-driven trust computations. We have also evaluated the performance of the proposed approach with existing approaches among several potential attacks. The extensive simulation outcomes show that EdgeTrust can recognize IoT's possible attacks to maintain a robust environment. In comparison, the proposed approach assigns a lower degree of trust, i.e., 0.25 and 0.18 in the self-promoting attack. Furthermore, EdgeTrust also identifies the good-mouthing instantly and maintains the lower trust degree, whereas, in the case of SGSQ-TM, malicious nodes regain the trustworthiness. Another notable challenge addressed is the lightweight approach that requires less energy consumption, which makes it suitable for the real-world scenario. In the future, the proposed mechanism can be extended by evaluating the storage challenges that the edge nodes may face and formulating a two-way approach to maintain hospital-side trust management.

Author Contributions: Conceptualization, K.A.A. and I.U.D.; methodology, A.A.; software, K.A.A.; validation, K.A.A., I.U.D., A.A. and H.A.K.; formal analysis, K.A.A. and J.J.P.C.R.; investigation, I.U.D. and J.J.P.C.R.; resources, A.A.; data curation, A.A., H.A.K. and J.J.P.C.R.; writing—original draft preparation, K.A.A.; writing—review and editing, I.U.D. and A.A.; visualization, A.A.; supervision, I.U.D.; project administration, A.A.; funding acquisition, A.A. All authors have read and agreed to the published version of the manuscript.

Funding: This work was supported by the Deanship of Scientific Research at King Saud University, Riyadh, Saudi Arabia, through the Vice Deanship of Scientific Research Chairs: Chair of Cyber Security.

Institutional Review Board Statement: Not applicable.

Informed Consent Statement: Not applicable.

Data Availability Statement: Not applicable.

Conflicts of Interest: The authors declare no conflict of interest.

References

1. Din, I.U.; Asmat, H.; Guizani, M. A review of information centric network-based internet of things: Communication architectures, design issues, and research opportunities. *Multimed. Tools Appl.* **2019**, *78*, 30241–30256. [CrossRef]
2. Din, I.U.; Guizani, M.; Rodrigues, J.J.; Hassan, S.; Korotaev, V.V. Machine learning in the Internet of Things: Designed techniques for smart cities. *Future Gener. Comput. Syst.* **2019**, *100*, 826–843. [CrossRef]
3. Gulzar, M.; Abbas, G. Internet of Things security: A survey and taxonomy. In Proceedings of the 2019 International Conference on Engineering and Emerging Technologies (ICEET), Lahore, Pakistan, 21–22 February 2019; pp. 1–6.
4. Yan, Z.; Zhang, P.; Vasilakos, A.V. A survey on trust management for Internet of Things. *J. Netw. Comput. Appl.* **2014**, *42*, 120–134. [CrossRef]
5. Qu, C.; Tao, M.; Yuan, R. A hypergraph-based blockchain model and application in Internet of Things-enabled smart homes. *Sensors* **2018**, *18*, 2784.
6. Guo, Y.; Wang, N.; Xu, Z.Y.; Wu, K. The internet of things-based decision support system for information processing in intelligent manufacturing using data mining technology. *Mech. Syst. Signal Process.* **2020**, *142*, 106630. [CrossRef]
7. Shafique, K.; Khawaja, B.A.; Sabir, F.; Qazi, S.; Mustaqim, M. Internet of things (IoT) for next-generation smart systems: A review of current challenges, future trends and prospects for emerging 5G-IoT scenarios. *IEEE Access* **2020**, *8*, 23022–23040. [CrossRef]
8. Diène, B.; Rodrigues, J.J.; Diallo, O.; Ndoye, E.H.M.; Korotaev, V.V. Data management techniques for Internet of Things. *Mech. Syst. Signal Process.* **2020**, *138*, 106564. [CrossRef]
9. Tseng, L.; Yao, X.; Otoum, S.; Aloqaily, M.; Jararweh, Y. Blockchain-based database in an IoT environment: Challenges, opportunities, and analysis. *Clust. Comput.* **2020**, *23*, 2151–2165. [CrossRef]

10. Ephzibah, E.; Dharinya, S.S.; Remya, L. Decision Making Models Through AI for Internet of Things. In *Internet of Things for Industry 4.0*; Springer: Berlin/Heidelberg, Germany, 2020; pp. 57–72.
11. Hui, H.; Zhou, C.; Xu, S.; Lin, F. A novel secure data transmission scheme in industrial internet of things. *China Commun.* **2020**, *17*, 73–88. [CrossRef]
12. Sicari, S.; Rizzardi, A.; Coen-Porisini, A. 5G in the Internet of Things era: An overview on security and privacy challenges. *Comput. Netw.* **2020**, *179*, 107345. [CrossRef]
13. Sharif, A.; Ouyang, J.; Yang, F.; Chattha, H.T.; Imran, M.A.; Alomainy, A.; Abbasi, Q.H. Low-cost inkjet-printed UHF RFID tag-based system for internet of things applications using characteristic modes. *IEEE Internet Things J.* **2019**, *6*, 3962–3975. [CrossRef]
14. Wu, F.; Wu, T.; Yuce, M.R. An internet-of-things (IoT) network system for connected safety and health monitoring applications. *Sensors* **2019**, *19*, 21. [CrossRef]
15. Singh, B. The Internet of Things: A Vision for Smart World. In *Advances in Signal Processing and Communication*; Springer: Berlin/Heidelberg, Germany, 2019; pp. 165–172.
16. Janjua, K.; Shah, M.A.; Almogren, A.; Khattak, H.A.; Maple, C.; Din, I.U. Proactive Forensics in IoT: Privacy-Aware Log-Preservation Architecture in Fog-Enabled-Cloud Using Holochain and Containerization Technologies. *Electronics* **2020**, *9*, 1172. [CrossRef]
17. Khan, M.A.; Din, I.U.; Jadoon, S.U.; Khan, M.K.; Guizani, M.; Awan, K.A. G-RAT I A novel graphical randomized authentication technique for consumer smart devices. *IEEE Trans. Consum. Electron.* **2019**, *65*, 215–223. [CrossRef]
18. Gong, X.; Feng, T.; Albettar, M. PEASE: A PUF-Based Efficient Authentication and Session Establishment Protocol for Machine-to-Machine Communication in Industrial IoT. *Electronics* **2022**, *11*, 3920. [CrossRef]
19. Qiu, J.; Tian, Z.; Du, C.; Zuo, Q.; Su, S.; Fang, B. A survey on access control in the age of internet of things. *IEEE Internet Things J.* **2020**, *7*, 4682–4696. [CrossRef]
20. Awan, K.A.; Ud Din, I.; Almogren, A.; Almajed, H. AgriTrust—A Trust Management Approach for Smart Agriculture in Cloud-based Internet of Agriculture Things. *Sensors* **2020**, *20*, 6174. [CrossRef]
21. Awan, K.A.; Din, I.U.; Almogren, A.; Almajed, H.; Mohiuddin, I.; Guizani, M. NeuroTrust-Artificial Neural Network-based Intelligent Trust Management Mechanism for Large-Scale Internet of Medical Things. *IEEE Internet Things J.* **2020**, *8*, 15672–15682. [CrossRef]
22. Almogren, A.; Mohiuddin, I.; Din, I.U.; Al Majed, H.; Guizani, N. FTM-IoMT: Fuzzy-based Trust Management for Preventing Sybil Attacks in Internet of Medical Things. *IEEE Internet Things J.* **2020**, *8*, 4485–4497. [CrossRef]
23. Haseeb, K.; Almogren, A.; Ud Din, I.; Islam, N.; Altameem, A. SASC: Secure and Authentication-Based Sensor Cloud Architecture for Intelligent Internet of Things. *Sensors* **2020**, *20*, 2468. [CrossRef]
24. García-García, L.; Jiménez, J.M.; Abdullah, M.T.A.; Lloret, J. Wireless technologies for IoT in smart cities. *Netw. Protoc. Algorithms* **2018**, *10*, 23–64. [CrossRef]
25. Tortorella, G.L.; Fogliatto, F.S.; Esposto, K.F.; Mac Cawley Vergara, A.; Vassolo, R.; Tlapa Mendoza, D.; Narayanamurthy, G. Measuring the effect of Healthcare 4.0 implementation on hospitals' performance. *Prod. Plan. Control* **2022**, *33*, 386–401. [CrossRef]
26. Galletly, C.L.; Barreras, J.L.; Lechuga, J.; Glasman, L.R.; Cruz, G.; Dickson-Gomez, J.B.; Brooks, R.A.; Ruelas, D.M.; Stringfield, B.; Espinoza-Madrigal, I. US public charge policy and Latinx immigrants' thoughts about health and healthcare utilization. *Ethn. Health* 2022, 1–18, *online ahead of print.*
27. Sony, M.; Antony, J.; McDermott, O. The Impact of Healthcare 4.0 on the Healthcare Service Quality: A Systematic Literature Review. *Hosp. Top.* 2022, 1–17. *online ahead of print.*
28. Gardas, B.B. Organizational hindrances to Healthcare 4.0 adoption: An multi-criteria decision analysis framework. *J. Multi-Criteria Decis. Anal.* **2022**, *29*, 186–195. [CrossRef]
29. Mahajan, H.B.; Rashid, A.S.; Junnarkar, A.A.; Uke, N.; Deshpande, S.D.; Futane, P.R.; Alkhayyat, A.; Alhayani, B. Integration of Healthcare 4.0 and blockchain into secure cloud-based electronic health records systems. *Appl. Nanosci.* 2022, 1–14. *online ahead of print.*
30. Azad, M.A.; Bag, S.; Hao, F.; Shalaginov, A. Decentralized self-enforcing trust management system for social Internet of Things. *IEEE Internet Things J.* **2020**, *7*, 2690–2703. [CrossRef]
31. Esposito, C.; Tamburis, O.; Su, X.; Choi, C. Robust Decentralised Trust Management for the Internet of Things by Using Game Theory. *Inf. Process. Manag.* **2020**, *57*, 102308. [CrossRef]
32. Lloret, J.; Parra, L.; Taha, M.; Tomás, J. An architecture and protocol for smart continuous eHealth monitoring using 5G. *Comput. Netw.* **2017**, *129*, 340–351. [CrossRef]
33. Kouicem, D.E.; Imine, Y.; Bouabdallah, A.; Lakhlef, H. A Decentralized Blockchain-Based Trust Management Protocol for the Internet of Things. *IEEE Trans. Dependable Secur. Comput.* **2020**, *19*, 1292–1306. [CrossRef]
34. Khan, Z.A. Using energy-efficient trust management to protect IoT networks for smart cities. *Sustain. Cities Soc.* **2018**, *40*, 1–15. [CrossRef]
35. Khattak, H.A.; Tehreem, K.; Almogren, A.; Ameer, Z.; Din, I.U.; Adnan, M. Dynamic pricing in industrial internet of things: Blockchain application for energy management in smart cities. *J. Inf. Secur. Appl.* **2020**, *55*, 102615. [CrossRef]
36. Siddiqua, A.; Shah, M.A.; Khattak, H.A.; Din, I.U.; Guizani, M. ICAFE: Intelligent congestion avoidance and fast emergency services. *Future Gener. Comput. Syst.* **2019**, *99*, 365–375. [CrossRef]

37. Caminha, J.; Perkusich, A.; Perkusich, M. A smart middleware to detect on-off trust attacks in the Internet of Things. In Proceedings of the 2018 IEEE International Conference on Consumer Electronics (ICCE), Las Vegas, NV, USA, 12–14 January 2018.
38. Chen, J.I.Z. Embedding the MRC and SC Schemes into Trust Management Algorithm Applied to IoT Security Protection. *Wirel. Pers. Commun.* **2018**, *99*, 461–477. [CrossRef]
39. Mukherjee, S.; Sharma, N. Intrusion detection using naive Bayes classifier with feature reduction. *Procedia Technol.* **2012**, *4*, 119–128. [CrossRef]
40. Rish, I. An empirical study of the naive Bayes classifier. In Proceedings of the IJCAI 2001 Workshop on Empirical Methods in Artificial Intelligence, Seattle, WA, USA, 4–6 August 2001; Volume 3, pp. 41–46.
41. Triantafyllou, A.; Sarigiannidis, P.; Lagkas, T.D. Network protocols, schemes, and mechanisms for internet of things (iot): Features, open challenges, and trends. *Wirel. Commun. Mob. Comput.* **2018**, *2018*, 5349894. [CrossRef]
42. Zinca, D.; Popa, M.O. Development of a ZettaJS driver for the ESP8266 IoT hardware. In Proceedings of the 2018 International Symposium on Electronics and Telecommunications (ISETC), Timişoara, Romania, 8–9 November 2018; pp. 1–4.
43. Elsayed, K.; Ibrahim, M.A.B.; Hamza, H.S. Service discovery in heterogeneous IoT environments based on OCF/IoTivity. In Proceedings of the 2019 15th International Wireless Communications & Mobile Computing Conference (IWCMC), Tangier, Morocco, 24–28 June 2019; pp. 1160–1165.
44. Gočal, P.; Macko, D. EEMIP: Energy-efficient communication using timing channels and prioritization in ZigBee. *Sensors* **2019**, *19*, 2246. [CrossRef]
45. Qureshi, K.N.; Iftikhar, A.; Bhatti, S.N.; Piccialli, F.; Giampaolo, F.; Jeon, G. Trust management and evaluation for edge intelligence in the Internet of Things. *Eng. Appl. Artif. Intell.* **2020**, *94*, 103756. [CrossRef]
46. Das, R.; Singh, M.; Majumder, K. SGSQoT: A community-based trust management scheme in Internet of Things. In Proceedings of the International Ethical Hacking Conference, Kolkata, India 31 March–1 April 2018; pp. 209–222.

MDPI
St. Alban-Anlage 66
4052 Basel
Switzerland
Tel. +41 61 683 77 34
Fax +41 61 302 89 18
www.mdpi.com

Electronics Editorial Office
E-mail: electronics@mdpi.com
www.mdpi.com/journal/electronics